# PARTITION AND ADSORPTION OF ORGANIC CONTAMINANTS IN ENVIRONMENTAL SYSTEMS

# PARTITION AND ADSORPTION OF ORGANIC CONTAMINANTS IN ENVIRONMENTAL SYSTEMS

Cary T. Chiou

WILEY-INTERSCIENCE

A JOHN WILEY & SONS, INC., PUBLICATION

***Library of Congress Cataloging-in-Publication Data Is Available***

ISBN 0-471-23325-0

Printed in the United States of America

10  9  8  7  6  5  4  3  2  1

# CONTENTS


**Preface**                                                                               ix

**1   Important Thermodynamic Properties**                                                1

    1.1   Introduction   1
    1.2   First Law of Thermodynamics   2
    1.3   Second Law of Thermodynamics   3
    1.4   Extensive and Intensive Properties   6
    1.5   Chemical Potential   6
    1.6   Chemical Potentials in Multiple Phases   7
    1.7   Change in Chemical Potential with Pressure   8
    1.8   Activity of a Substance   9
    1.9   Vapor–Liquid and Vapor–Solid Equilibria   10

**2   Fundamentals of the Solution Theory**                                               14

    2.1   Introduction   14
    2.2   Raoult's Law   14
    2.3   Henry's Law   18
    2.4   Flory–Huggins Theory   19
    2.5   Variation of Activity Coefficient with Concentration   21
    2.6   Molar Heat of Solution   22
    2.7   Cohesive Energy Density and Solubility Parameter   27

**3   Interphase Partition Equations**                                                    30

    3.1   Partition between Two Separate Phases   30
    3.2   Partition between an Organic Solvent and Water   31
    3.3   Partition between a Macromolecular Phase and Water   32
    3.4   Temperature Dependence of Partition Coefficient   33
    3.5   Concentration Dependence of Partition Coefficient   36

**4   Fundamentals of the Adsorption Theory**                                             39

    4.1   Introduction   39
    4.2   Langmuir Adsorption Isotherm   41
    4.3   Freundlich Equation   43
    4.4   BET Multilayer Adsorption Theory   43

4.5   Polanyi Adsorption Potential Theory   45
4.6   Surface Areas of Solids   48
4.7   Isosteric Heat of Adsorption   50

**5  Contaminant Partition and Bioconcentration**      **53**

5.1   Introduction   53
5.2   Octanol–Water Systems   54
5.3   Heptane–Water Systems   59
5.4   Butanol–Water Systems   62
5.5   Substituent Contributions to Partition Coefficients   63
5.6   Lipid–Water Systems   68
      5.6.1   Solubility of Solutes in Lipids   68
      5.6.2   Lipid–Water Partition Coefficient   72
5.7   Correlations of Partition Coefficients   77
5.8   Bioconcentration of Organic Contaminants   80

**6  Adsorption of Vapors on Minerals and Other Solids**      **86**

6.1   Introduction   86
6.2   Nitrogen Isotherm and Solid Surface Area   86
6.3   Micropore Volume   90
6.4   Improper Surface-Area Measurement   92
6.5   Adsorption of Water and Organic Vapors   100

**7  Contaminant Sorption to Soils and Natural Solids**      **106**

7.1   Introduction   106
7.2   Background in Sorption Studies   107
      7.2.1   Influences of Mineral Matter, Organic Matter, and Water   107
      7.2.2   Soils as a Dual Sorbent for Organic Compounds   109
7.3   Sorption from Water Solution   112
      7.3.1   General Equilibrium Characteristics   112
      7.3.2   Effect of Soil Organic Matter versus Sediment Organic Matter   124
      7.3.3   Effect of Contaminant Water Solubility   133
      7.3.4   Behavior of PAHs versus Other Nonpolar Contaminants   138
      7.3.5   Estimation of Sorption Coefficients for Nonpolar Contaminants   145
      7.3.6   Sorption to Previously Contaminated Soils   146
      7.3.7   Deviations from Linear Sorption Isotherms   149
      7.3.8   Influence of Dissolved and Suspended Natural Organic Matter   168
      7.3.9   Influence of Surfactants and Microemulsions   178

7.4  Sorption from Organic Solvents   192

    7.4.1  Effect of Solvent Polarity   192

    7.4.2  Effects of Temperature, Moisture, and Contaminant Polarity   195

7.5  Sorption from Vapor Phase   200

    7.5.1  General Aspects of Vapor Sorption   200

    7.5.2  Influence of Moisture on Vapor Sorption   203

7.6  Influence of Sorption on Contaminant Activity   210

**8  Contaminant Uptake by Plants from Soil and Water**   **214**

8.1  Introduction   214

8.2  Background in Plant-Uptake Studies   215

8.3  Theoretical Considerations   216

8.4  Uptake by Small Plant Roots from Water   220

8.5  Uptake by Plant Seedlings from Soil   223

8.6  Uptake by Root Crops from Different Soils   226

8.7  Effect of Plant Composition   228

8.8  Contaminant Levels in Aquatic Plants and Sediments   229

8.9  Time Dependence of Contaminants in Plants   231

**Bibliography**   **235**

**Index**   **249**

# PREFACE

The concern for the presence of a wide variety of contaminants in the environment calls for development and assemblage of information about their behavioral characteristics so that appropriate strategies can be adopted to either prevent or minimize their adverse impacts on human welfare and natural resources. This information is especially warranted for toxic chemicals that persist for extended periods of time in the environment. When chemicals enter the environment, they are usually not confined to a specific location but rather are in dynamic motion either within a medium or across the adjacent media. The propensity for a contaminant to move into and distribute itself between the media (or phases) is determined by its physical and chemical properties and environmental factors and variables. The quantity of a contaminant in a given medium and the state of its existence affect its environmental impact. It is therefore important to understand what drives a contaminant from one medium to another and the manner and extent that a contaminant associates with the different media or phases within a local environmental system.

This book is essentially a monograph that depicts the processes by which nonionic organic contaminants are sorbed to natural biotic and abiotic substances. The book focuses on physical principles and system parameters that affect the contaminant uptake by soil from water, air, and other media, by fish from water, and by plants from soil and water. Since contaminant uptake by natural organic substances is often predominantly by a partition interaction, the partition characteristics in several solvent–water model mixtures are treated in some detail to elucidate the relevant physicochemical parameters. When addressing these subjects, the author has relied heavily on the views drawn from his published studies and on those derived from other supporting literature sources. At the risk of appearing immodest, the author has made no attempt to give equal weight to all views on the subject, preferring instead to present a coherent point of view that accounts for many observed contaminant-uptake phenomena. This book is intended to be a good starting point for beginning researchers in the field who might otherwise have difficulties in making sense of the often conflicting and confusing literature.

The book is written primarily for graduate students and beginning professionals in environmental science and engineering in the hope that it will facilitate their research on contaminant sorption to soils and biotic species. Senior scientists may also find the discussion on certain aspects of the sorption process to be beneficial. A great emphasis has been placed on the principles underlying the contaminant sorption to these media and the related medium-contaminant properties. Our intent is to derive from a range of laboratory and field measurements some relatively simple views and rules that can guide us toward a sufficiently accurate account of the activity and fate of contaminants in the environment. In Chapters 1 through 4 of the book we provide requisite backgrounds in thermodynamics and theories of solution and adsorption to assist students and junior professionals to comprehend the discussion in subsequent chapters on sorption-related thermodynamic properties. As we will see in Chapters 5 through 8, nonionic contaminants are sorbed to natural substances usually either by a partition process (a solution phenomenon) or by an adsorption process (a surface phenomenon), or by both in some situations.

It would not have been possible for the author to complete this book without invaluable contributions of his co-workers at Oregon State University (Corvallis, Oregon) and the U.S. Geological Survey (Denver, Colorado) and without the continuous inspiration of Professor Milton Manes, his former research adviser at Kent State University (Kent, Ohio) and the co-author of several research papers. The author thanks the National Institute of Environmental Health Sciences, the U.S. Environmental Protection Agency, and the National Science Foundation for their supports of his earlier research at Oregon State University (1976–1983) and the U.S. Geological Survey for continuous support of his research (1983–date). The author is also indebted to the encouragement from many of his colleagues to write this book and to their assistance during the book's preparation. Finally, the author thanks the U.S. Geological Survey for the granting of official time to prepare the book and for financial assistance in the drafting of the illustrative figures and graphs.

*U.S. Geological Survey*                                                CARY T. CHIOU
*Denver, Colorado*
*February 2002*

# 1 Important Thermodynamic Properties

## 1.1 INTRODUCTION

In environmental systems, one is keenly interested in the transfer of a chemical (contaminant) from one phase (or medium) to another and in the manner it distributes itself between phases at equilibrium. In most cases, contaminants are transported through mobile water or atmosphere into other natural biotic or abiotic phases or media. Depending on the material properties of individual phases and on variable environmental factors, such as temperature and humidity, the manner by which a contaminant is retained by individual natural phases can vary widely. For most organic contaminants, particularly electrically neutral species, the way a contaminant is retained by a biotic or abiotic matter falls mainly into either or both of two categories: The contaminant adheres only onto the surface of a natural material, or it dissolves into the latter's molecular network. Although these different modes of action are not readily distinguishable to our eyes, they are consequential to the extent of contaminant uptake and to the activity and fate of the contaminant in its local environment.

It is important to understand the terms *system*, *phase*, and *medium* as they are referred to in the context above. A *system* is defined as a physical domain enclosed by a real or imaginary boundary that separates it from its surroundings. The content of a system may be simple or complex, ranging from a single vapor, liquid, or solid to a multicomponent and heterogeneous mixture of considerable complexity. In heterogeneous systems, there exist molecularly homogeneous regions, which we refer to as *phases*. Examples of phases in a heterogeneous system are the organic solvent and water phases in their partially miscible mixtures and the vapor and liquid phases of a volatile liquid in a partially filled vessel or a subsurface space. The term *medium* is less precise than the term *phase*, although they are sometimes used interchangeably. The former refers to matter that is apparently uniform in its macroscopic appearance but is not well characterized, such as a soil sample composed of many finely divided mineral and organic phases or a plant-matter sample composed of many constituents or phases (e.g., water, cellulose, and lipids) in its composition.

Whether mass transfer occurs for any component across phases or the component at the time is at equilibrium between phases at constant temperature and pressure (where there is no net exchange of mass) is governed by the equality or inequality of its chemical potentials with the (various) phases. The chemical potentials being referred to are the molar Gibbs free energies of the component in individual phases. There is a natural tendency of a chemical to come to a state of equilibrium between all contacted phases, where the chemical potential gradient across phase boundaries is zero. The chemical potentials are derived from the first and second laws of thermodynamics. In the derivation of Gibbs free energy, the reader will also be introduced to two other important thermodynamic properties, enthalpy (heat) and entropy, by which one can distinguish a surface process from a solution process, as shown later. For a more detailed treatment of the thermodynamic quantities and their relationships, the reader is directed to a physical chemistry textbook.

## 1.2    FIRST LAW OF THERMODYNAMICS

The first law of thermodynamics is a consequence of the principle of conservation of energy: that is, that heat, kinetic energy, potential energy, and electrical energy are different forms of energy that can be interconverted but can be neither created nor destroyed. Consider any system enclosed in a vessel that can change its volume and exchange heat with its surroundings but is impervious to the passage of matter. We postulate a property called the *internal energy* of the system, $E$. We will be concerned with the change in $E$ and not with its absolute value. If the system absorbs an amount of heat $q$ with no other changes, the conservation of energy requires that its internal energy increase by the amount of $q$; conversely, the internal energy will decrease by the amount of $q$ if an amount of heat $q$ is released to its surroundings. Similarly, if the system does work $w$ on its surroundings with no other changes, its internal energy will decrease by the amount of $w$. If the system both exchanges heat and does work, the change in internal energy is then

$$\Delta E = q - w \qquad (1.1)$$

where $q$ is here taken as positive for heat absorbed by the system and $w$ as positive for work done by the system. The first law also implies that $E$ is a *state function*: that its magnitude is solely dependent on its state variables (e.g., temperature, pressure, and volume). For any series of processes that end with a return to the original state variables, $\Delta E = 0$.

For a constant-pressure system involving only the work of expansion and contraction (i.e., no electrical work), $w$ equals $P\Delta V$, where $P$ is the (constant) pressure and $\Delta V$ is the (finite) change in volume. In this case, the change in $E$ is therefore

$$\Delta E = q - P\Delta V \qquad (1.2)$$

If one defines a new state function, $H$, called *enthalpy*, as

$$H = E + PV \tag{1.3}$$

then the change in $H$ at constant pressure will be

$$\Delta H = \Delta E + P\Delta V \tag{1.4}$$

A comparison of Eqs. (1.2) with (1.4) leads to

$$\Delta H = q \quad \text{for a constant-pressure process} \tag{1.5}$$

The enthalpy is therefore a useful state function for describing the heat exchange at constant pressure.

## 1.3   SECOND LAW OF THERMODYNAMICS

We first begin with the concept of a *reversible process* in thermodynamics. In addition to the usual sense of a reversible process, the condition of thermodynamic reversibility for any process is that it proceeds at all times infinitesimally close to equilibrium, so that its direction can be reversed by an infinitesimally small change in one or more of the state variables. A close approximation to a reversible process is the freezing of water in a vessel maintained below but very close to the equilibrium freezing point (which is 0°C at 1 atmosphere); the process can be reversed by raising the temperature very slightly above the freezing point. Conversely, the freezing process of supercooled water can be carried out irreversibly by seeding it with an ice crystal. In the reversible expansion of a gas against a resistance that is close to the gas pressure at all times, the differential work is $P\,dV$ and the overall work is $\int P\,dV$. By contrast, in the extreme case of the gas expanding into a vacuum, the work is zero.

The most useful statement of the second law of thermodynamics is described in terms of a state function called the *entropy* ($S$), which is a measure of the degree of randomness or disorder in a system. For a system undergoing a change in state, the change in entropy is such that

$$dS = dq/T \quad \text{for an infinitesimal reversible process} \tag{1.6}$$

$$dS > dq/T \quad \text{for an infinitesimal spontaneous process} \tag{1.7}$$

where $T$ is the thermodynamic temperature [Kelvin (K)]. For a reversible process, $dq = T\,dS$. By relating Eq. (1.6) to the first law, one finds for a reversible process in a closed system that involves only the $P$–$V$ work (i.e., no electrical work) that

$$dE = T\,dS - P\,dV \tag{1.8}$$

For any other process, $dq \neq T\,dS$ and $dw \neq P\,dV$. However, the difference between $T\,dS$ and $P\,dV$ (i.e., $dE$) is a state function. Therefore, Eq. (1.8) holds for all processes, whether or not reversible.

According to the second law of thermodynamics, the criterion for whether a process is taking place reversibly (i.e., at equilibrium) or spontaneously within a completely isolated system (i.e., the one at constant volume and internal energy) is given as

$$(dS)_{E,V} \geq 0 \tag{1.9}$$

that is, the overall entropy change of the system is zero for an equilibrium process but increases for a spontaneous process. The fact that $(dS)_{E,V}$ can never be less than zero is a consequence of the second law.

Chemical processes of most interest usually take place at constant temperature and pressure. A new criterion is therefore required to indicate whether a process is reversible or spontaneous under this condition. If we now allow a process to take place initially in an isolated system and then adjust the temperature by reversible absorption (or emission) of heat and adjust the pressure by reversible expansion (or contraction) at constant temperature, the entropy change from the adjustment will be $dq/T = (dE + P\,dV)/T$. The change in entropy of the system, which is no longer an isolated system, after this adjustment will be

$$(dS)_{T,P} = (dS)_{E,V} + dE/T + P\,dV/T \tag{1.10}$$

Substituting Eq. (1.9) into Eq. (1.10) gives

$$-T(dS)_{E,V} = dE + P\,dV - T(dS)_{T,P} \leq 0 \tag{1.11}$$

The quantities on both sides will therefore be negative for spontaneous processes, zero for equilibrium processes, and never positive.

One can express the right side of Eq. (1.11) by defining a new state function, $G$, the Gibbs function or Gibbs free energy, as

$$G = E + PV - TS = H - TS \tag{1.12}$$

At constant temperature and pressure, one gets

$$(dG)_{T,P} = dE + P\,dV - T\,dS \tag{1.13}$$

From Eqs. (1.11) and (1.13) the condition that

$$(dG)_{T,P} \leq 0 \tag{1.14}$$

becomes the criterion for any infinitesimal process within a closed system (i.e., where no mass transfer occurs across the system boundary) to take place at equilibrium [i.e., $(dG)_{T,P} = 0$] or spontaneously [i.e., $(dG)_{T,P} < 0$] at constant temperature and pressure. For a single-component system, $dG$ is a function of temperature and pressure (or volume). For a complex mixture, $dG$ depends also on the composition, as will be seen.

If a phase transition (e.g., from liquid to vapor) takes place in a closed single-component system at constant $T$ and $P$, the transition can thus be carried out at equilibrium with any phase-mass ratio as long as both phases coexist in finite amounts. In this case, $dG/d\lambda$, or $\Delta G$, is equal to 0, where $\Delta G$ corresponds to a finite phase transition and $\lambda$ is the progress variable. In a closed multicomponent system where a chemical reaction takes place or a component distributes between phases at fixed $T$ and $P$, usually only one composition can satisfy the condition for equilibrium (i.e., $dG/d\lambda = \Delta G = 0$).

For simple systems without mass and composition changes, one can thus write

$$dE = T\,dS - P\,dV \tag{1.8}$$

and

$$dG = T\,dS - P\,dV + P\,dV + V\,dP - T\,dS - S\,dT$$

or

$$dG = V\,dP - S\,dT \tag{1.15}$$

In a closed system where a change in state or a chemical reaction takes place at constant temperature, one finds from Eq. (1.12) an important relation as follows:

$$\Delta G = \Delta H - T\,\Delta S \tag{1.16}$$

Thus, the reduction in free energy of a closed system at constant temperature is favored by a decrease in system enthalpy or by an increase in system entropy. However, chemical processes seldom occur with emission of heat (i.e., $\Delta H < 0$) coupled with an increase in $\Delta S$. In some special cases, the process may proceed with $\Delta H = 0$ and $T\,\Delta S > 0$, such as the expansion and mixing of ideal gases or the formation of an ideal solution, or with $\Delta H < 0$ and $T\,\Delta S \simeq 0$, such as chemical reactions in which the moles of reactants equal the moles of products. Frequently, chemical processes occur with opposing effects of $\Delta H$ and $T\,\Delta S$, in which one outweighs the other.

To illustrate how either $\Delta H$ or $T\,\Delta S$ may act as the main driving force for a spontaneous process, let us consider two physical processes, *vaporization* and *adsorption*, at constant temperature in a closed system. When a fraction of a liquid in excess quantity is being evaporated into a fixed vacuum space, the heat absorbed by the system to evaporate the liquid (i.e., $\Delta H > 0$) increases virtually linearly with the mass of liquid evaporated, whereas the rate of increase in system entropy (i.e., $\Delta S > 0$) is relatively large at first but decreases as the vapor density increases. Here the unfavorable endothermic heat of evaporation is outbalanced by the more favorable entropy increase until the system reaches equilibrium, at which point $\Delta H = T\Delta S$ and $\Delta G = 0$. Conversely, when a vapor is adsorbing onto a previously evacuated surface, the exothermic heat of adsorption (i.e., $\Delta H < 0$) is relatively large initially but decreases rapidly when more vapor is adsorbed (because the adsorption sites are usually energetically heterogeneous, as discussed in Chapter 4). The system entropy decreases (i.e., $\Delta S < 0$) in a similar fashion but at a different rate. Thus the system reaches equilibrium at some point, where $\Delta H = T\Delta S$ and $\Delta G = 0$. In this case, the unfavorable entropy loss is outbalanced by the more favorable decrease in enthalpy before the system reaches equilibrium.

## 1.4  EXTENSIVE AND INTENSIVE PROPERTIES

Extensive thermodynamic properties are those whose magnitudes are related to the sizes (or the moles) of the chemical species present. Examples are

$$G, H, V, E, S \quad \text{or} \quad \Delta G, \Delta H, \Delta V, \Delta E, \Delta S$$

Intensive properties are those whose magnitudes are not a function of their sizes or masses. Examples are $T$, $P$, $\rho$ (density), and the partial molar quantities of the extensive properties.

For any extensive property $Y$ at constant $T$ and $P$ in a multiple-component system, the differential change of the property is thus

$$dY = (\partial Y/\partial n_1)\,dn_1 + (\partial Y/\partial n_2)\,dn_2 + (\partial Y/\partial n_3)\,dn_3 + \cdots \tag{1.17}$$

or

$$dY = \overline{Y}_1\,dn_1 + \overline{Y}_2\,dn_2 + \overline{Y}_3\,dn_3 + \cdots \tag{1.18}$$

where the partial molar quantity, $\overline{Y}_i$, is an intensive thermodynamic property.

## 1.5  CHEMICAL POTENTIAL

The chemical potential of a substance in a phase serves as a measure of its escaping tendency. We already know that when two phases in a system are at equilibrium, they must be at the same $T$ and $P$. When the transfer of a substance between two phases is allowed, an additional requirement for equilibrium is that the chemical potentials of the substance must be the same in the two phases. For a system involving a change in the quantity of its components, due, for example, to chemical reactions or transfer of mass to and out of the system, the previous differential equations are adjusted to take into account the changes in the moles ($n_1, n_2, n_3$, etc.) of individual components. By extending Eqs. (1.8) and (1.15), one now obtains

$$dE = T\,dS - P\,dV + (\partial E/\partial n_1)_{V,S,n_j}\,dn_1 + \cdots + (\partial E/\partial n_k)_{V,S,n_j}\,dn_k \qquad (1.19)$$

and

$$dG = V\,dP - S\,dT + (\partial G/\partial n_1)_{T,P,n_j}\,dn_1 + \cdots + (\partial G/\partial n_k)_{T,P,n_j}\,dn_k \qquad (1.20)$$

Thus, for a reversible process involving a change in individual-component mass in a phase,

$$dE = T\,dS - P\,dV + \sum \mu_i\,dn_i \qquad (1.21)$$

and

$$dG = V\,dP - S\,dT + \sum \mu_i\,dn_i \qquad (1.22)$$

in which the chemical potential or the molar Gibbs function of component $i$ is defined as

$$\mu_i = (\partial E/\partial n_i)_{V,S,n_j} = (\partial G/\partial n_i)_{T,P,n_j} \qquad (1.23)$$

Since the chemical potential of a substance is an intensive property, the difference in its values between regions in a phase or between phases of a system determines the direction of mass transfer (from the one of higher potential to the one of lower potential), just as the temperature gradient determines the direction of heat flow. The usefulness of the chemical potential as a criterion for equilibrium of a substance between phases is illustrated below.

## 1.6  CHEMICAL POTENTIALS IN MULTIPLE PHASES

Consider a closed system consisting of two separate phases, A and B, to which an organic compound (solute) is added at constant temperature and pressure, as shown in Figure 1.1. The solute $i$ will then distribute itself between phases A and B, to arrive eventually at some stable concentrations when the system reaches the state of equilibrium. Here one may express the change in Gibbs free energy of the entire system as

$$\Delta G_i = \Delta G_{i,A} + \Delta G_{i,B} \quad \text{or} \quad dG_i = dG_{i,A} + dG_{i,B} \qquad (1.24)$$

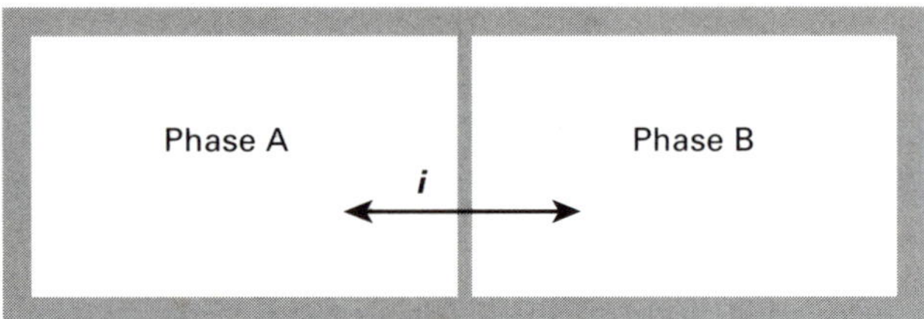

**Figure 1.1**   Distribution of a component ($i$) between two separate phases, A and B, at constant temperature and pressure.

At constant $T$ and $P$, then

$$dG_i = \mu_{i,A}\, dn_{i,A} + \mu_{i,B}\, dn_{i,B} \qquad (1.25)$$

When component $i$ is at the state of equilibrium between phases A and B, $dG_i = 0$. To maintain equilibrium, any infinitesimal increase of component $i$ in phase A must be accompanied by an equal amount of loss of component $i$ in phase B, that is,

$$dn_{i,A} = -dn_{i,B} \qquad (1.26)$$

thus,

$$\mu_{i,A} = \mu_{i,B} \qquad (1.27)$$

Equation (1.27) defines the state of equilibrium for component $i$ between any two phases at constant temperature and pressure. A similar operation can be carried out for a component in a multiple-phase system through a series of steps that allow the transfer of mass between only two phases at a time. This leads to the conclusion

$$\mu_{i,A} = \mu_{i,B} = \mu_{i,C} = \mu_{i,D} \cdots \qquad (1.28)$$

We shall see later that Eq. (1.27) serves as the criterion for the distribution (e.g., partition) relations of organic contaminants between water and other phases of environmental interest (e.g., the soil organic matter in sorption and the fish lipid in bioconcentration).

## 1.7   CHANGE IN CHEMICAL POTENTIAL WITH PRESSURE

For 1 mole of a component in a system with no mass change, Eq. (1.15) gives

$$d\mu_i = \overline{V}_i\, dP - \overline{S}_i\, dT \qquad (1.29)$$

and at fixed $T$,

$$\int d\mu_i = \int \overline{V}_i\, dP \tag{1.30}$$

$$\mu_i - \mu_i^{\circ} = \int_{P^{\circ}}^{P} \overline{V}_i\, dP \tag{1.31}$$

where $\mu_i$ is the chemical potential at $P$ and $\mu_i^{\circ}$ is the chemical potential at a reference pressure $P^{\circ}$. For solid or liquid substances, $\overline{V}_i$ does not vary much with $P$ and may be treated as constant; hence,

$$\mu_i = \mu_i^{\circ} + \overline{V}_i(P - P^{\circ}) \tag{1.32}$$

For ideal gases, $\overline{V}_i = RT/P_i$, where $R$ is the gas constant (8.31 J/mol·K); one gets

$$\mu_i - \mu_i^{\circ} = \int_{P^{\circ}}^{P} (RT/P_i)\, dP_i \tag{1.33}$$

or

$$\mu_i = \mu_i^{\circ} + RT \ln(P_i/P_i^{\circ}) \tag{1.34}$$

In our treatment of gases, it is common and convenient to set $P_i^{\circ} = 1$ atmosphere (atm) as the reference state of a gas at temperature $T$. In this case, $\mu_i^{\circ}$ is the reference chemical potential of gas $i$ at 1 atm pressure and temperature $T$, and $\mu_i$ is the chemical potential of gas $i$ at $P_i$ (atm) and $T$. With $P_i^{\circ} = 1$ atm, Eq. (1.34) is thus reduced to

$$\mu_i = \mu_i^{\circ} + RT \ln P_i \tag{1.35}$$

If $P_i^{\circ} \neq 1$ atm, $P_i$ in Eq. (1.35) is simply the dimensionless ratio of $P_i$ to $P_i^{\circ}$. If the gas behaves nonideally (when under high pressure), $\mu_i = \mu_i^{\circ} + RT \ln f_i$ is used instead, where $f_i$ is the fugacity of vapor $i$ (i.e., the vapor pressure corrected for deviation from the ideal-gas law). From Eq. (1.35), the differential change in $\mu_i$ with $P_i$ at constant temperature is therefore

$$d\mu_i = RT\, d\ln P_i = RT\, d\ln f_i \tag{1.36}$$

## 1.8  ACTIVITY OF A SUBSTANCE

By Raoult's law convention, the activity of a substance at temperature $T$ is the ratio of its fugacity or vapor pressure to that of the fugacity or vapor pressure of the substance at some reference state at $T$, that is,

$$a_i = f_i/f_i^{\circ} \simeq P_i/P_i^{\circ} \tag{1.37}$$

where $f_i^\circ$ is the reference fugacity of substance $i$ at $P_i^\circ$ and $T$ and $f_i$ is the fugacity of substance $i$ at $P$ and $T$. To a good approximation, except at extremely high vapor pressure, $f_i^\circ = P_i^\circ$ and $f_i = P_i$. The most useful and convenient reference states for gases, liquids, and solids are chosen as follows:

$$\text{For gases:} \qquad f_i^\circ = 1 \text{ atm of the gas} \qquad (1.38)$$

$$\text{For liquids:} \qquad f_i^\circ = f_i^\circ \text{ of the pure liquid} \qquad (1.39)$$

$$\text{For solids:} \qquad f_i^\circ = f_i^\circ \text{ of the pure supercooled liquid} \qquad (1.40)$$

By Eqs. (1.37) and (1.39), one sees that $a_i = 1$ for a pure liquid at $T$. Contrarily, one sees with Eq. (1.40) that $a_i < 1$ for a pure solid at $T$ below its melting point, because the vapor pressure of the solid is less than that of its supercooled liquid, as shown later in Figure 1.2. From the relations above, one also gets

$$d\mu_i = RT\, d\ln P_i = RT\, d\ln f_i = RT\, d\ln a_i \qquad (1.41)$$

Therefore, an alternative way of stating the equilibrium of a chemical between any two separate phases is that the activity or fugacity (or partial pressure) of the chemical is the same in the two phases.


## 1.9  VAPOR–LIQUID AND VAPOR–SOLID EQUILIBRIA

It was indicated earlier that when a chemical species in two phases (A and B) reaches equilibrium,

$$\Delta\overline{G}_i = \mu_{i,A} - \mu_{i,B} = 0 \quad \text{or} \quad d\mu_{i,A} = d\mu_{i,B}$$

For a pure liquid in equilibrium with its vapor, one therefore gets

$$d\mu_l(\text{liquid}) = d\mu_v(\text{vapor}) \qquad (1.42)$$

or

$$\overline{V}_l\, dP - \overline{S}_l\, dT = \overline{V}_v\, dP - \overline{S}_v\, dT \qquad (1.43)$$

that is,

$$dP/dT = (\overline{S}_v - \overline{S}_l)/(\overline{V}_v - \overline{V}_l) \qquad (1.44)$$

With

$$\Delta \overline{S} = \overline{S}_v - \overline{S}_l = \Delta \overline{H}_{\text{evap}}/T \tag{1.45}$$

where $\Delta \overline{H}_{\text{evap}}$ is the molar heat of evaporation of the liquid, one gets

$$dP/dT = \Delta \overline{H}_{\text{evap}}/T(\overline{V}_v - \overline{V}_l) \tag{1.46}$$

which is known as the Clapeyron equation. Because $\overline{V}_v \gg \overline{V}_l$, Eq. (1.46) can be reduced further to

$$dP/dT = \Delta \overline{H}_{\text{evap}}/T\overline{V}_v \tag{1.47}$$

If the ideal-gas law holds, $P\overline{V}_v = RT$, then

$$d\ln P/dT = \Delta \overline{H}_{\text{evap}}/RT^2 \quad \text{or} \quad d\ln P/d(1/T) = -\Delta \overline{H}_{\text{evap}}/R \tag{1.48}$$

Integration of Eq. (1.48) on the assumption of constant $\Delta \overline{H}_{\text{evap}}$ gives

$$\ln P = -\Delta \overline{H}_{\text{evap}}/RT + \text{constant} \tag{1.49}$$

or

$$\log P = -\Delta \overline{H}_{\text{evap}}/2.303RT + \text{constant} \tag{1.50}$$

For the solid–vapor equilibrium, one gets a similar expression:

$$\log P = -\Delta \overline{H}_{\text{sub}}/2.303RT + \text{constant} \tag{1.51}$$

where $\Delta \overline{H}_{\text{sub}}$ is the molar heat of sublimation of the solid,

$$\Delta \overline{H}_{\text{sub}} = \Delta \overline{H}_{\text{evap}} + \Delta \overline{H}_{\text{fus}} \tag{1.52}$$

Equation (1.50) or (1.51) is called the *Clausius–Clapeyron equation*. It enables one to determine the heat associated with liquid–vapor or solid–vapor transition from the *P–T* data more conveniently than by direct calorimetry. This heat serves as a useful reference for comparison with the heat involved when the vapor is transferred to a substrate or phase where it may either condense onto the surface (as in adsorption) or disperse into the matrix (as in partition).

Given in Table 1.1 are the vapor pressures of some liquids as a function of temperature (0 to 100°C). The vapor–pressure data at temperatures considerably higher than the normal boiling points of the liquids are excluded. Over this small-to-moderate temperature range, the *P–T* data of the liquids are reasonably well represented by Eq. (1.50), as illustrated in Figure 1.2, showing that $\Delta \overline{H}_{\text{evap}}$ is not very sensitive to temperature. A similar plot for solid compounds with the vapor pressure data below the melting points would lead to a similar conclusion that the $\Delta \overline{H}_{\text{sub}}$ is essentially constant over a small-to-

**TABLE 1.1.  Vapor Pressures of Some Liquids as a Function of Temperature**[a]

| | Vapor Pressure, $P°$ (mmHg) | | | | | | | |
|---|---|---|---|---|---|---|---|---|
| $t$ (°C) | CT | TCE | TEE | BEN | $o$-XYL | $m$-DCB | HEX | OCT |
| 0 | 33 | 21 | 4.2 | | 1.3 | 0.4 | 45 | 2.8 |
| 10 | 56 | 36 | 7.9 | 44 | 3.6 | 0.9 | 75 | 5.4 |
| 20 | 91 | 58 | 14 | 73 | 4.9 | 1.6 | 120 | 10.5 |
| 30 | 143 | 94 | 25 | 120 | 8.8 | 3.0 | 185 | 17 |
| 40 | 216 | 140 | 40 | 183 | 15 | 5.3 | 275 | 31 |
| 50 | 317 | 215 | 64 | 273 | 25 | 9.0 | 400 | 50 |
| 60 | 451 | 317 | 97 | 390 | 41 | 15 | 570 | 78 |
| 70 | 622 | 450 | 145 | 550 | 63 | 24 | 790 | 118 |
| 80 | 843 | 610 | 207 | 750 | 95 | 37 | 1050 | 175 |
| 90 | | 840 | 290 | 1040 | 140 | 56 | | 253 |
| 100 | | | 400 | | 197 | 83 | | 354 |

[a] CT, carbon tetrachloride; TCE, trichloroethylene; TEE, tetrachloroethylene; BEN, benzene; $o$-XYL, $o$-xylene; $m$-DCB, $m$-dichlorobenzene; HEX, $n$-hexane; OCT, $n$-octane.

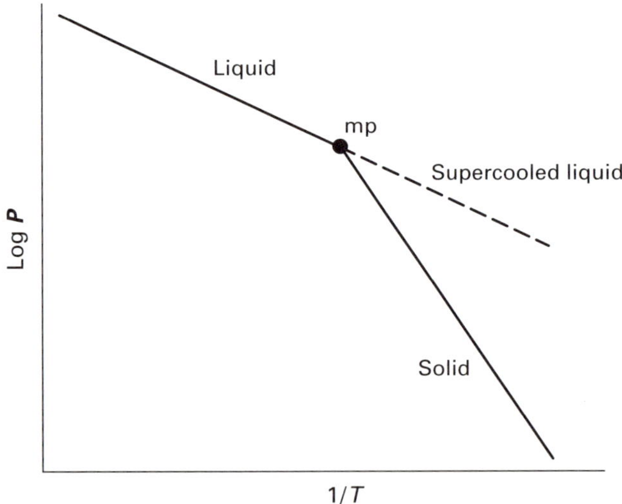



**Figure 1.2**  Clausius–Clapeyron plot of the vapor pressure (log $P$) of a substance against the reciprocal of absolute temperature (1/$T$) below the critical temperature. The intersection of the liquid and solid lines is the melting point (mp) of the substance.

moderate temperature range. If the vapor pressure data at temperatures below and above the melting points are available, the plot of log $P$ versus 1/$T$ should then display a discontinuity (because $\Delta \overline{H}_{\text{sub}} > \Delta \overline{H}_{\text{evap}}$) at the melting point, as illustrated schematically in Figure 1.2. By extrapolation of the liquid line across the melting point to lower temperatures, one obtains the vapor pressure ($P$) of the supercooled liquid. The heat of fusion of a solid compound can be obtained with the relation $\Delta\overline{H}_{fus} = \Delta\overline{H}_{sub} - \Delta\overline{H}_{evap}$.

The observation that $\Delta\overline{H}_{evap}$ (or $\Delta\overline{H}_{sub}$) is not very sensitive to temperature below the normal boiling point of a substance may be understood on the basis that the powerful molecular forces that keep the substance in a liquid or solid state are not significantly affected by the thermal energy of the molecules below the boiling point. As the temperature increases significantly toward the critical point of the substance (at which the properties of the gas and the liquid coalesce to form a single phase), $\Delta\overline{H}_{evap}$ will then decrease very sharply and become zero at the critical temperature.

# 2 Fundamentals of the Solution Theory

## 2.1 INTRODUCTION

In natural systems, the solubilities of organic contaminants in water and other phases play a crucial role in the behavior and fate of the compounds. The solubility affects not only the limit to which a substance can be solubilized by a solvent or a phase, but also dictates the distribution pattern of the substance between any two solvents or phases of interest. Water is apparently the most important natural solvent, not only because it is a huge medium to hold various contaminants, but also because it is a common medium through which contaminants are transported to other media. Depending on specific local environments, natural organic substances such as mineral oils, biological lipids, soil organic matter, and plant organic matter play vital roles in extracting and sequestering these contaminants, thereby mediating their environmental impact and fate.

As recognized, the water solubilities of organic compounds vary much more widely with their structures and compositions than do their corresponding solubilities in an organic-solvent phase. For liquid substances (i.e., solutes), the solubility in a solvent (or medium) is determined by the degree of solute–solvent compatibility. For solid substances, the solubility is also affected by the energy required to overcome the solid-to-liquid transition (called the *melting-point effect*). These features suggest immediately that both the potential level of contamination and the distribution pattern may vary widely for the various sources or types of organic compounds. To understand the solubility and partition behavior of organic compounds in natural systems, it is essential that one capture the essentials of the relevant solution theory.

## 2.2 RAOULT'S LAW

Raoult (1887, 1888) recognized that the addition of a small amount of solutes to a solvent does not radically change any extensive property of the solvent, because it changes the solvent mole fraction only slightly. On the other hand, the properties of a solute may change much more substantially as it goes from a pure substance to one in dilute solution. One would therefore expect the ideal-solution approximation to apply more widely and closely to the solvent

14

than to solutes in dilute solution. When the solute and solvent are similar substances that mix with a minimal thermal effect, the activity or partial pressure of the solute at a given mole fraction may also be approximated, usually with less accuracy, by applying the ideal-solution assumption. When the behavior of the solute deviates significantly from this ideal state, it is expressed by an activity coefficient, which is attributed to a mixing thermicity. In essence, Raoult's law gives no account of the effect of the molecular-size disparity between the components in a solution on individual component activities.

By *Raoult's law*, the partial pressure of component $i$ in solution may be expressed in the following form:

$$P_i = P_i^\circ x_i \gamma_i \quad \text{or} \quad a_i = P_i/P_i^\circ = x_i \gamma_i \tag{2.1}$$

where $P_i$ is the partial pressure of component $i$ at mole fraction $x_i$, $P_i^\circ$ the reference-state vapor pressure of pure substance $i$ at the same temperature $T$, $\gamma_i$ the activity coefficient of component $i$ at $x_i$, and $a_i$, as stated earlier, is the activity of component $i$ at $x_i$. Here the reference-state vapor pressure for a liquid or a solid substance at temperature $T$ is simply the saturation vapor pressure of the pure liquid or the supercooled liquid at $T$. For the solvent, in which $x_i$ is close to 1, the solvent $\gamma_i$ should approach 1 according to the law, as is generally found when the solute and solvent have similar molecular sizes. If the solute behaves ideally, then $\gamma_i = 1$ and hence $a_i = x_i$. For a solid solute, one can thus calculate its ideal mole fraction solubility in a solvent based on the calculated activity of the pure solid, $a_i^s = P_i^s/P_i^\circ$, where $P_i^s$ is the vapor pressure of the solid at $T$. Although the model holds when the solute and solvent have similar sizes and compositions, it does not hold, as shown later, if one of them is a macromolecular substance, even in the absence of a thermal effect.

To the extent that Raoult's law is obeyed, a graphic illustration of the behavior of a component $(i)$ in a solution is depicted in Figure 2.1, in which the partial pressure $(P_i)$ or fugacity $(f_i)$ of the component is plotted against its mole fraction $(x_i)$ in solution. It is assumed here that component $i$ is a liquid completely miscible with the solvent at system temperature $T$. We shall consider the case for a solid solute later. In Figure 2.1, the straight line between the origin and the $P_i^\circ$ in the ordinate (at $x_i = 1$) is the ideal-solution line for component $i$ (i.e., where $\gamma_i = 1$ at all $x_i$). The upper curve is for a nonideal system, where component $i$ exhibits a positive deviation from ideality (i.e., $\gamma_i > 1$). The lower curve is for another nonideal system, where the component $i$ exhibits a negative deviation from ideality (i.e., $\gamma_i < 1$). If $\gamma_i > 1$, as for most systems, the compatibility between molecule $i$ and other molecules is considered to be less than that between $i$ molecules; if $\gamma_i < 1$, the reverse is the case. The latter condition applies for rare systems where specific interactions (e.g., complexation) occur between $i$ and other components.

As depicted in Figure 2.1, whereas $\gamma_i = 1$ applies over all $x_i$ if the solution is ideal, one sees that $\gamma_i$ always approaches 1 as $x_i \to 1$ if component $i$ is completely miscible, whether the solution is ideal or not at other $x_i$. In other words,

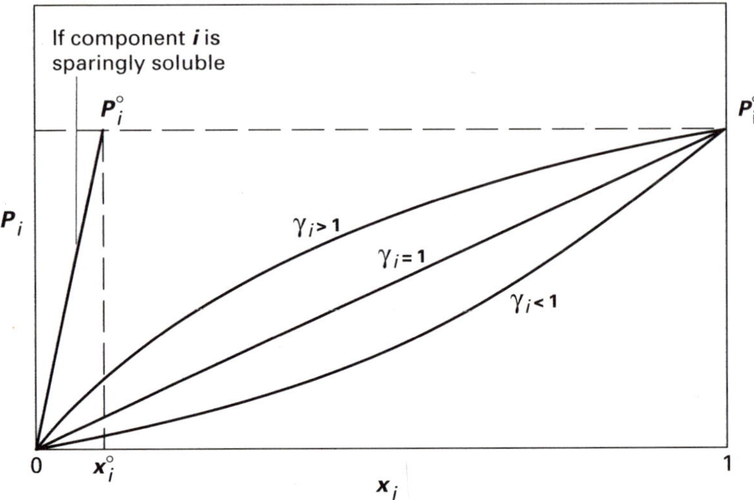


**Figure 2.1**   Relation between the partial pressure and mole fraction of a liquid solute at a system temperature according to Raoult's law.

$P_i$ approaches $P_i^o$ and $a_i$ approaches 1 as $x_i \to 1$ for a liquid substance that is completely miscible with the solvent. If a liquid is completely miscible with a solvent and exhibits a positive deviation from ideality (i.e., $\gamma_i > 1$), the $\gamma_i$ should be relatively small in magnitude (say, $5 > \gamma_i > 1$), because a higher $\gamma_i$ would force a phase separation, as with a partially miscible system. Although the $\gamma_i$ of a component can be either greater or smaller than 1 depending on the system involved, it cannot undergo a transition from greater than 1 to smaller than 1 with a change in its concentration in a given system.

If a liquid is only partially miscible with a solvent (i.e., they exhibit large mutual incompatibility), the relation between $P_i$ and $x_i$ will end at $x_i < 1$. An example is given in Figure 2.1 for a sparingly soluble liquid substance in a solvent with $x_i^o \ll 1$, where $x_i^o$ is the mole fraction solubility. In this case, $P_i$ rises rapidly and reaches its maximum at $x_i = x_i^o$. At saturation (i.e., at $x_i = x_i^o$), $P_i$ is equal to $P_i^o$, as depicted by the dashed line. In this case, by Eq. (2.1), one gets

$$x_i^o \gamma_i^o = 1 \qquad (2.2)$$

or

$$\gamma_i^o = 1/x_i^o \qquad (2.3)$$

where $\gamma_i^o$ is the activity coefficient of a sparingly soluble liquid at saturation. As seen, the $\gamma_i$ value is inversely proportional to the mole fraction solubility and has no theoretical upper boundary, since the lower limit of $x_i^o$ is zero. For small $x_i^o$ values, $\gamma_i$ would be relatively independent of its concentration (i.e.,

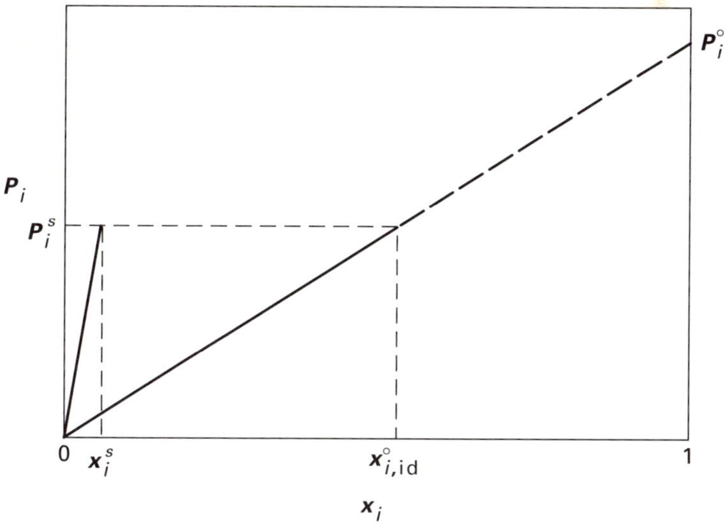



**Figure 2.2**   Relation between the partial pressure and mole fraction of a solid solute at a system temperature according to Raoult's law.

$\gamma_l \simeq \gamma_i^\circ \cong$ constant), as shown later. Thus, for liquid substances that are sparingly soluble in a solvent at $T$, reasonably accurate $\gamma_i$ values can be obtained readily from the solubility data.

We now consider the relation between $P_i$ and $x_i$ for a solid substance in solvents in which its melt exhibits unequal compatibilities with the solvents, as depicted in Figure 2.2.

For a solid substance at $T$, the $P_i$–$x_i$ relation cannot be extended to $x_i = 1$ because the activity of a solid will be always less than 1 [see Eqs. (1.37) to (1.40)]. Therefore, an excess solid phase will be formed in solid–solvent equilibria at the point of saturation. Figure 2.2 depicts two contrasting systems, one in which the solid melt forms an ideal solution with a solvent and the other in which the solution of the solid melt exhibits a large positive deviation from ideality with a different solvent. In the former case, a linear relation between $P_i$ and $x_i$, with $\gamma_i = 1$, exists between $x_i = 0$ and $x_i = x_{i,\mathrm{id}}^\circ$, where $x_{i,\mathrm{id}}^\circ$ is the ideal mole fraction solubility of the solid substance at temperature $T$. From Eq. (2.1) one obtains at saturation,

$$x_{i,\mathrm{id}}^\circ = a_i^s = P_i^s \big/ P_i^\circ \tag{2.4}$$

where $a_i^s$ is the activity of the pure solid substance and $P_i^s$ is its saturation vapor pressure at $T$. The dashed line, which extends to $x_i = 1$, is for the corresponding supercooled liquid at $T$.

If the solution is nonideal with $\gamma_i > 1$, the solid solubility (designated as $x_i^s$) will be less than its ideal solubility ($x_{i,\mathrm{id}}^\circ$), and the relation between $P_i$ and $x_i$

terminates at $x_i = x_i^s$. If $x_i^s \ll 1$, the $P_i$–$x_i$ relation is expected to be largely linear over the entire $x_i$ range; that is, $\gamma_i$ is essentially independent of the concentration. Since at the point of saturation the activity of a solid substance at $T$ is the same in any system, whether its solution is ideal or not, one gets

$$a_i^s = x_{i,\mathrm{id}}^\circ = x_i^s \gamma_i^\circ \tag{2.5}$$

and

$$\gamma_i^\circ = x_{i,\mathrm{id}}^\circ \big/ x_i^s = a_i^s \big/ x_i^s \tag{2.6}$$

As shown later, the value of $a_i^s$ of a solid substance can be calculated from its molar heat of fusion ($\Delta \overline{H}_{\mathrm{fus}}$) and melting point ($T_m$), if they are known; this allows the calculation of $\gamma_i^\circ$ (at saturation) from the measured solubility ($x_i^s$). Again, if $x_i^s$ is very small, the $\gamma_i$ value for the dissolved solid at any concentration below saturation is practically equal to $\gamma_i^\circ$.

## 2.3 HENRY'S LAW

Whereas Raoult's law applies well for a component (generally, the solvent) when its mole fractions is close to 1, Henry's law applies to components at high dilution. *Henry's law* can be expressed in a number of forms, such as

$$P_i = k_i x_i \quad \text{or} \quad a_i = k_i^* x_i \tag{2.7}$$

where $P_i$, $a_i$, and $x_i$ are as defined before and $k_i$ and $k_i^*$ are Henry's constants. The value of $k_i$ or $k_i^*$ depends on the solvent type. By reference to Raoult's law with a given solvent, one finds that

$$k_i = P_i^\circ \gamma_i^\infty \quad \text{and} \quad k_i^* = \gamma_i^\infty \tag{2.8}$$

where $\gamma_i^\infty$ is the Raoult's activity coefficient of substance $i$ at infinite dilution (i.e., $x_i \rightarrow 0$). The linear Henry's law is thus limited to $x_i \ll 1$ such that $\gamma_i$ is practically the same as $\gamma_i^\infty$. The $P_i^\circ \gamma_i^\infty$ term in Eq. (2.8) may be considered as a hypothetical vapor pressure of the pure substance according to Henry's law, which is obtained by a linear extrapolation of $P_i$ with a constant slope of $\gamma_i^\infty$ from infinite dilution to $x_i = 1$.

If the activity of a substance at low concentrations is not sufficiently linear with respect to its concentration, a Henry's law activity coefficient ($\eta_i$) is added to the right of Eq. (2.7), similar to the correction for deviation from the ideal Raoult's law:

$$P_i = k_i x_i \eta_i \quad \text{or} \quad a_i = k_i^* x_i \eta_i \tag{2.9}$$

For systems involving no specific interaction between components, $\eta_i$ usually does not deviate greatly from 1, which is analogous to the limiting case of Raoult's law that $\gamma_i$ approaches 1 as $x_i$ approaches 1.

Raoult's law and Henry's law have their respective advantages in describing solution and partition processes, depending on the system involved. If a substance of interest is either completely miscible with the solvent or has a very high solubility in the solvent, Henry's law is preferred to account for the behavior of the substance in the dilute range, and Raoult's law is undoubtedly preferred in the high concentration range; this minimizes the effort to characterize the system over the entire range of concentration. If the substance of interest has a limited solubility in the solvent, especially if the solubility is very small, both Raoult's law and Henry's law are readily applicable, but Raoult's law allows for a rapid determination of $\gamma_i$ or $\gamma_i^\infty$ from the solubility data.

In addition to the discussion above, the term *Henry's law* has been used in a broader sense to refer to any linear relation that exists for a dilute substance between the solution and a neighboring phase, regardless of whether the process involved with the other phase is a solution or a surface phenomenon.

## 2.4  FLORY–HUGGINS THEORY

Whereas Raoult's law accounts for the behavior of solutions of small molecules of comparable size, the Flory–Huggins theory provides a more accurate treatment for systems where the difference in molecular size between the components is considerable, such as for common contaminants in a polymeric or macromolecular substance. The inability of Raoult's law to deal with the latter system has to do with its adoption of the mole fraction as the weighting basis for the activity of a substance. This criterion gives a satisfactory measure of the component activity if the components are comparable in molecular size, in which case the deviation from Raoult's law as accounted for by the activity coefficient gives a reasonable assessment of the molecular incompatibility.

Small molecules behave more like rigid bodies because they have more rigid molecular segments, which prevent the molecules from assuming a large number of configurations. Large molecules, typically polymers, contain many relatively flexible repeating units or molecular segments (like a string of beads) that enable them to take on a large number of spatial orientations. For this reason, these segments may interact relatively freely with each other and with other molecular species. In this sense, a macromolecule behaves as if it consists of many independent small molecules when it interacts with ordinary small molecules. Therefore, the mole fraction concept, as adopted by Raoult's law, is not an effective measure of the component activity in a solution of a macromolecular substance. The Flory–Huggins theory (see Flory, 1953) offers a more accurate and rigorous treatment of the chemical activity of a component in a macromolecular solution in terms of its volume fraction.

The inability of Raoult's law to account for the solution behavior of a macromolecular substance is illustrated in the next example. Consider a binary-component system in which the two components (1 and 2) have similar structures and compositions (e.g., styrene and polystyrene) and their molecuclar weights are 100 and 10,000 daltons, respectively. Assume that the solution is made 10% by weight of component 1 and 90% by weight of component 2. The mole fractions calculated for components 1 and 2 are thus $x_1 = 0.917$ and $x_2 = 0.083$. Since in this case $x_1$ is close to 1, Raoult's law would suggest that $\gamma_1$ should approach 1 and thus $P_1$ should be close to $P_1^\circ$. However, the measured $P_1$ in this case is only about $0.1 P_1^\circ$ instead. To reconcile this deviation on the basis of Raoult's law, one would be forced to assume that $\gamma_1 = 0.11$ (i.e., $\gamma_1 < 1$), but this assumption cannot be justified because there is no evidence for a specific interaction between the two molecularly similar components.

Consider a binary solution consisting of a small molecule with a molar volume of $\overline{V}_1$ and a linear polymer of a molar volume of $\overline{V}_2$, in which $\overline{V}_2 >> \overline{V}_1$. By a statistical treatment of the number of spatial configurations that component 2 may assume in mixing with component 1, Flory (1941) and Huggins (1942) developed a thermodynamic expression for the free energy of mixing between components 1 and 2. The chemical activities of components 1 and 2 are given as

$$\ln a_1 = \ln \phi_1 + (1 - \overline{V}_1/\overline{V}_2)\phi_2 + \chi_1 \phi_2^2 \tag{2.10}$$

and

$$\ln a_2 = \ln \phi_2 - (\overline{V}_2/\overline{V}_1 - 1)\phi_1 + \chi_1(\overline{V}_2/\overline{V}_1)\phi_1^2 \tag{2.11}$$

where $a_i$ is the activity of component $i$, $\phi_i$ is the volume fraction (where $\phi_1 + \phi_2 = 1$), and $\chi_1$ is the Flory–Huggins interaction parameter for component 1 [i.e., the sum of its excess enthalpic ($\chi_H$) and entropic ($\chi_S$) contributions to its incompatibility with component 2]. The $\chi_H$ term accounts for the heat of mixing, similar to the $\ln \gamma$ term in Raoult's law. For systems with completely linear and flexible polymer segments, the entropy of mixing for components 1 and 2 is given by the first two terms to the right in Eqs. (2.10) and (2.11). The $\chi_S$ term corrects for the entropy loss upon mixing when the polymer suffers certain restriction on its orientation. Thus, $\chi_S$ is approximately 0 if the polymer segments are highly flexible to adopt a large number of spatial orientations. In Eq. (2.11), the $\chi_1(\overline{V}_2/\overline{V}_1)$ term may be viewed as the $\chi_2$ term for component 2. As seen, if there is no molecular-size disparity between the two components, (i.e., $\overline{V}_1 = \overline{V}_2$), Eqs. (2.10) and (2.11) are then reduced to Raoult's law, since in this case $x_1 \simeq \phi_1$, $x_2 \simeq \phi_2$, and $\chi = \ln \gamma$.

As seen later, Eq. (2.10) offers a more general account of the activity of an organic solute with natural organic matter and biological lipids, where a moderate-to-large molecular-size disparity is observed. If $\overline{V}_1/\overline{V}_2 \simeq 0$, Eq. (2.10) then becomes

$$\ln a_1 = \ln \phi_1 + \phi_2 + \chi_1 \phi_2^2 \qquad (2.12)$$

If the pure component 1 is a liquid at temperature $T$ and has a limited solubility in a high-molecular-weight polymeric or macromolecular substance (component 2), at the point of saturation Eqs. (2.10) and (2.12) become

$$\ln \phi_1^{\circ} + (1 - \overline{V}_1/\overline{V}_2)\phi_2 + \chi_1 \phi_2^2 = 0 \qquad (2.13)$$

and

$$\ln \phi_1^{\circ} + \phi_2 + \chi_1 \phi_2^2 = 0 \qquad \text{if } \overline{V}_1/\overline{V}_2 \simeq 0 \qquad (2.14)$$

where $\phi_1^{\circ} = 1 - \phi_2$ is the volume fraction solubility of the liquid at temperature $T$.

If the pure component 1 is a solid at $T$, the corresponding equations are

$$\ln \phi_1^{\circ} + (1 - \overline{V}_1/\overline{V}_2)\phi_2 + \chi_1 \phi_2^2 = \ln a_1^s \qquad (2.15)$$

and

$$\ln \phi_1^{\circ} + \phi_2 + \chi_1 \phi_2^2 = \ln a_1^s \qquad \text{if } \overline{V}_1/\overline{V}_2 \simeq 0 \qquad (2.16)$$

where $a_1^s$ is the activity of pure component 1 as a solid at temperature $T$, as defined before. We shall later make use of Eqs. (2.10) to (2.16) to account for the solubility and partition behaviors of organic compounds with some macromolecular natural organic substances, including biological lipids that are only moderately large in molecular size.

## 2.5   VARIATION OF ACTIVITY COEFFICIENT WITH CONCENTRATION

For a nonideal solution, as noted, the activity coefficient of a substance ($\gamma_i$) is a function of its concentration ($x_i$). The relation between $\gamma_i$ and $x_i$ for a binary-component solution was derived by van Laar (1910, 1913) and extended by Carlson and Colburn (1942):

$$\log \gamma_1 = A \big/ \left(1 + Ax_1/Bx_2\right)^2 \qquad (2.17)$$

and

$$\log \gamma_2 = B \big/ \left(1 + Bx_2/Ax_1\right)^2 \qquad (2.18)$$

where $\gamma_1$ and $\gamma_2$ are the activity coefficients of components 1 and 2 at mole fractions $x_1$ and $x_2$ and $A$ and $B$ are defined as

$$A = \log \gamma_1^\infty \qquad (\text{at } x_1 \to 0) \qquad\qquad (2.19)$$

and

$$B = \log \gamma_2^\infty \qquad (\text{at } x_2 \to 0) \qquad\qquad (2.20)$$

Thus, in the limit of $x_2 \to 0$ (i.e., $x_1 \to 1$),

$$\log \gamma_2 \to \log \gamma_2^\infty \quad \text{and} \quad \log \gamma_1 \to 0 \quad (\text{i.e., } \gamma_1 \to 1) \qquad (2.21)$$

Similarly, as $x_1 \to 0$ (i.e., $x_2 \to 1$),

$$\log \gamma_1 \to \log \gamma_1^\infty \quad \text{and} \quad \log \gamma_2 \to 0 \quad (\text{i.e., } \gamma_2 \to 1) \qquad (2.22)$$

Equation (2.21) or (2.22) is simply Raoult's law, which must be satisfied. If $x_2$ is small ($\ll 1$),

$$\log \gamma_2 \simeq B \simeq \log \gamma_2^\infty \simeq \text{constant} \qquad\qquad (2.23)$$

Therefore, for a substance at dilution, Henry's law [i.e., Eq. (2.7)] is also satisfied. If the $x_i$ of a substance is small in a solution, the $\log \gamma_i$ varies with $(1 + x_i)^{-2}$ according to Eqs. (2.17) and (2.18). In this case, the variation of $\gamma_i$ with $x_i$ will not be substantial. The van Laar equations are best suited for systems that exhibit positive deviations from Raoult's law. Whereas $\log \gamma_1$ and $\log \gamma_2$ vary with the solution composition, their values cannot change in sign with composition (i.e., the numerical value of $\gamma_1$ or $\gamma_2$ cannot undergo a transition from above 1 to less than 1).

As noted with the Flory–Huggins theory, the relations above hold mainly for systems where components 1 and 2 have comparable molecular sizes. For systems where the size disparity is large, the $\log \gamma$ term should be replaced by $\chi/2.303$ and the mole fraction ($x$) by the volume fraction ($\phi$) in the van Laar equations. However, the conclusion regarding the variability of $\chi$ with concentration would remain the same as that of $\log \gamma$ with concentration.

## 2.6  MOLAR HEAT OF SOLUTION

For solid and liquid solutes that have limited solubility in a given solvent, the partial molar heat of solution for the solute may be calculated from the temperature dependence of the solute solubility. To derive this property, we begin with the partial molar free energy (i.e., the chemical potential) of the solute in solution with respect to some standard state at constant pressure ($P$) and temperature ($T$). One recalls from Eq. (1.27) that when a substance with two or more phases is brought to equilibrium, the chemical potentials of the substance in all phases must be the same. Therefore, for solute $i$ at equilibrium between the vapor and solution phases,

$$\mu_i(\text{sol}) = \mu_i(\text{vap}) \tag{2.24}$$

where $\mu_i(\text{sol})$ is the chemical potential of solute $i$ in solution and $\mu_i(\text{vap})$ is the chemical potential in the vapor phase. By Eq. (1.35), one arrives at

$$\mu_i(\text{vap}) = \mu_i^\circ + RT \ln P_i \tag{2.25}$$

where $\mu_i^\circ$ is the chemical potential of the vapor $i$ at 1 atm and $T$.

If we choose Raoult's law to express the dissolved solute activity and assume that the solute is a liquid at $T$, then $P_i = P_i^\circ x_i \gamma_i$, where $P_i^\circ$ is the vapor pressure of pure liquid (or supercooled liquid) $i$. Thus, for liquid or supercooled-liquid solutes at temperature $T$, Eq. (2.25) can be written as

$$\mu_i(\text{vap}) = \mu_i^\circ + RT \ln P_i^\circ + RT \ln x_i \gamma_i \tag{2.26}$$

or

$$\mu_i(\text{vap}) = \mu_i^*(\text{liq}) + RT \ln x_i \gamma_i \tag{2.27}$$

where $\mu_i^*(\text{liq})$ is recognized as the chemical potential of pure liquid $i$ at $T$. Hence, the chemical potential of solute $i$ in solution can be related to its concentration as

$$\mu_i = \mu_i(\text{sol}) = \mu_i^*(\text{liq}) + RT \ln x_i \gamma_i \tag{2.28}$$

or alternatively,

$$\mu_i = \mu_i^*(\text{liq}) + RT \ln a_i = \mu_i^*(\text{liq}) + RT \ln(P_i / P_i^\circ) \tag{2.29}$$

In light of the fact that the supercooled-liquid state of a solid substance is metastable, it is also desirable to express the chemical potential of the dissolved solid with the pure solid as the reference state. The relation between the chemical potentials of the solid, $\mu_i^*(\text{sld})$, and its supercooled liquid, $\mu_i^*(\text{liq})$, at temperature $T$ is given as

$$\mu_i^*(\text{liq}) = \mu_i^*(\text{sld}) + RT \ln(P_i^\circ / P_i^s) \tag{2.30}$$

In Eq. (2.30), since $P_i^\circ > P_i^s$ one finds that $\mu_i^*(\text{liq}) > \mu_i^*(\text{sld})$, as the supercooled-liquid state is unstable. Here $\mu_i^*(\text{liq}) - \mu_i^*(\text{sld}) = \Delta G_{i(\text{fus})}$ is called the molar free energy of fusion of the solid at $T$. Substitution of Eq. (2.30) into (2.28) gives an alternative expression for a dissolved solid solute as

$$\mu_i = \mu_i^*(\text{sld}) + RT \ln[(x_i \gamma_i)(P_i^\circ / P_i^s)] = \mu_i^*(\text{sld}) + RT \ln(P_i / P_i^s) \tag{2.31}$$

The partial molar free-energy change for converting 1 mole of pure liquid $i$ into its solution having a mole fraction concentration of $x_i$ is therefore

$$\Delta \overline{G}_{i(sol)} = \mu_i - \mu_i^*(liq) = RT \ln a_i = RT \ln x_i \gamma_I \qquad (2.32)$$

A similar expression is obtained for converting 1 mole of pure solid $i$ into its solution at $x_i$:

$$\Delta \overline{G}_{i(sol)} = \mu_i - \mu_i^*(sld) = RT \ln[(x_i \gamma_i)(P_i^\circ / P_i^s)] \qquad (2.33)$$

The partial molar entropic change for the solution of solute $i$ (whether solid or liquid) may then be obtained through Eq. (1.22) from the derivative of $\Delta \overline{G}_{i(sol)}$ with $T$ at constant external pressure ($P$) and solution composition ($x_i$):

$$\Delta \overline{S}_{i(sol)} = -\left[\frac{\partial(\Delta \overline{G}_{i(sol)})}{\partial T}\right]_{P,x_i} = \left[\frac{\partial(\Delta \overline{G}_{i(sol)})}{\partial \ln x_i}\right]_{P,T}\left[\frac{\partial \ln x_i}{\partial T}\right]_{\Delta \overline{G}_i,P} \qquad (2.34)$$

If the excess solid or liquid solute in contact with its solution is essentially pure (i.e., if there is no significant amount of the dissolved solvent), the first term on the right of Eq. (2.34) may be evaluated by reference to Eq. (2.32) or (2.33) as

$$\left[\frac{\partial(\Delta \overline{G}_{i(sol)})}{\partial \ln x_i}\right]_{P,T} = RT\left[\frac{\partial \ln(x_i \gamma_i)}{\partial \ln x_i}\right]_{P,T} \qquad (2.35)$$

If the solute has a limited solubility in the solvent, the derivative in Eq. (2.35) is essentially 1 because the value of $\gamma_i$ is largely independent of $x_i$. Thus Eq. (2.34) is reduced to

$$\Delta \overline{S}_{i(sol)} = RT\left[\frac{\partial \ln x_i}{\partial T}\right]_{\Delta \overline{G}_i,P} \qquad (2.36)$$

For solutes having a limited solubility in the solvent, the solution process ceases at the point of equilibrium as the solute concentration reaches saturation (i.e., at $x_i = x_i^\circ$). At this point the chemical potential of the solute in solution equals that in the excess-solute phase (i.e., $\Delta \overline{G}_{i(sol)} = 0$), such that $\Delta \overline{H}_{i(sol)} = T \Delta \overline{S}_{i(sol)}$. The molar heat of solution at the point of equilibrium is therefore

$$\Delta \overline{H}_{i(sol)} = RT^2\left[\frac{\partial \ln x_i^\circ}{\partial T}\right]_P \qquad (2.37)$$

If $\Delta \overline{H}_{i(sol)}$ is relatively invariant over a range of temperature, the integration of Eq. (2.37) yields

$$\ln x_i^\circ = -\Delta \overline{H}_{i(\text{sol})} \big/ RT + \text{constant} \qquad (2.38)$$

Equation (2.37) or (2.38) is often referred to as the *van't Hoff equation* and is used extensively to obtain the molar heat of solution of a solute in a partially miscible solvent from a plot of $\ln x_i^\circ$ versus $1/T$, which gives a slope of $-\Delta \overline{H}_{i(\text{sol})}/R$. In dilute systems, the overall heat of solution results mainly from the solute heat of solution. If the $x_i^\circ$ term in Eq. (2.37) is small, it can be replaced by more convenient alternative forms (e.g., by molar concentration or weight percent). Although Eq. (2.38) is derived with the application of Raoult's law for solute activity in solution, in which the Flory–Huggins model provides a more accurate account of the solute activity with certain solvents as mentioned, the equation remains valid as long as the solute of interest exhibits a limited solubility in mass or volume fraction in the solvent.

As noted with Eq. (2.3) for a liquid solute with a small solubility, where $x_i^\circ \simeq 1/\gamma_i$, the $\Delta \overline{H}_{i(\text{sol})}$ term represents the "excess heat" required to disperse a mole of the liquid solute into solution. For a solid solute with a small solubility [see Eq. (2.5) with $x_i^\circ = x_i^s$], where $x_i^\circ \simeq a_i^s/\gamma_i$, the $\Delta \overline{H}_{i(\text{sol})}$ term is the sum of the molar excess heat of solution of the supercooled-liquid solute and the molar heat of fusion ($\Delta \overline{H}_{i(\text{fus})}$) of the solid. This accounts for the fact that the molar heat of solution for a solid solute is generally greater than that for a liquid solute if they have comparable structures and sizes (e.g., solid *p*-dichlorobenzene versus liquid *o*-dichlorobenzene at room temperature). Although the molar heat of solution of a solute depends strongly on solute–solvent polarities, it is generally less than the corresponding heat of vaporization because the van der Waals forces of attraction between solute and solvent offset part of the energy needed to break apart solute molecules. The heat of solution of a solute with a solvent serves as a useful reference to be compared with the heat effects associated with the transfer of the solute from that solvent into other phases of the system where the solute may be taken up by either surface adsorption or phase partition.

Since water is probably the most important medium for contaminant transfer to other natural phases, it is of considerable interest to determine the heats of solution in water of contaminants from their water solubility–temperature relations. For example, by Eq. (2.38), Friesen and Webster (1990) determined the heats of solution in water ($\Delta \overline{H}_w$) of 1,2,3,7-tetrachlorodibenzo-*p*-dioxin (T₄CDD), 1,2,3,4,7-pentachlorodibenzo-*p*-dioxin (P₅CDD), 1,2,3,4,7,8-hexachlorodibenzo-*p*-dioxin (H₆CDD), and 1,2,3,4,6,7,8-heptachlorodibenzo-*p*-dioxin (H₇CDD) from their measured water solubilities in the temperature range 7 to 41°C, where all the compounds exist as solids. The water solubility data are presented in Table 2.1, and a plot of $\ln x^\circ$ versus $1/T$ is shown in Figure 2.3. The van't Hoff plot yields virtually straight lines, meaning that the $\Delta \overline{H}_w$ values of the four solid compounds are relatively temperature independent over this temperature range. The $\Delta \overline{H}_w$ values calculated for T₄CDD, P₅CDD, H₆CDD, and H₇CDD are 39.8, 47.5, 45.5, and 42.2 kJ/mol, respectively. If the van't Hoff plot does not produce a straight line, the $\Delta \overline{H}_w$ value at a

**TABLE 2.1. Solubilities in Water of 1,2,3,7-Tetrachlorodibenzo-*p*-dioxin (T$_4$CDD), 1,2,3,4,7-Pentachlorodibenzo-*p*-dioxin (P$_5$CDD), 1,2,3,4,7,8-Hexachlorodibenzo-*p*-dioxin (H$_6$CDD), and 1,2,3,4,6,7,8-Heptachlorodibenzo-*p*-dioxin (H$_7$CDD) as a Function of Temperature**

| Temp. (°C) | Solubility in Water, $S_w$ (mol/L) | | | |
|---|---|---|---|---|
| | T$_4$CDD | P$_5$CDD | H$_6$CDD | H$_7$CDD |
| 7.0 | $(7.56 \pm 0.20) \times 10^{-10}$ | $(1.42 \pm 0.01) \times 10^{-10}$ | $(5.91 \pm 0.05) \times 10^{-12}$ | $(2.20 \pm 0.09) \times 10^{-12}$ |
| 11.5 | $(8.12 \pm 0.11) \times 10^{-10}$ | $(1.88 \pm 0.01) \times 10^{-10}$ | $(7.98 \pm 0.15) \times 10^{-12}$ | $(2.69 \pm 0.01) \times 10^{-12}$ |
| 17.0 | $(12.5 \pm 3.6) \times 10^{-10}$ | $(2.44 \pm 0.01) \times 10^{-10}$ | $(10.7 \pm 0.4) \times 10^{-12}$ | $(3.04 \pm 0.06) \times 10^{-12}$ |
| 21.0 | $(14.9 \pm 2.1) \times 10^{-10}$ | $(3.45 \pm 0.08) \times 10^{-10}$ | $(12.5 \pm 1.2) \times 10^{-12}$ | $(5.40 \pm 0.77) \times 10^{-12}$ |
| 26.0 | $(22.6 \pm 1.0) \times 10^{-10}$ | $(4.63 \pm 003) \times 10^{-10}$ | $(20.2 \pm 0.4) \times 10^{-12}$ | $(6.03 \pm 0.18) \times 10^{-12}$ |
| 41.0 | $(43.3 \pm 5.4) \times 10^{-10}$ | $(12.8 \pm 0.1) \times 10^{-10}$ | $(48.6 \pm 1.4) \times 10^{-12}$ | $(14.9 \pm 0.5) \times 10^{-12}$ |

*Source*: Data from Friesen and Webster (1990).

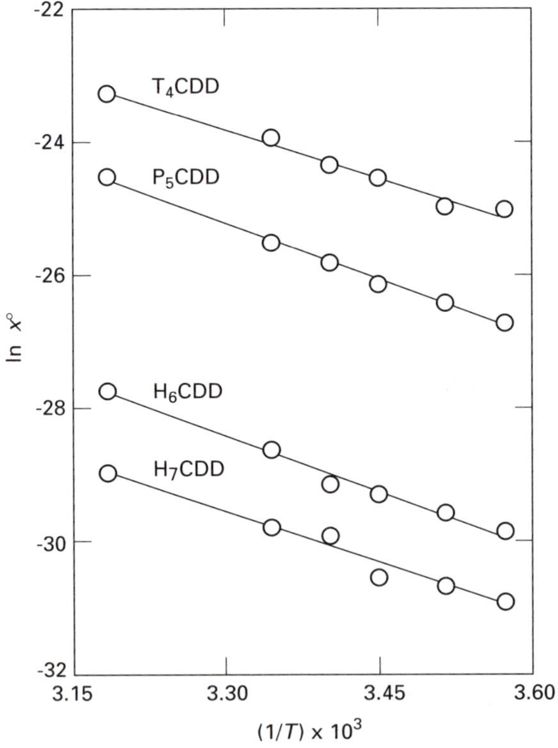


**Figure 2.3** Temperature dependence of the water solubilities of four polychlorinated dibenzo-*p*-dioxins. [Data from Friesen and Webster (1990). Reproduced with permission.]

temperature of interest is obtained from the derivative of $\ln x^\circ$ with $1/T$ at that particular temperature. Heats of solution are discussed in more detail in Chapter 3.

Let us consider the associated molar entropies of solution in water $(\Delta \bar{S}_w)$ of the chlorinated dibenzo-$p$-dioxins in the example above. It is important to note that whereas the $\Delta \bar{H}_w$, or $\Delta \bar{H}_{i(\text{sol})}$ in general, for a liquid or solid substance that exhibits a limited solubility in water (or a solvent) is practically independent of the solute concentration $(x)$, the $\Delta \bar{S}_w$, or $\Delta \bar{S}_{i(\text{sol})}$, varies with $x$ on its path toward saturation $(x^\circ)$ [see Eq. (2.36)]. However, at $x = x^\circ$, $\Delta \bar{G}_w$ is zero according to Eq. (2.33) (where $x^\circ \gamma^\circ = P^s/P^\circ$), and one obtains under this condition $\Delta \bar{S}_w = \Delta \bar{H}_w/T$ over the temperature range studied (Chiou and Manes, 1990). For example, the $\Delta \bar{S}_w$ values calculated at $T = 299\,\text{K}$ for solid $T_4CDD$, $P_5CDD$, $H_6CDD$, and $H_7CDD$ at the points of solid–water equilibria are 113, 159, 152, and 141 J/K·mol, respectively; the $\Delta \bar{S}_w$ values at other temperatures can be calculated similarly. The finding that both $\Delta \bar{H}_w$ and $\Delta \bar{S}_w$ values are positive is much expected, since the solubilization of nonionic organic solutes in water or a solvent is favored by the entropic effect and commonly disfavored by the enthalpic effect.


## 2.7  COHESIVE ENERGY DENSITY AND SOLUBILITY PARAMETER

The sum of the various attractive forces that hold the molecules of a substance in a liquid or solid state is called the *cohesive energy*. The magnitude of this energy is not only a function of the molecular makeup but also of the molecular size. Types of cohesive forces that operate in uncharged liquids and solids include the *induced dipole–induced dipole force* (also called the *London force*), the *dipole–dipole force* (the *Debye force*), the *dipole–induced dipole force* (the *Keesom force*), and the *H-bonding force*. With the possible exception of the H-bonding force, these molecular forces are frequently lumped together as the *van der Waals forces*. The London force, also referred to as the *dispersion force*, originates from the momentary distortion of electrons around nuclei and is thus operative in all molecules. This molecular force is temperature independent and is the sole attractive force for nonpolar substances. On the other hand, the involvement of dipolar and H-bonding forces for a substance requires the presence of polar and H-bonding groups in its molecular structure. The energy of evaporation per unit volume of a liquid or a supercooled liquid, called the *cohesive energy density*, is a critical parameter in determining its compatibility with other liquid species. The *cohesive energy density* (CED) is defined as

$$\text{CED} = \Delta \bar{E}_{\text{int}}/\bar{V} \tag{2.39}$$

where $\Delta \bar{E}_{\text{int}}$ is the internal energy per mole of the liquid and $\bar{V}$ is the molar volume of the liquid at a given system temperature. Thus CED has units of

**TABLE 2.2. Solubility Parameters for Selected Compounds at Room Temperature**

| Compound | $\delta$ (cal/cm$^3$)$^{0.5}$ | Compound | $\delta$ (cal/cm$^3$)$^{0.5}$ |
|---|---|---|---|
| Aliphatic hydrocarbons | | ethylene dibromide | 9.7 |
| *n*-pentane | 7.1 | trichloroethylene | 9.2 |
| *n*-hexane | 7.3 | tetrachloroethylene | 9.3 |
| *n*-heptane | 7.4 | chlorobenzene | 9.5 |
| *n*-octane | 7.5 | bromobenzene | 9.9 |
| cyclopentane | 8.1 | *o*-dichlorobenzene | 10.0 |
| cyclohexane | 8.2 | Alcohols | |
| Aromatic hydrocarbons | | methanol | 14.5 |
| benzene | 9.2 | ethanol | 12.7 |
| toluene | 8.9 | *n*-propanol | 11.9 |
| ethylbenzene | 8.8 | *n*-butanol | 11.4 |
| *o*-xylene | 9.0 | benzyl alcohol | 12.1 |
| *m*-xylene | 8.8 | cyclohexanol | 11.4 |
| *p*-xylene | 8.8 | *n*-octanol | 10.3 |
| *n*-propylbenzene | 8.6 | ethylene glycol | 14.6 |
| styrene | 9.3 | glycerol | 16.5 |
| naphthalene | 9.9 | Ketones | |
| phenanthrene | 9.8 | acetone | 9.9 |
| anthracene | 9.9 | methyl ethyl ketone | 9.3 |
| Halogenated carbons | | acetophenone | 10.6 |
| methylene dichloride | 9.7 | Nitrogen compounds | |
| ethylene dichloride | 9.8 | aniline | 10.3 |
| chloroform | 9.3 | pyridine | 10.7 |
| carbon tetrachloride | 8.6 | quinoline | 10.8 |
| 1,1,1-trichloroethane | 8.5 | | |

*Source*: Data from compilations of Hildebrand et al. (1970) and Barton (1975).

energy/volume and is commonly expressed in cal/cm$^3$. $\Delta \overline{E}_{int}$ is related to the molar enthalpy of the liquid, which is numerically equal to the molar heat of evaporation of the liquid ($\Delta \overline{H}_{evap}$), such that

$$\Delta \overline{E}_{int} = \Delta \overline{H}_{evap} - RT \tag{2.40}$$

where $R$ is the gas constant and $T$ is the system temperature. Since at room temperature the $RT$ term is usually small relative to $\Delta \overline{H}_{evap}$, except for liquids of very small molecular sizes, $\Delta \overline{E}_{int}$ is approximately equal to $\Delta \overline{H}_{evap}$. In theory, for any two liquids to be miscible or sufficiently compatible with each other in forming a solution, their CEDs must be close to each other. Conversely, if the two liquids differ markedly in their CEDs, the solution as formed will then deviate considerably from being ideal (or athermal).

A related parameter, $\delta$, which is the square root of CED, has been widely used to account for the compatibility of a liquid (or a supercooled liquid) with others. The $\delta$ parameter, developed initially by Hildebrand (Hildebrand and Scott, 1964), is termed the *solubility parameter* of a substance:

$$\delta = (\text{CED})^{1/2} \tag{2.41}$$

The $\delta$ values at room temperature for some hydrocarbons, halogenated hydrocarbons, and compounds with polar groups are presented in Table 2.2. As noted in Table 2.2, the $\delta$ values for aliphatic hydrocarbons (e.g., $n$-pentane to $n$-octane) are comparable in their magnitudes but are significantly smaller than for aromatic hydrocarbons (e.g., benzene, naphthalene, and phenanthrene). This is because aromatic compounds with labile $\pi$ electrons are more polarizable, thus promoting the molecular attraction by London forces. Meanwhile, the $\delta$ values for small polar liquids, such as alcohols, ketones, and nitrogen-containing aromatics, are considerably higher than for aliphatic liquids, because the polar or H-bonding force adds to the London force, the net effect being more pronounced for small molecules than for large molecules. It is also worth noting that aliphatic hydrocarbons substituted with halogens (except F) show a significant increase in $\delta$. This may be reasoned on the basis that the large halogen atoms (e.g., Cl and Br) of the substituted molecules contain many labile outer-shell electrons, making the compounds more easily polarizable as well as enabling them to form dipole moments if their electron clouds become unevenly oriented.

Strictly speaking, the concept of solubility parameters as a criterion for compatibilities of two components in a solution is followed strictly only when the same molecular forces are operative for two components. It thus works well either among nonpolar liquids or among those polar liquids with the same or similar polar functional groups that respond with the same principal molecular forces. Finally, for macromolecules or polymers, it is difficult to determine CED or $\delta$ directly from Eq. (2.40) or (2.41) because of their nondetectable vapor pressures. In this case, the CED or $\delta$ values are usually estimated from their solution properties with suitable solvents. As many polymers or macromolecules often possess large polar and apolar domains, the estimated $\delta$ values depend strongly on the polarity of the solvent used (Barton, 1975). Therefore, the $\delta$ values reported for polymers may fall into a range rather than being discrete values.

# 3 Interphase Partition Equations

## 3.1 PARTITION BETWEEN TWO SEPARATE PHASES

In Chapter 2 we have conveniently expressed the chemical potential of a component in solution at temperature $T$ and constant pressure in terms of its concentration and its pure-liquid or supercooled-liquid reference chemical potential at $T$. To establish the equilibrium partition coefficient of an organic solute (contaminant) between any two separable solvent phases, one equates the chemical potential of the solute in one phase with that in the other. Let us designate the two separate phases of interest as A and B. By Eq. (1.41), the equality in chemical potential of the solute in phases A and B requires that the solute activities in the two phases be identical at equilibrium (i.e., $a_{i,A} = a_{i,B}$), or

$$(x_i \gamma_i)_A = (x_i \gamma_i)_B \tag{3.1}$$

where $x_i$ and $\gamma_i$ are as defined earlier. Thus, the partition coefficient of solute $i$ on the basis of its mole fractions in phases A and B is then

$$K_{i,AB}^* = x_{i,A} / x_{i,B} = \gamma_{i,B} / \gamma_{i,A} \tag{3.2}$$

To the extent that $\gamma_{i,A}$ and $\gamma_{i,B}$ may vary with $x_{i,A}$ and $x_{i,B}$, respectively, $K_{i,AB}^*$ may then vary with $x_{i,A}$ and $x_{i,B}$. If the solute is present at low concentrations in both phases A and B, as is commonly the case, $K_{i,AB}^*$ will be practically invariant because $\gamma_{i,B}$ and $\gamma_{i,A}$ should be essentially constant.

The partition coefficient of a solute is expressed more frequently as the ratio of the solute molar concentrations rather than the respective mole fractions in the two phases involved, because the former can be measured more readily and finds more practical utility. If the solute of interest is dilute, the solute mole fraction and the solute molar concentration are linearly related to each other such that

$$x_{i,A} = C_{i,A} \overline{V}_A \quad \text{and} \quad x_{i,B} = C_{i,B} \overline{V}_B \tag{3.3}$$

where $C_{i,A}$ is the molar concentration of solute $i$ in phase A (mol/L), $C_{i,B}$ the molar concentration of solute $i$ in phase B, $\overline{V}_A$ the molar volume of the phase A solvent (L/mol), and $\overline{V}_B$ the molar volume of the phase B solvent. Substituting Eq. (3.3) into (3.2) gives

$$K_{i,AB} = C_{i,A}/C_{i,B} = (\gamma_{i,B}/\gamma_{i,A})(\overline{V}_B/\overline{V}_A) \qquad (3.4)$$

In natural aquatic systems, contaminants are usually present at subsaturated levels in water, and thus one is largely interested in the partition coefficients of contaminants at low concentrations between an organic phase and water.

The expressions for $K_{i,AB}^*$ and $K_{i,AB}$ in Eqs. (3.2) and (3.4) have so far been simplified with the assumption that the two solvent phases between which the solute partitions (i.e., dissolves) are completely immiscible to each other. Thus, the solubility behavior of the solute with the two separable phases is assumed to be the same as that with the two pure solvents. Although this assumption holds as a good approximation for a number of systems, such as mixtures of water and a highly water-insoluble aliphatic hydrocarbon, it is not practical for many systems in which the two solvent phases are mutually soluble to a significant extent. A more general expression for the solute partition coefficient should be written as

$$K_{i,AB} = C_{i,A}^*/C_{i,B}^* = (\gamma_{i,B}^*/\gamma_{i,A}^*)(\overline{V}_B^*/\overline{V}_A^*) \qquad (3.5)$$

where the associated terms are labeled with a superscript asterisk to take into account the change of the property of one solvent by the saturated amount of the other solvent as a result of their mutual saturation. Thus, $C_{i,A}^*$ is the concentration of the solute in solvent A–rich phase, which contains a saturated amount of solvent B, $\gamma_{i,A}^*$ is the activity coefficient of the solute at a given concentration in solvent A–rich phase with a saturated amount of solvent B, and $\overline{V}_A^*$ is the molar volume of the solvent A–rich phase. In general, if the solvent–solvent (or phase–phase) mutual saturation is not substantial, the change in molar volume is usually less significant than the change in solute activity coefficient. Further, the solvent mutual-saturation effect on solute activity coefficient varies with the extent of solute solubility, while the effect on the solvent molar volume is identical for all solutes. As expected, if the solvent–solvent mutual saturation effect is insignificant, Eq. (3.5) simplifies to Eq. (3.4). It must be kept in mind, however, that the derivations of Eqs. (3.1) to (3.5) are theoretically rigorous only to the extent that the solute solubilities in the two solvents are well represented by Raoult's law: namely, that there is no large molecular-size disparity between the solute and the solvent of interest.


## 3.2   PARTITION BETWEEN AN ORGANIC SOLVENT AND WATER

Since we are commonly interested in the partition behavior of organic solutes between water and partially water-miscible organic solvents, we consider first Eq. (3.5) for solutes with a solvent–water mixture in terms of the associated parameters. In the text that follows, the subscript $s$ is used to refer to quantities associated with the organic-solvent phase and the subscript $w$ to quantities associated with the water phase. By Eq. 3.5 and omitting the subscript $i$, one finds that

$$\log K_{sw} = \log \gamma_w^* - \log \gamma_s^* + \log \overline{V}_w^* - \log \overline{V}_s^* \tag{3.6}$$

If the solute has a low solubility in water, one obtains from Eqs. (2.2), (2.5), and (3.3) the molar solubility of the solute in water ($S_w$) as

$$S_w^{(l)} = 1/\gamma_w \overline{V}_w \qquad \text{for liquid solutes} \tag{3.7}$$

and

$$S_w^{(s)} = a^s/\gamma_w \overline{V}_w \qquad \text{for solid solutes} \tag{3.8}$$

where $a^s$ is the activity of the pure solid substance at $T$ (i.e., $a^s = P^s/P^\circ$) as defined by Eq. (2.4). From Eqs. (3.7) and (3.8), the supercooled liquid solubility of a solid, $S_w^{(l)}$, is related to the solid solubility $S_w^{(s)}$ as follows:

$$S_w^{(l)} = S_w^{(s)}/a^s = S_w^{(s)}(P^\circ/P^s) \tag{3.9}$$

Substituting Eq. (3.7) into (3.6), one obtains for liquid (or supercooled-liquid) solutes,

$$\log K_{sw} = -\log S_w - \log \overline{V}_s^* - \log \gamma_s^* - \log(\gamma_w/\gamma_w^*) + \log(\overline{V}_w^*/\overline{V}_w) \tag{3.10}$$

In Eq. (3.10), the $S_w$ value for a solid solute is that of the supercooled liquid, as determined according to Eq. (3.9). The melting-point effect that affects the solid solubility in a single phase (e.g., water) does not affect the partition coefficient ($K_{sw}$) because the effect cancels out in solute partition between any two separable phases. The $\gamma_w/\gamma_w^*$ term corrects for the effect of the dissolved organic solvent in water on the solute water solubility. The value of $\gamma_w/\gamma_w^*$ is usually greater than 1 and is called the *solubility enhancement factor*. In solvent–water systems, where the solvent has a limited solubility in water, the change in the molar volume of water due to solvent saturation is not substantial (i.e., $\overline{V}_w^*/\overline{V}_w \simeq 1$). Under this condition, Eq. (3.10) is further reduced to

$$\log K_{sw} = -\log S_w - \log \overline{V}_s^* - \log \gamma_s^* - \log(\gamma_w/\gamma_w^*) \tag{3.11}$$

## 3.3   PARTITION BETWEEN A MACROMOLECULAR PHASE AND WATER

The solute partition coefficient at dilution between an amorphous polymeric or macromolecular organic substance and water ($K_{pw}$) cannot be represented by Eq. (3.6) or (3.11). This is because the solubility of common organic solutes in a macromolecular phase, as expressed by Eqs. (2.13) and (2.15), is underestimated by Raoult's law. The more rigorous expression for $K_{pw}$ is obtained by equating the activity of the solute in water [Eq. (2.1)] with that in the polymeric organic phase [Eq. (2.10)]. With a dilute-solution approximation [Eq. (3.3)] for the solute in water, this gives

$$\log K_{pw} = \log \gamma_w - \log \overline{V} + \log \overline{V}_w^* - [(1 - \overline{V}/\overline{V}_p) + \chi]/\,2.303 - \log(\gamma_w/\gamma_w^*)$$

$$(3.12)$$

where $\overline{V}$ is the molar volume of the solute and $\overline{V}_p$ is the molar volume of the polymeric or macromolecular substance. Substituting Eq. (3.7) into (3.12) with $\overline{V}_w^* = \overline{V}_w$, one obtains for liquid (or supercooled-liquid) solutes,

$$\log K_{pw} = -\log S_w \overline{V} - [(1 - \overline{V}/\overline{V}_p) + \chi]/2.303 - \log(\gamma_w/\gamma_w^*) \qquad (3.13)$$

and, if $\overline{V}/\overline{V}_p \simeq 0$,

$$\log K_{pw} = -\log S_w \overline{V} - (1 + \chi)/2.303 - \log(\gamma_w/\gamma_w^*) \qquad (3.14)$$

It is often more convenient to express the concentration of the solute in a polymeric or macromolecular phase on a mass-to-mass basis when the condition $\overline{V}/\overline{V}_p \simeq 0$ suffices. The $K_{pw}$ in Eq. (3.14), expressed as a mass-to-mass concentration ratio, is therefore

$$\log K_{pw} = -\log S_w \overline{V} - \log \rho - (1 + \chi)/2.303 - \log(\gamma_w/\gamma_w^*) \qquad (3.15)$$

where $\rho$ is the density (g/mL) of the macromolecular organic phase and the density of the water phase is assumed to be 1. The superiority of Eq. (3.14) or (3.15) to (3.11) for the partition of liquid and supercooled-liquid solutes from water into a macromolecular phase is illustrated later.

## 3.4　TEMPERATURE DEPENDENCE OF PARTITION COEFFICIENT

Temperature affects the partition coefficient of an organic solute between two separated solvent phases. Since the partition coefficient of a solute ($K$) is a function of its relative solubilities in the two solvents, the dependence of $K$ on temperature reflects the net temperature effect of the solute solubilities with the solvents.

One may start with Eq. (3.11) for a solute at dilution in a solvent–water mixture. The solute is considered to have a limited solubility in water. Taking a derivative of $\log K_{sw}$ with $T$ gives

$$\frac{d \log K_{sw}}{dT} = -\frac{d \log S_w}{dT} - \frac{d \log \overline{V}_s^*}{dT} - \frac{d \log \gamma_s^*}{dT} - \frac{d \log (\gamma_w/\gamma_w^*)}{dT} \qquad (3.16)$$

For simplicity, the second and fourth terms on the right of Eq. (3.16) may be approximated as zero because the molar volume of a liquid (solvent) is not very sensitive to $T$ and because the solubility enhancement factor ($\gamma_w/\gamma_w^*$) will not vary strongly with $T$ if the solvent has a small solubility in water.

We consider first the effect of $T$ on $\log S_w$. For a liquid or a supercooled-liquid solute at $T$, one finds according to Eq. (3.7) that

$$\frac{d\log S_w}{dT} = -\frac{d\log\gamma_w}{dT} = \frac{\Delta\overline{H}_w^{\,ex}}{2.303RT^2} \tag{3.17}$$

in which $\Delta\overline{H}_w^{\,ex}$ is called by Hildebrand et al. (1970) the molar *excess heat* of mixing for the solute with a solvent (water). (Note that in an ideal solution, where $\gamma = 1$, $\Delta\overline{H}^{\,ex}$ is therefore zero.)

Although the $S_w$ term for a solid solute in partition equilibrium is that of the corresponding supercooled liquid, it is of interest to illustrate the different temperature dependence of the solid solubility with temperature. By Eq. (3.8) with $a^s = P^s/P^\circ$, one obtains

$$\frac{d\log S_w^{(s)}}{dT} = -\frac{d\log\gamma_w}{dT} + \frac{d\log(P^s/P^\circ)}{dT} = \frac{\Delta\overline{H}_w^{\,ex}}{2.303RT^2} + \frac{\Delta\overline{H}_{fus}}{2.303RT^2} \tag{3.18}$$

in which $\Delta\overline{H}_{fus} = \Delta\overline{H}_{sub} - \Delta\overline{H}_{evap}$ according to Eq. (1.52). Thus, the molar heat of solution ($\Delta\overline{H}_w$) for a solute in water may be expressed as:

$$\Delta\overline{H}_w = \begin{cases} \Delta\overline{H}_w^{\,ex} & \text{for liquid solutes} \tag{3.19} \\[2mm] \Delta\overline{H}_w^{\,ex} + \Delta\overline{H}_{fus} & \text{for solid solutes} \tag{3.20} \end{cases}$$

Since $\Delta\overline{H}_{fus}$ is always positive and $\Delta\overline{H}_w^{\,ex}$ is normally positive for a liquid solute that exhibits a limited solubility in water (i.e., when $\gamma_i \gg 1$), the solubility of a solid solute in water (or other solvents) is usually much more sensitive to $T$ than that of a similar liquid solute.

The derivative of $\log\gamma_s^*$ with $T$ is similar to that of $\log\gamma_w$ with $T$:

$$\frac{d\log\gamma_s^*}{dT} = \frac{-\Delta\overline{H}_s^{\,ex}}{2.303RT^2} = \frac{-\Delta\overline{H}_s}{2.303RT^2} \tag{3.21}$$

in which $\Delta\overline{H}_s^{\,ex}$ is the molar excess heat of mixing of the liquid or supercooled-liquid solute with the organic-solvent phase (which is saturated by a certain amount of water) at $T$ and $\Delta\overline{H}_s$ is the molar heat of solution of the solute at $T$. Again, for solid solutes, $\Delta\overline{H}_s = \Delta\overline{H}_s^{\,ex} + \Delta\overline{H}_{fus}$. Thus, by combining Eqs. (3.16), (3.17), and (3.21), one obtains

$$\frac{d\log K_{sw}}{dT} = \frac{\Delta\overline{H}_s^{\,ex} - \Delta\overline{H}_w^{\,ex}}{2.303RT^2} = \frac{\Delta\overline{H}_s - \Delta\overline{H}_w}{2.303RT^2} \tag{3.22}$$

and thus,

$$\Delta\overline{H}_{sw} = \Delta\overline{H}_{s} - \Delta\overline{H}_{w} \qquad (3.23)$$

where $\Delta\overline{H}_{sw}$ is the enthalpy (heat) of partition when a mole of the solute is transferred from water to an organic phase of interest at equilibrium.

It is evident from derivations above that the heat associated with the equilibrium of a solute (whether it is a liquid or a solid substance) between an organic phase and water is the difference of the solute's molar heats of solution in the organic phase and water. For solid solutes, the heat of fusion ($\Delta\overline{H}_{fus}$), which affects the solid solubility in a single phase (e.g., water), has no net effect on the solute partition coefficient ($K_{sw}$) or on the heat of solute partition ($\Delta\overline{H}_{sw}$).

For most low-polarity organic compounds with a limited solubility in water, both $\Delta\overline{H}_{w}$ and $\Delta\overline{H}_{s}$ are positive with $\Delta\overline{H}_{w} > \Delta\overline{H}_{s}$, and therefore the $K_{sw}$ value would exhibit an exothermic heat that is smaller in magnitude than the reverse heat of solution of the solute in water ($-\Delta\overline{H}_{w}$). In other words, while the $K_{sw}$ will normally decrease and the $S_{w}$ increase with a temperature rise, the extent of variation would be much smaller for $K_{sw}$ than for $S_{w}$. The opposing heat effects (i.e., the temperature dependencies) between $K_{sw}$ and $S_{w}$ are often greatly magnified for solid solutes because $\Delta\overline{H}_{fus}$ is part of $\Delta\overline{H}_{w}$, but not of $\Delta\overline{H}_{sw}$, as described by Eqs. (3.20) and (3.23). For most solutes in organic–water mixtures, the $\Delta\overline{H}_{sw}$ values are normally less than 12 kJ/mol in exothermicity. The estimated variation in $K_{sw}$ for a solute with a temperature rise from 20°C to 25°C is therefore less than 10%.

If the organic phase of interest is macromolecular in nature, in which Eq. (3.13) or (3.14) defines more properly the solute partition coefficient, one may derive relations identical to those of Eqs. (3.17) to (3.23) by assuming that the molar volume terms are largely invariant with temperature and by substituting $\chi/2.303$ for $\log\gamma_{s}^{*}$. Thus, the relations of Eqs. (3.17) to (3.23) hold for solutes of a limited solubility in water at dilution, much independent of the relative molecular sizes of the solute and the organic phase.

For solid substances of interest, if one knows the molar heat of fusion ($\Delta\overline{H}_{fus}$) and the melting point ($T_{m}$), the activity of the solid can then be determined, through Eq. (3.18), as

$$\int d\ln(P^{s}/P^{\circ}) = \int_{T}^{T_{m}} (\Delta\overline{H}_{fus}/RT^{2})\,dT \qquad (3.24)$$

with the boundary condition that $P^{s}/P^{\circ} = 1$ at $T = T_{m}$. If one assumes that $\Delta\overline{H}_{fus}$ is practically constant between $T$ and $T_{m}$, one gets the important equation

$$\ln a^{s} = \ln\frac{P^{s}}{P^{\circ}} = \frac{-\Delta\overline{H}_{fus}}{R}\frac{T_{m}-T}{TT_{m}} \qquad (3.25)$$

where the term $\Delta\overline{H}_{fus}/T_{m} = \Delta\overline{S}_{fus}$ is the molar entropy of fusion of the solid substance (at $T = T_{m}$, the solid and its melt are at equilibrium, thus $\Delta\overline{G}_{fus} = 0$). The relation between the $P^s$ of a solid substance and the $P^\circ$ of the corresponding supercooled liquid at temperature $T$ is illustrated graphically in Figure 1.2. Once the $a^s$ of a solid is determined by Eq. (3.25), it then enables one to calculate the supercooled liquid solubility at $T$ from the measured solid solubility by Eq. (3.9).

## 3.5  CONCENTRATION DEPENDENCE OF PARTITION COEFFICIENT

In derivations of the partition coefficients for solutes between an organic phase and water, we have assumed that the solutes are present at dilution in both solvent phases. For solutes that are sparingly soluble in water, this assumption is closely met in the water phase and similarly met in the organic phase if the concentration is kept low. If the solute of interest is a solid with a relatively high melting point (say, $t_m \gg 100°C$), the maximum solute concentration in any solvent will always be low in absolute magnitude and hence the dilute-solution approximation will always hold, largely independent of the solute concentration. For liquid solutes or other solid solutes with low melting points that are relatively soluble in the organic phase of a solvent–water mixture (or a macromolecular phase–water mixture), the partition coefficients may be concentration dependent if the concentration is more than 10 to 20% in the organic phase. Although it is not common to measure the partition coefficient at such high concentrations, it is of interest to consider the possible concentration dependence of the partition coefficient, since the result may be of value to the characterization of the association of an organic solute with a natural organic substrate.

To evaluate the dependence of the partition coefficient ($K_{sw}$) on concentration, one makes use of an *isotherm* that relates the solute concentration in the organic–solvent phase ($C_s$) to that in water ($C_w$) over a wide range of $C_w$ at a given temperature. If the dilute-solution approximation holds for the entire concentration range, as for solutes (liquids or solids) that exhibit small solubilities in both solvents, the relation between $C_s$ and $C_w$ by Eq. (3.11), or that between $C_p$ and $C_w$ by Eq. (3.13), should be virtually linear from $C_w = 0$ to $C_w = S_w$, as depicted in Figure 3.1.

On the other hand, if the solute is very soluble in the organic phase but does not behave nearly ideally, the isotherm will not have a linear shape but will instead exhibit a moderate concave-upward curvature at the high $C_w$ region, as depicted in Figure 3.2. In this case the expression for partition coefficient is more sophisticated than that given by Eq. (3.11) or (3.13). The main cause for the nonlinear partition coefficient is the change in solute activity coefficient in the organic phase ($\gamma_s^*$ or the equivalent $\chi_H$) with solute concentration. This may be expected on the grounds that when the solute concentration reaches an appreciable level in an organic phase, the composition of the organic phase is modified significantly, such that it becomes appreciably

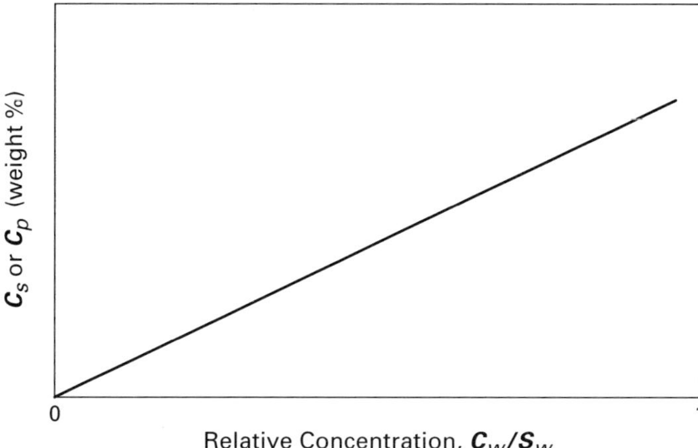

**Figure 3.1**  Linear isotherm for the weak partition of a solute from water into an organic solvent or a polymer. Here the highest $C_s$ or $C_p$ value is presumably less than 10% by weight.

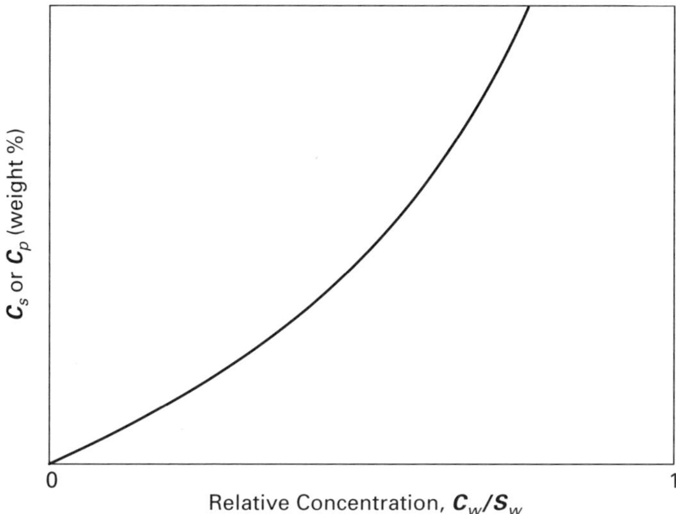

**Figure 3.2**  Concave-upward isotherm for the strong partition of a solute into an organic solvent or a polymer. Here the highest $C_s$ or $C_p$ value is presumably much greater than 10 to 20% by weight.

similar to that of the solute and hence becomes more compatible with the solute. As a result, $\gamma_s^*$ or $\chi_H$ decreases progressively with increasing $C_s$. This nonlinear effect would not be remarkable, since $\gamma_s^*$ or $\chi_H$ is not expected to change sharply with concentration, as solutes highly soluble in a solvent (or an organic phase) must have properties closely similar to those of the solvent, rendering $\gamma_s^*$ or $\chi_H$ small. If the solute forms an ideal solution (or close to it) with the organic phase, the partition isotherm will then be linear because $\gamma_s^*$ is essentially 1, or $\chi_H$ is close to zero, at all concentrations.

# 4 Fundamentals of the Adsorption Theory

## 4.1 INTRODUCTION

*Adsorption* is a surface phenomenon that is characterized by the concentration of a chemical species (adsorbate) from its vapor phase or from a solution onto or near the surfaces or pores of a solid (adsorbent). This surface excess occurs in general when the attractive energy of a substance with the solid surface (i.e., the adhesive work) is greater than the cohesive energy of the substance itself (Manes, 1998). The adsorptive uptake is amplified if the solid material has a high surface area. If the adsorption occurs by London–van der Waals forces of the solid and adsorbate, it is called *physical adsorption*. If the forces leading to adsorption are related to chemical bonding forces, the adsorption is referred to as *chemisorption*. However, the distinction between physical adsorption and chemisorption is not always sharp. For example, the adsorption of polar vapors onto polar solids may fall under either classification, depending on the adsorption energy. From a thermodynamic point of view, the concentration of a substance from a dilute vapor phase or solution onto a solid surface corresponds to a reduction in freedom of motion of molecules and thereby to a loss in system entropy. As such, the adsorption process must be exothermic to the extent that the negative $\Delta H$ is greater in magnitude than the associated negative $T \Delta S$ to maintain a favorable free-energy driving force (i.e., for $\Delta G$ to be negative). For more detailed discussions on the thermodynamic aspect of the adsorption process, see Adamson (1967), Gregg and Sing (1982), and Manes (1998).

When a vapor is adsorbed onto a previously unoccupied solid surface or its pore space, the amount of the vapor adsorbed is proportional to the solid mass. The vapor uptake also depends on temperature ($T$), the equilibrium partial pressure of the vapor ($P$), and the nature of the solid and vapor. For a vapor adsorbed on a solid at a fixed temperature, the adsorbed quantity per unit mass of the solid ($Q$) is then only a function of $P$. The relation between $Q$ and $P$ at a given temperature is called the *adsorption isotherm*. $Q$ is frequently presented as a function of the relative pressure, $P/P^\circ$, where $P$ is normalized to the saturation vapor pressure ($P^\circ$) of the adsorbate at temperature $T$. The normalized isotherm is often more useful, as it enables one to assess readily the net adsorption heats and other characteristics of vapors over a range of temperatures. For adsorption of solutes from solution, one constructs similar isotherm forms by relating $Q$ with $C_e$ (the equilibrium concentration) or with the relative concentration, $C_e/C_s$, where $C_s$ is the solubility of the solute.

Except for rare cases where the microscopic structure of a solid surface is nearly uniform, the surfaces of most solids are heterogeneous, with the result that adsorption energies are variable. The adsorption sites are taken up sequentially, starting from the highest-energy sites to the lowest-energy sites, with increasing partial pressure or solute concentration. Thus the net (differential) molar heat of adsorption decreases with increasing adsorption and vanishes when the vapor pressure or solute concentration reaches saturation. Adsorption isotherms are typically nonlinear because of the energetic heterogeneity and the limited active sites or surfaces of the solid. Since a given site or a surface of the solid cannot be shared by two or more different kinds of adsorbates, the adsorption process is necessarily competitive, which is in contrast to a partition process. The surface area or porosity of the solid is usually the principal factor affecting the amount of vapor adsorption; therefore, a powerful adsorbent must have a large surface area. Adsorption of a solute from solution is subject to competition by the solvent and other components in the solution. Therefore, a powerful adsorbent for single vapors is not necessarily a strong adsorbent for solutes from solution.

A number of adsorption isotherms have been recorded for vapors on a wide variety of solids. Brunauer (1945) grouped the isotherms into five principal classes, types I to V, as illustrated in Figure. 4.1. Type I is characterized by Langmuir-type adsorption (see below), which shows a monotonic approach to a limiting value that corresponds theoretically to the completion of a surface monolayer. Type II is perhaps most common for physical adsorption on relatively open surfaces, in which adsorption proceeds progressively from submonolayer to multilayer; the isotherm exhibits a distinct concave-downward curvature at some low relative pressure ($P/P°$) and a sharply rising curve at high $P/P°$. The point B at the knee of the curve signifies completion of an adsorbed monolayer. It forms the basis of the Brunauer–Emmett–Teller (BET) model for surface-area determination of a solid from the assumed monolayer capacity, described below.

A type III isotherm signifies a relatively weak gas–solid interaction, as exemplified by the adsorption of water and alkanes on nonporous low-polarity solids such as polytetrafluroethylene (Teflon) (Graham, 1965; Whalen, 1968; Gregg and Sing, 1982). In this case, the adsorbate does not effectively spread on the solid surface. Type IV and V isotherms are characteristic of vapor adsorption by capillary condensation into small adsorbent pores, in which the adsorption reaches an asymptotic value as the saturation pressure is approached. Adsorption of organic vapors on activated carbon is typically type IV, whereas adsorption of water vapor on activated carbon is type V (Manes, 1998), as shown later. The shape of the adsorption isotherm of a solute from solution depends sensitively on the competitive adsorption of the solvent and other components and may deviate greatly from that of its vapor on the solid.

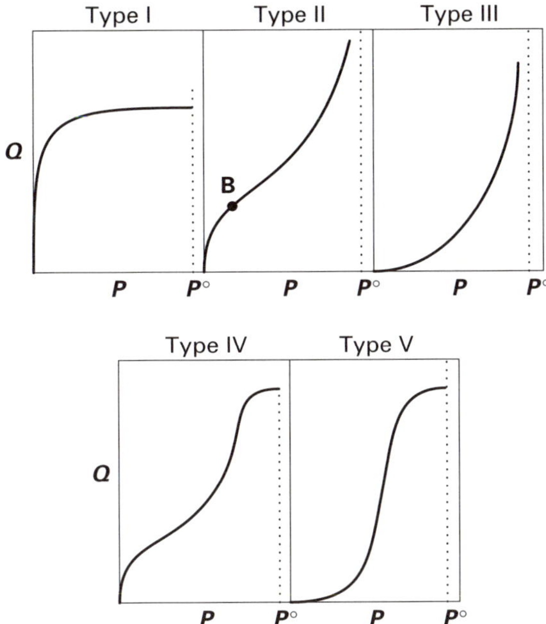


**Figure 4.1**  The five types of adsorption isotherms according to the classification of Brunauer (1945).

One notes with interest the similarity in shape of type III adsorption isotherm and a special partition isotherm, as depicted in Figure 3.2, when a solute partitions very favorably from water (or other media) into an organic phase (or solvent). Whereas the isotherm in Figure 3.2 is for the solute partition from water into an organic phase, a similar partition isotherm arises when the vapor of a liquid partitions strongly into an organic substance. Illustrative examples of such a vapor partition phenomenon are benzene, toluene, xylene, and carbon tetrachloride on rubber, polystyrene, and polyiosbutylene (Eichinger and Flory, 1968a,b). A practical means to distinguish a type III vapor adsorption isotherm from a similarly shaped vapor partition isotherm is that the vapor partition should display a very high uptake capacity, usually more than 10% by weight at $P/P° = 0.5$, while a type III vapor adsorption exhibits a very low capacity, usually far less than 1% by weight, at $P/P° = 0.5$.

## 4.2   LANGMUIR ADSORPTION ISOTHERM

Langmuir (1918) considered the adsorption of gases or vapors on a plane surface that contains a fixed number of identical active sites. From a kinetic consideration, the rate of vapor desorption from the occupied sites is set equal to the rate of adsorption on the unoccupied sites at equilibrium:

$$k_d\theta = k_a P(1-\theta) \tag{4.1}$$

where $\theta$ is the fraction of the total sites occupied by the vapor at an equilibrium partial pressure $P$, $k_d$ the desorption rate constant, and $k_a$ the adsorption rate constant. Therefore,

$$\theta = \frac{k_a P}{k_d + k_a P} = \frac{(k_a/k_d)P}{1+(k_a/k_d)P} \tag{4.2}$$

Since the amount $Q$ of vapor adsorbed by a unit mass of the solid is proportional to $\theta$, one gets an adsorption isotherm as

$$Q = \frac{Q_m bP}{1+bP} \tag{4.3}$$

where $Q_m$ is the limiting (monolayer) adsorption capacity (i.e., when the surface is covered with a complete monolayer of the adsorbed vapor) and $b = k_a/k_d$ is related to the heat of adsorption per unit mass (or per mole) of the vapor, which is considered to be independent of the adsorbed amount.

As seen, at low $P$, where $bP \ll 1$, $Q$ is proportional to $P$ (i.e., $Q = kP$), where $k$ is a constant, and the relation between $Q$ and $P$ is therefore linear. At high $P$, $bP \gg 1$, $Q$ approaches $Q_m$ asymptotically and the isotherm is concave toward the $P$ axis. The linear relation between $Q$ and $P$ at low $P$ may be referred to as the *Henry region*. The general shape of the Langmuir-type isotherm falls under Brunauer's classification of type I. Examples of systems that closely meet Eq. (4.3) are the adsorption of relatively inert vapors of nitrogen, argon, methane, and carbon dioxide on plane (open) surfaces of mica and glass at liquid air or liquid nitrogen temperature (Langmuir, 1918).

Although Eq. (4.3) is intended originally only for vapor adsorption, a similar form is frequently adapted to fit the adsorption data of a substance (solute) from a solution, in which case the $P$ term in Eq. (4.3) is replaced by the equilibrium solute concentration. The constant $Q_m$ and $b$ in the Langmuir equation may be determined by rewriting the equation as

$$\frac{1}{Q} = \frac{1}{Q_m bP} + \frac{1}{Q_m} \tag{4.4}$$

By Eq. (4.4), a plot of $1/Q$ versus $1/P$ gives a slope of $1/Q_m b$ and an intercept of $1/Q_m$. From the slope and intercept values, $Q_m$ and $b$ can be calculated.

Although the adsorption data of many vapors or solutes on solids conform to the general shape of the Langmuir equation, this is not necessarily a proof that the system complies with the Langmuir model. For most solids, the adsorption sites are energetically heterogeneous, and this energetic heterogeneity along with site limitations may give rise to a *Langmuir-shape isotherm*.

In other words, the $b$ constant in Eq. (4.3), which is related to the molar heat of adsorption, varies with the range of $P$ in many of these systems. By contrast, the nonlinearity in the original Langmuir derivation is attributed to the degree of site saturation (i.e., to an entropic effect) rather than to an energetic factor. Thus, unless the observed nonlinearity is proven to be truly entropic in nature, the isotherm is more appropriately referred to as a *Langmuir-type isotherm*, or simply a type I isotherm.

## 4.3  FREUNDLICH EQUATION

The Freundlich equation was developed mainly to allow for an empirical account of the variation in adsorption heat with concentration of an adsorbate (vapor or solute) on an energetically heterogeneous surface. It has the general form

$$Q = K_f C^n \tag{4.5}$$

where $Q$ is the amount adsorbed per unit mass of the solid (adsorbent); $C$ is the vapor or solute concentration at equilibrium; $K_f$ is the Freundlich constant, equal to the adsorption capacity at $C = 1$; and $n$ is an exponent related to the intrinsic heat of vapor or solute adsorption. The $n$ value is in principle less than 1, because the adsorption isotherm is commonly concave to the $C$ axis, and varies with the extent of adsorption (i.e., with $Q$). Depending on the adsorbent, the constancy of $n$ may apply to a narrow or wide range of $C$. It can be determined from the slope of the plot of log $Q$ versus log $C$ over a specific range.

Unlike the Langmuir model, the Freundlich equation does not approach (arithmetic) linearity at low $C$, nor does it approach a limiting (fixed) adsorption capacity as $C$ reaches saturation. These features are opposed to the general adsorption characteristics. Basically, the Freundlich equation with its adjustable parameters offers a simple mathematical tool rather than a physical model to account for the energetic heterogeneity of adsorption at different regions of the isotherm. Interpretation of the temperature effect on adsorption by Freundlich equation is generally difficult. This is because the vapor or solute concentration ($C$) can be increased by increasing the temperature while the adsorbed mass ($Q$) usually decreases with increasing temperature. For many applications, however, the Freundlich equation is quite mathematically convenient.

## 4.4  BET MULTILAYER ADSORPTION THEORY

The Brunauer–Emmett–Teller (BET) theory (Brunauer et al., 1938) was formulated to deal with submonolayer-to-multilayer vapor adsorption on a solid.

The model sets a theoretical basis for calculating the surface area of the solid. The theory was derived on the assumptions that (1) the Langmuir equation applies to each adsorbed layer (i.e., the surface has uniform and localized sites so that there is no interference in adsorption between neighboring sites); (2) the adsorption and desorption occur only onto and from the exposed layer surfaces; (3) at solid–vapor equilibrium, the rate of adsorption onto the $i$th layer is balanced by the rate of desorption from the $(i + 1)$th layer; and (4) the molar heat of adsorption for the first layer is considered to be higher than for the succeeding layers, the latter assumed to be equal to the heat of liquefaction of the vapor. These considerations lead to an isotherm of the form

$$\frac{Q}{Q_m} = \frac{Cx}{(1-x)[1+(C-1)x]} \tag{4.6}$$

where $Q$ is the amount of vapor adsorbed at relative vapor pressure $x = P/P^\circ$, $P$ the equilibrium pressure of the vapor, $P^\circ$ the saturation pressure of the vapor at the system temperature, $Q_m$ the (statistical) monolayer capacity of the adsorbed vapor on the solid, and $C$ is a constant related to the difference between the heat of adsorption in the first layer and the heat of liquefaction of the vapor. Equation (4.6) may be transformed into

$$\frac{x}{Q(1-x)} = \frac{(C-1)x}{CQ_m} + \frac{1}{CQ_m} \tag{4.7}$$

A plot of $x/[Q(1-x)]$ versus $x$ should yield a straight line (usually, at $0.05 < x < 0.30$), with a slope of $(C-1)/CQ_m$ and an intercept of $1/CQ_m$, from which $C$ and $Q_m$ can be determined. The linear relation of $x/[Q(1-x)]$ versus $x$ usually does not go beyond $x > 0.30$, much because the multilayer adsorption does not proceed indefinitely as the theory contends. Once $Q_m$ is determined, and if the molecular area of the vapor is known, the surface area of the solid (adsorbent) can then be calculated. The magnitude of $C$ accounts for the curvature of an adsorption isotherm; a large $C$ ($>> 1$) produces a highly concave-downward shape at low $x$, and a small $C$ ($<< 1$) leads to a concave-upward shape at low $x$.

Generally speaking, the BET model accounts satisfactorily for multilayer adsorption of vapors on surfaces that are not highly heterogeneous (i.e., if the surface area of the solid is small to moderate in magnitude). This is because the model assumes that the solid surface has uniform energetic sites with a constant adsorption energy and that the molar heat of adsorption beyond the first layer is all the same, both of which are not well satisfied in vapor adsorption on microporous solids. The BET model, with an inert gas as the adsorbate, has proven to be the best available analytical method for surface-area determination of solids. Nitrogen ($N_2$) gas at its boiling point (77 K) is the most commonly used adsorbate, with which the $Q_m$ of $N_2$ on a solid is obtained;

the surface area is then calculated along with the assumed $N_2$ molecular area of $16.2 \times 10^{-20}\,m^2$.

## 4.5  POLANYI ADSORPTION POTENTIAL THEORY

If adsorption is highly energetically heterogeneous, as with high-surface-area microporous solids such as activated carbon and silica gel, the adsorption data exhibit serious deviations from the Langmuir model or the BET model. This is because the force field within a pore space (adsorption space) of a microporous material that attracts a molecule varies considerably with the location. The Polanyi adsorption potential theory (Polanyi, 1916) has long been recognized as the most powerful model for dealing with vapor adsorption on energetically heterogeneous solids (Brunauer, 1945). The basic Polanyi model has been extended to a wide range of vapor- and liquid-phase systems by Manes and co-workers (Manes, 1998), and will therefore be referred to as *Polanyi–Manes model*. The model relates a wide variety of both vapor- and liquid-phase data to each other, and in particular, it correlates liquid-phase with vapor-phase adsorption. For a detailed account of the extended model, see Manes (1998).

The Polanyi theory considers that for a molecule located within the attractive force field of a microporous solid, there exists an (attractive) adsorption potential ($\varepsilon$) between the molecule and the solid surface. This attraction derives from the induced dipole–induced dipole force (i.e., the London force) of the molecule and surface atoms, which is short range in nature. The potential $\varepsilon$ at a particular location within the adsorption space may be viewed as the energy required to remove the molecule from that location to a point outside the attractive force field of the solid. Thus, the magnitude of $\varepsilon$ for an adsorbate depends on its proximity to the solid surface. It is highest in the narrowest pore (or in the narrowest portion of a pore) because the adsorbate is close to more solid material. A series of equipotential surfaces are formed by connecting the points in adsorption space with the same $\varepsilon$, as shown schematically in Figure 4.2.

When a vapor is placed within an attractive force field of a solid, two opposing thermodynamic effects occur. The system energy is minimized by vapor concentration into the region of the lowest potential energy, but the system entropy is reduced by this concentration. The impact of these two effects at a constant temperature on the molar free energy is given by

$$d\overline{G} = -d\varepsilon + \overline{V}\,dP \tag{4.8}$$

where $-d\varepsilon$ is the differential potential energy change per mole of the vapor, $\overline{V}$ the molar volume of the vapor, and $dP$ the differential change in vapor partial pressure. At adsorption equilibrium, $d\overline{G} = 0$, and the reduction in potential energy offsets the loss in entropy:

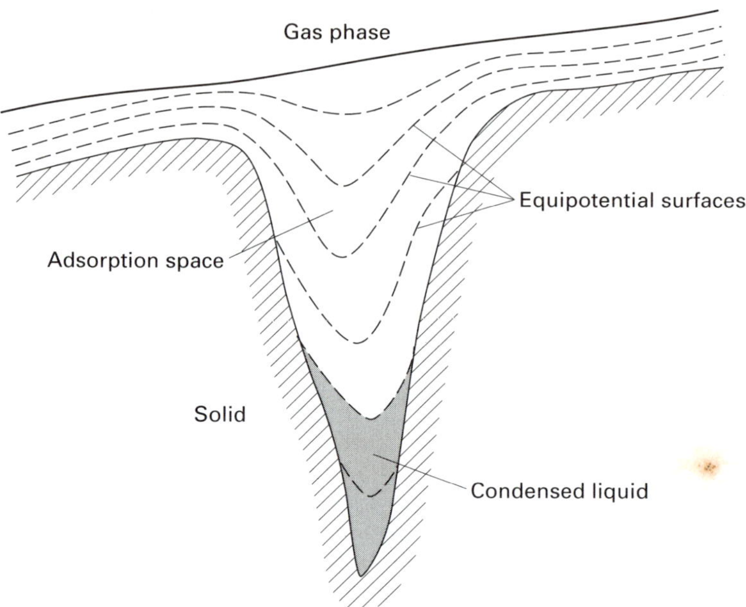


**Figure 4.2**  Rough schematic model for a region of the porous carbon surface (pore) showing the equipotential surfaces corresponding to successively lower values of the adsorption potential with increasing pore size. The vapor liquefies wherever the adsorption potential required to concentrate it to saturation is equaled or exceeded.

$$d\varepsilon = \overline{V}\,dP \tag{4.9}$$

and

$$\varepsilon = \int \overline{V}\,dP \tag{4.10}$$

According to the Polanyi theory, a vapor will condense to form a liquid or liquidlike adsorbate if $\varepsilon$ (taken as positive) at an equipotential surface is greater than or equal to the work required to concentrate the vapor from ambient pressure $P$ (where $\varepsilon = 0$) to its saturation pressure $P°$ at the equipotential surface. If the vapor follows the ideal-gas law, Eq. (4.10) becomes

$$\varepsilon = RT \ln(P°/P) \tag{4.11}$$

Thus, if a porous solid (adsorbent) is exposed to increasing partial pressure of a vapor, condensation takes place beginning with the region of the highest potential (or in the finest pore) and then with the region of progressively lower adsorption potential until all adsorption space is filled as the ambient pressure becomes saturated (i.e., as the adsorption potential becomes zero).

For a vapor at a given equilibrium $P/P°$ or a given $\varepsilon$, which corresponds to a given equipotential surface inside the adsorption space, the volume enclosed by the equipotential surface and the solid surface is the adsorbed volume. The net molar heat of adsorption at the equilibrium potential surface is $-\varepsilon$. If the vapor is condensed as a liquidlike adsorbate, the total molar heat of adsorption is $-(\varepsilon + \Delta\overline{H}_{evap})$, where $\Delta\overline{H}_{evap}$ is the molar heat of evaporation of the liquid. If the vapor is condensed as a solid adsorbate, the total molar heat of adsorption is $-(\varepsilon + \Delta\overline{H}_{sub})$, where $\Delta\overline{H}_{sub}$ is the molar heat of sublimation of the adsorbate.

For vapor adsorption on a relatively inert porous solid (e.g., activated carbon) that involves primarily London forces (i.e., in the absence of chemisorption or specific interaction), the adsorption potential ($\varepsilon$) is independent of temperature. A direct consequence of this temperature independence and of the vapor condensation is that a plot of the total adsorbed liquid (or solid) volume ($\phi$) against $\varepsilon$ at that volume (called a *characteristic curve*) is temperature invariant and depends only on the vapor and the solid structure. Thus, once the characteristic curve is obtained for a vapor on a porous solid from its adsorption data at one temperature, it can be used in a reverse manner to construct the isotherm at a different temperature. The Polanyi model postulates no specific mathematical form for the characteristic curve, which is fixed instead by the structure of the porous solid.

If there is no molecular sieving involved in vapor adsorption, the Polanyi model expects the characteristic curves for all vapor adsorbates on a chemically inert porous solid to have a common shape and a common limiting adsorbate volume (at $\varepsilon = 0$). For any adsorbed volume, the adsorption potentials of different vapor adsorbates are related to each other by constant characteristic factors. Therefore, all characteristic curves on a given solid can be made to collapse into a single curve by appropriate divisors of the individual adsorption potentials for any given adsorbed volumes. The most effective and convenient divisors are found to be the liquid molar volumes ($\overline{V}$) of the vapor adsorbates (Dubinin and Timofeyev, 1946). The resulting plot of the adsorbed volume versus $\varepsilon/\overline{V}$ for a vapor adsorbate is called a *correlation curve* (Lewis et al., 1950). As shown by Polanyi and Manes, correlation curves provide a basis to predict the adsorption of a solute from solution on an inert porous solid from the respective vapor isotherms of the pure solute and solvent.

If a solute in solution is partially miscible with the solvent, the basic Polanyi model expects that the solute condense into the adsorption space as a liquid or a solid phase, depending on the state of the pure solute at the system temperature. Therefore, the critical difference between vapor-phase and liquid-phase adsorption is that the vapor condenses in a hitherto unoccupied space, whereas the liquid or solid solute condenses to displace an equal volume of the solvent. According to Polanyi, the adsorption potential of a partially miscible solute can thus be expressed as

$$\varepsilon_{sl} = \varepsilon_s - \varepsilon_l(\overline{V}_s/\overline{V}_l) = RT \ln(C_s/C_e) \qquad (4.12)$$

where $\varepsilon_s$ is the (molar) adsorption potential of the solute, $\varepsilon_l$ the adsorption potential of the solvent, $\varepsilon_{sl}$ the adsorption potential of the solute from solution, $\overline{V}_s$ and $\overline{V}_l$ the respective molar volumes of the solute and solvent, $C_s$ the solute solubility in the solvent, and $C_e$ the solute concentration in the solvent at equilibrium. Equation (4.12) may be further converted to give

$$\varepsilon_{sl}/\overline{V}_s = \varepsilon_s/\overline{V}_s - \varepsilon_l/\overline{V}_l = (RT/\overline{V}_s)\ln(C_s/C_e) \tag{4.13}$$

As seen, the net adsorption potential density of the solute ($\varepsilon_{sl}/\overline{V}_s$) is simply the difference between the potential densities of pure solute ($\varepsilon_s/\overline{V}_s$) and solvent ($\varepsilon_l/\overline{V}_l$). Thus one may in principle predict the adsorption of a partially miscible solute from solution from established or estimated correlation curves of the pure solute and solvent. Equation (4.13) has been found most successful for partially miscible liquid solutes in solution, in which the effective molar volume of the liquid adsorbate ($\overline{V}_s$) is practically the same as the molar volume of the pure liquid. For solid solutes, the effective adsorbate molar volume may well exceed that of the pure substance, because packing of the condensed solid crystallite into fine-pore adsorption spaces may be hindered significantly by crystalline structure; therefore, for solid solutes, adjustment of molar volumes for packing efficiency is often required. Manes (1998) extended the Polanyi theory to a wide range of vapor and solution systems, including single and multiple vapors and solutes that are either completely or partially miscible to each other.

In adsorption from solution, the net heat of adsorption for a partially miscible solute ($\varepsilon_{sl}$) is usually smaller than that of its single vapor-phase adsorption ($\varepsilon_s$) because of the energy required to displace the solvent, as depicted by Eq. (4.12). In such systems (i.e., where the solute separates out as a liquid or a solid phase in adsorption space), the total molar heat of adsorption is $-(\varepsilon_{sl} + \Delta\overline{H}_{sol})$, where $\Delta\overline{H}_{sol}$ is the molar heat of solution of the solute. One may recall from the discussion in Chapter 3 that the $\Delta\overline{H}_{sol}$ for solid solutes includes the associated heats of fusion ($\Delta\overline{H}_{fus}$).

## 4.6  SURFACE AREAS OF SOLIDS

The surface area of a solid (adsorbent) plays a fundamental role in the physical adsorption of vapors. The BET method with appropriate adsorbate gases has become a universal method for determining the solid surface area. Suitable vapor adsorbates must be chemically inert, not subject to molecular sieving by the solid pore, and confined only to the exterior of the solid (i.e., no vapor penetration into the interior network). The use of an inert vapor as the adsorbate is to eliminate any specific interaction (or reaction) with either solid surface or its interior network. Prevention of molecular sieving is accomplished by the use of small adsorbates. Measurement at low temperature ensures that the adsorbate solubilization into the solid matrix is minimized. Although $N_2$ vapor at its normal boiling point (ca. 77 K) is used most frequently as the adsorbate, the choice is by no means restricted to $N_2$, and use of a wide variety of other inert vapor adsorbates (e.g., krypton) should yield similar results. The surface area, considered as the solid–vapor or solid–vacuum interfacial area, which is external to solid material, is assumed to predate the experiment and to be unchanged by the experiment. The surface area is therefore a property of the solid; that is, within the precision of the measurement method, it should be independent of the choice of any suitable adsorbates used.

For highly porous solids, the term *internal surface* is frequently used to refer to the surface associated with the walls of pores that have narrow openings, which extend inward from the granule surface to the interior of the granule. On the other hand, the term *external surface* is used to refer to the surface from all prominences and those cracks that are wider than they are deep (Gregg and Sing, 1982). It is understood that the internal surface is restricted to open-ended pores and does not apply to sealed-off pores (i.e., those having no openings to the exterior of the granule). Although these two kinds of surfaces are somewhat operational in their definitions, it is understood that the internal surface is nonetheless *external* to the material and accessible to gases, as is measured in surface-area determination. Thus, as long as the adsorbate does not penetrate the field of force that exists between the atoms, ions, or molecules inside the solid, it is considered to be on the external surface, despite the fact that it may adsorb on the solid's internal surface (Brunauer, 1945). For a highly microporous solid such as activated carbon, one may then say that the solid has a very high surface area, as determined by the BET method, because it has a large internal surface.

It is unfortunate that the term *internal surface* practiced in soil science literature gives a confusing implication to the surface area. The confusion initiates from the use of the amounts of some polar solvents (e.g., water and ethylene glycol) retained by a unit mass of the soil or mineral sample under certain evacuating conditions for determining the *total surface area* of the sample. By taking the surface area from the BET method using an inert gas (e.g., $N_2$) as the external surface area, the difference between the thus determined total surface area and the external surface area is considered to be the internal surface area of the sample. Evidently, the analytical method leading to the internal surface area does not comply with the accepted criterion in surface area determination, and consequently, this internal surface area is mainly an artifact of the method. As mentioned earlier, appropriate adsorbates for surface-area determination must be chemically inert so that they neither alter the structure of the solid nor penetrate the molecular network of the solid. Because some polar solvents can potentially alter the solid structure, such as by a solvation process with some clay minerals, or penetrate the soil organic matter by dissolution, the resulting internal surface area is often a measure of phenomena other than physical adsorption. As pointed out by

Brunauer (1945), when an adsorbate penetrates the interior of the solid, it either dissolves in the solid to form a solution or reacts with the solid to form a new compound. We shall see in Chapter 6 that the uptake of polar vapors or liquids by soils and certain minerals is eminently consistent with this expectation.

## 4.7  ISOSTERIC HEAT OF ADSORPTION

Because the adsorbent surface is commonly energetically heterogeneous, the exothermic heat of adsorption of a vapor (or a solute) usually varies with the amount adsorbed. To account for the variation in adsorption heat, the isosteric heats of adsorption at some fixed adsorbate loadings are determined from the equilibrium vapor pressures (or solute concentrations) of the isotherms at different temperatures with the aid of the Clausius–Clapeyron equation. Although the concept of isosteric heat is originally intended for adsorption systems, it has been extended to nonadsorption systems (e.g., partition) to elucidate whether a concentration-dependent heat effect occurs with the system. For adsorption of a vapor by an solid, consider the adsorption isotherms at $T_1$, $T_2$, and $T_3$ (in K) in Figure 4.3, with $T_1 < T_2 < T_3$, where the amount of vapor uptake ($Q$) at each temperature is plotted against the equilibrium partial pressure ($P$). Similarly, for adsorption of a solute from solution, one considers the isotherms at different temperatures in which the solute uptake ($Q$) is plotted against the equilibrium solute concentration ($C_e$).

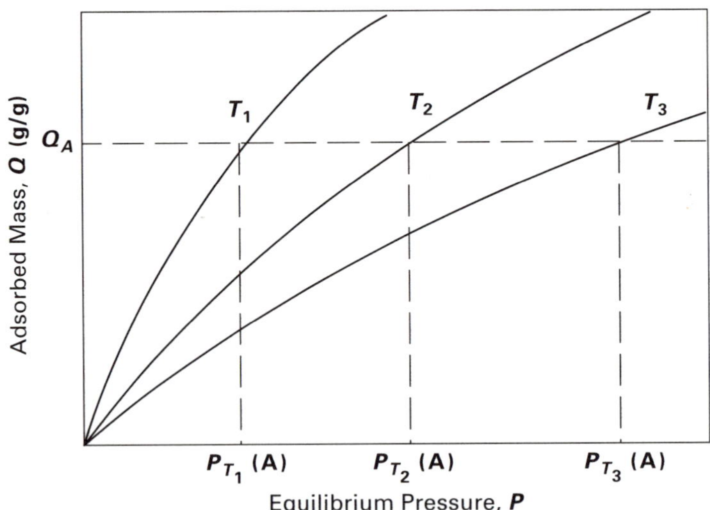


**Figure 4.3**  Schematic drawing showing the equilibrium pressures ($P$) of a vapor at three temperatures ($T_1 < T_2 < T_3$) with a fixed adsorbate mass ($Q_A$) on an adsorbent.

The isosteric-heat data describe how sensitively the molar heat of adsorption of a vapor or a solute varies with the amount adsorbed by a solid. To determine the isosteric heat of adsorption at a given $Q$ (say, $Q_A$ in Figure 4.3), one accounts for the variation of $P$ (or $C_e$) with $T$ at a fixed $Q$ using the general form of the Clausius–Clapeyron equation:

$$\frac{d \log P}{dT} = \frac{\Delta \overline{H}_d}{2.303RT^2} = \frac{-\Delta \overline{H}_a}{2.303RT^2} \tag{4.14}$$

or

$$\frac{d \log P}{d(1/T)} = \frac{-\Delta \overline{H}_d}{2.303R} \tag{4.15}$$

Similarly,

$$\frac{d \log C_e}{dT} = \frac{\Delta \overline{H}_d}{2.303RT^2} = \frac{-\Delta \overline{H}_a}{2.303RT^2} \tag{4.16}$$

or

$$\frac{d \log C_e}{d(1/T)} = \frac{-\Delta \overline{H}_d}{2.303R} \tag{4.17}$$

where $\Delta \overline{H}_d$ is the molar heat of desorption and $\Delta \overline{H}_a$ is the molar heat of adsorption ($\Delta \overline{H}_d = -\Delta \overline{H}_a$). By repeating the calculations for $\Delta \overline{H}_a$ at other fixed $Q$, the dependence of $\Delta \overline{H}_a$ on $Q$ can then be determined. For vapor or solute adsorption, the $\Delta \overline{H}_a$ should have the largest negative value (i.e., the molar exothermic heat) at the lowest $Q$ and hence the smallest negative value at the highest $Q$. As stated before, if the adsorbed vapor forms a condensed phase on the adsorbent, the $\Delta \overline{H}_a$ should be more exothermic than the molar heat of vapor condensation (i.e., $-\Delta \overline{H}_{\text{evap}}$ or $-\Delta \overline{H}_{\text{sub}}$). Similarly, if the adsorbed solute displaces the solvent to form a separate phase on the adsorbent surface, $\Delta \overline{H}_a$ should be more exothermic than the reverse molar heat of solute solution (i.e., $-\Delta \overline{H}_{\text{sol}}$). When the adsorption reaches the maximum on an adsorbent, the net adsorption heat is zero and thus $\Delta \overline{H}_a$ is equal to the heat of adsorbate condensation. In systems where the adsorption energy is not high enough to condense the vapor into a separate phase or to condense the solute by displacing the solvent, the adsorption will be weak. In this case, the thermicity of adsorption would be small and notably less exothermic than the heat of adsorbate condensation. However, as long as a net adsorption occurs, the system will nevertheless exhibit an exothermic effect, despite the fact that it may be very small. When adsorption is weak, the isotherm usually assumes a relatively linear shape over the entire range. For strong adsorption, which normally involves adsorbate condensation, the isotherm develops a marked concave-downward shape at low $P$ or $C_e$.

Because the sorption of organic compounds to many natural solids may be dictated by processes other than adsorption (e.g., by a partition interaction), the isosteric plot of the isotherms provides useful heat data for the undergoing process. For example, in a typical partition process of an organic solute from water to a partially miscible organic phase, the isotherm is usually highly linear over a wide concentration range, and therefore the molar isosteric heat of sorption is largely constant, independent of solute concentrations. This unique characteristic enables one to distinguish an uptake by partition from that by adsorption for a contaminant of interest. As we will find out later, ordinary soils act as a dual sorbent in uptake of organic compounds, where either adsorption on soil minerals or partition into soil organic matter may predominate the soil uptake, depending on the system condition. The detected isosteric heat for the system helps to pinpoint the dominant mechanism.

# 5 Contaminant Partition and Bioconcentration

## 5.1 INTRODUCTION

The partition of organic compounds in a partially miscible solvent–water system has been an important subject in chemistry as the basis for solvent extraction of solutes from water. The application of partition coefficients to biochemical systems began about a century ago with Meyer (1899) and Overton (1901), who showed that the relative narcotic activities of drugs were well correlated with their oil–water partition coefficients. The usefulness of the partition coefficient as a system parameter for assessing the biochemical activity of an organic compound or a drug has been greatly extended by the work of Fujita et al. (1964), Hansch and Fujita (1964), Hansch (1969), Leo and Hansch (1971), and Leo et al. (1971). Leo and Hansch (1971) reviewed the partition characteristics of organic compounds in a variety of solvent–water systems and for practical reasons considered the octanol–water system to be the most appropriate reference for assessing the relative lipophilicity of organic solutes with biological components. Hansch (1969) showed, for example, that the partition coefficients of organic solutes between protein and water could be correlated successfully with their octanol–water partition coefficients, thus providing an assessment of the binding of small organic molecules with biological macromolecules.

The utility of partition coefficients to estimate the distribution of organic contaminants in environmental systems has meanwhile become increasingly evident since the 1970s. This development stemmed primarily from the observations that the potential of an organic contaminant to concentrate from water into aquatic organisms (e.g., fish) may be correlated successfully with its octanol–water partition coefficient (Neely et al., 1974; Lu and Metcalf, 1975; Chiou et al., 1977; Könemann and van Leeuwen, 1980; Oliver and Nimii, 1983, Banerjee et al., 1984; Chiou, 1985). Similar empirical correlations with octanol–water partition coefficients were also found for soil/sediment–water distribution coefficients of selected groups of organic contaminants (Chiou et al., 1979; Karickhoff et al., 1979; Kenaga and Goring, 1980; Means et al., 1980; Briggs, 1981; Schwarzenbach and Westall, 1981). Although the contaminant distribution between water and natural organic substrates may often be more complicated than a simple partition, these results manifest that an important driving force for contaminant distribution in natural aquatic systems is conceptually analogous to that of a solvent–water partition process. Data from laboratory and field studies on contaminant bioconcentration into fish and their correlations with solvent–water partition coefficients are presented later in this chapter. It is helpful to begin with an appreciation of the general and specific features of the solute partitioning in relation to solute and solvent properties.

## 5.2  OCTANOL–WATER SYSTEMS

Among current studies of the partition effects of nonionic organic compounds in various solvent–water mixtures, the partition coefficients in octanol–water mixtures have received the utmost attention because of the observed correlations between the octanol–water partition coefficients and the partition effects with natural organic substances and biological components. Part of the reasons for the success of $n$-octanol as a surrogate for natural organic matter and/or biological components has to do with the polar-to-nonpolar balance of the molecule through its hydrophilic OH and lipophilic alkyl chain that mimics to some extent the overall polarity of the natural organic matter and of biological materials. Here the term *polarity* is used to refer in a general sense to the ability of molecules to engage in hydrogen bonding and/or polar interactions as opposed to nonspecific dispersion (i.e., induced dipole–induced dipole) interactions. From the solubility-model standpoint, the octanol–water system is the one in which the partition behavior of most organic solutes with the solvent (octanol) follow closely the criterion of Raoult's law [Eqs. (3.10) and (3.11)] because the molecular-size disparity between the solute and octanol is generally not very significant.

To elucidate the relative effects of the factors that affect the octanol–water partition coefficient ($K_{ow}$), we recall Eq. (3.11), with changes in the subscripts for the related parameters, for solutes at low concentrations in both octanol and water phases, as follows:

$$\log K_{ow} = -\log S_w - \log \overline{V}_o^* - \log \gamma_o^* - \log \left( \gamma_w / \gamma_w^* \right) \tag{5.1}$$

in which $S_w$, $\gamma_w$, and $\gamma_w^*$ are as defined earlier, $\overline{V}_o^*$ is the molar volume of the water-saturated octanol, and $\gamma_o^*$ is the solute activity coefficient (Raoult's law) in the water-saturated octanol. Equation (5.1) was derived by Chiou et al. (1982b) on the assumption that the molar volume of octanol-saturated water ($\overline{V}_w^*$) is the same as the molar volume of pure water ($\overline{V}_o = 0.018\,L/mol$), because the solubility of octanol in water at room temperature is relatively small, $4.5 \times 10^{-3}\,M$. The solubility of water in octanol is $2.3\,M$, and thus $\overline{V}_o^*$ is computed as $0.12\,L/mol$ (instead of $\overline{V}_o = 0.157\,L/mol$ for pure octanol) on the basis of volume additivity based on component mole fractions. The $S_w$ value for a solid compound is that of the corresponding supercooled liquid at room temperature (25°C), as defined earlier by Eq. (3.9). A list of the measured

$K_{ow}$ and $S_w$ values for a number of substituted aromatic compounds at room temperature is given in Table 5.1.

Consider first the relation between $\log K_{ow}$ and $\log S_w$ for different organic solutes; the solutes have relatively small $S_w$ values (i.e., large $\gamma_w$ values) which span over several orders of magnitude, as shown in Table 5.1. By contrast, the solutes are usually very soluble in (i.e., highly compatible with) most organic solvents. If the solutes form ideal solutions in water-saturated octanol and if the solute solubility is the same in water and in octanol–saturated water, the last two terms in Eq. (5.1) drop out, and what remains is a linear plot of $\log K_{ow}$ versus $\log S_w$, with a slope of $-1$ and an intercept of $-\log \overline{V}_o^*$. The intercept in this case is essentially constant for all solutes in dilute solution. If one sets $K_{ow}^\circ$ as the partition coefficient from the *ideal line*, defined as

$$\log K_{ow}^\circ = -\log S_w - \log \overline{V}_o^* \qquad (5.2)$$

then the difference between $\log K_{ow}^\circ$ and $\log K_{ow}$ for a solute with a given $\log S_w$ expresses the effects of $\log \gamma_o^*$ and $\log (\gamma_w/\gamma_w^*)$ on $\log K_{ow}$. As noted in Eq. (3.10), the term $\gamma_w/\gamma_w^*$ expresses the extent of solute solubility enhancement in water by the dissolved organic solvent (in this case, octanol).

According to Eqs. (5.1) and (5.2), $\log K_{ow}^\circ - \log K_{ow} = \log \gamma_o^* + \log (\gamma_w/\gamma_w^*)$ must be satisfied if the measured values are accurate and if Raoult's law is valid. For *p,p′*-DDT and hexachlorobenzene (HCB), two highly insoluble solutes, the supercooled $\log S_w$ (mol/L) are $-6.74$ and $-5.57$ at 25°C, respectively, and their ($\log K_{ow}^\circ - \log K_{ow}$) values are 1.30 and 0.99. The respective experimental $\gamma_o^*$ values, based on measured solute solubilities in water-saturated octanol, are 7.8 and 5.4, or $\log \gamma_o^* = 0.89$ and 0.73 at 24 to 25°C; the respective experimental $\gamma_w/\gamma_w^*$ values, based on measured solubilities in octanol-saturated water and pure water, are 2.8 and 1.9, or $\log (\gamma_w/\gamma_w^*) = 0.45$ and 0.27 at 24 to 25°C. Thus, the data substantiate the expectation well. The results show that the relative effects of the terms on the right of Eq. (5.1) on $K_{ow}$ are, in decreasing order, water solubility ($S_w$), compatibility with water-saturated octanol ($\gamma_o^*$), and the influence of dissolved octanol on water solubility ($\gamma_w/\gamma_w^*$). The major effect of $S_w$ is evidenced by the small solubility (or the large $\gamma_w$) of relatively nonpolar organic solutes in water. The effect of $\gamma_o^*$, which increases with decreasing $S_w$, is less than 10 for practically all solutes. The effect of octanol saturation in water on solute water solubility ($\gamma_w/\gamma_w^*$), which also increases with decreasing $S_w$, is significant only for extremely water-insoluble solutes (as liquids or supercooled liquids). We shall see later that the solubility-enhancement effect for low-$S_w$ solutes by a dissolved organic substance is influenced not only by the concentration of the dissolved organic substance but also profoundly by its molecular weight, polar-group content, and molecular conformation.

In light of the fact that $S_w$ is the dominant factor in determining the magnitude of $K_{ow}$, a linear correlation should exist between these two parameters. A plot of $\log K_{ow}$ versus $\log S_w$ for compounds in Table 5.1 is shown in

**TABLE 5.1. Octanol–Water Partition Coefficients ($K_{ow}$) and Liquid or Supercooled-Liquid Solubilities in Water ($S_w$) of Substituted Aromatic Compounds at Room Temperature**[a]

| Compound | $\log S_w$ (mol/L) | $\log K_{ow}$ |
|---|---|---|
| Aniline | −0.405 | 0.90 |
| o-Toluidine | −0.817 | 1.29 |
| m-Toluidine | −0.853 | 1.40 |
| N-Methylaniline | −1.28 | 1.66 |
| N,N-Dimethylaniline | −2.04 | 2.31 |
| o-Chloroaniline | −1.53 | 1.90 |
| m-Chloroaniline | −1.37 | 1.88 |
| Benzene | −1.64 | 2.13 |
| Toluene | −2.25 | 2.69 |
| Ethylbenzene | −2.84 | 3.15 |
| n-Propylbenzene | −3.30 | 3.68 |
| Isopropylbenzene | −3.38 | 3.66 |
| 1,3,5-Trimethylbenzene | −3.09 | 3.42 |
| t-Butylbenzene | −3.60 | 4.11 |
| Fluorobenzene | −1.80 | 2.27 |
| Chlorobenzene | −2.36 | 2.84 |
| Bromobenzene | −2.55 | 2.99 |
| Iodobenzene | −2.78 | 3.25 |
| o-Xylene | −2.72 | 2.77 |
| m-Xylene | −2.73 | 3.20 |
| p-Xylene | −2.73 | 3.15 |
| Diphenylmethane | −4.07 | 4.14 |
| o-Dichlorobenzene | −2.98 | 3.38 |
| m-Dichlorobenzene | −3.04 | 3.38 |
| p-Dichlorobenzene | (−3.03) | 3.39 |
| 1,2,4-Trichlorobenzene | −3.57 | 4.02 |
| Biphenyl | (−3.88) | 4.09 |
| 2-PCB | (−4.57) | 4.54 |
| 3-PCB | −5.16 | 4.95 |
| Naphthalene | (−3.08) | 3.36 |
| 2-Methylnaphthalene | (−3.69) | 4.11 |
| Phenanthrene | (−4.48) | 4.57 |
| Anthracene | (−4.63) | 4.54 |
| Pyrene | (−5.24) | 5.18 |
| Hexachlorobenzene | (−5.57) | 5.50 |
| p,p'-DDT | (−6.74) | 6.36 |

*Source*: Data from Chiou et al. (1982b).

[a] Values at 20 to 25°C. Numbers in parentheses are the supercooled liquid solubilities of the solid solutes, estimated according to Eqs. (3.9) and (3.25) with $\Delta \bar{S}_{fus} = 56.5$ J/mol·K.

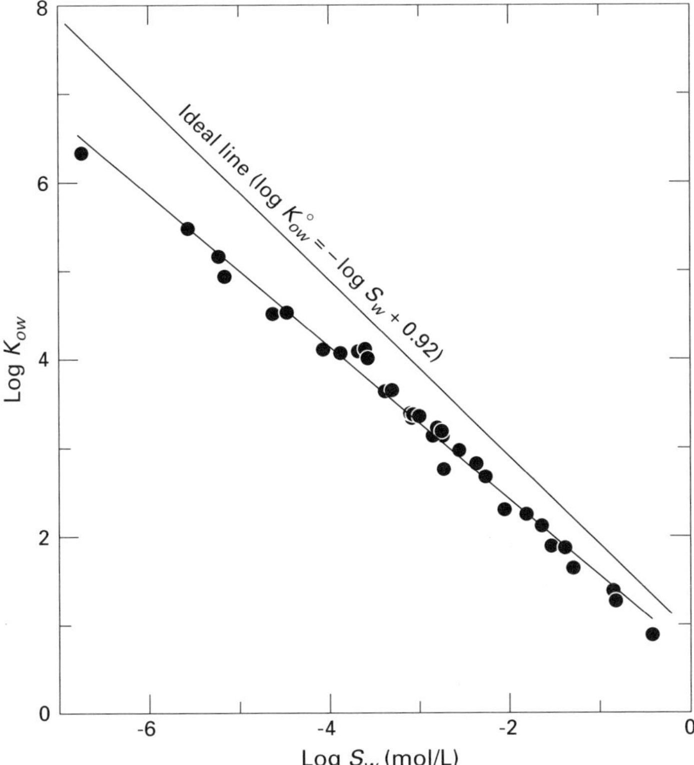


**Figure 5.1** Ideal line for solutes in octanol–water systems and the measured octanol–water partition coefficients ($K_{ow}$) of substituted aromatic compounds in Table 5.1 in relation to their water solubilities ($S_w$). [Data from Chiou et al. (1982b). Reproduced with permission.]

Figure 5.1 along with the ideal line for the octanol–water system. The regression of $\log K_{ow}$ with $\log S_w$ for 36 compounds gives

$$\log K_{ow} = -0.862 \log S_w + 0.710 \tag{5.3}$$

which extends over six orders of magnitude in $S_w$ and five orders of magnitude in $K_{ow}$ with a correlation coefficient $r^2 = 0.99$. As one will see later, the high correlation between $\log K_{ow}$ and $\log S_w$ for various solutes is closely related to the molecular properties of octanol.

The significance in the use of supercooled-liquid $S_w$ values for solid solutes for the $\log K_{ow}$–$\log S_w$ correlation of a group of solid and liquid solutes is underscored by the following examples. Consider, for instance, the $S_w$ and $K_{ow}$ values of phenanthrene and pyrene, both being solid at room temperature. If one were to use their respective solid–solute $\log S_w^{(s)}$ of –5.14 and –6.18 mol/L rather than their supercooled-liquid $\log S_w^{(l)}$ of $-4.48$ and $-5.24\,\text{mol/L}$, in the $\log K_{ow}$–$\log S_w$ plot, the resulting data points would shift toward the left by 0.66 and 0.94 unit, respectively, which would then separate these data points from the rest and reduce the overall $\log K_{ow}$–$\log S_w$ correlation. Another vivid example of the melting-point effect on $S_w$, and thus on the $\log K_{ow}$–$\log S_w$ correlation, is illustrated by the $S_w$ and $K_{ow}$ data of $m$-dichlorobenzene and $p$-dichlorobenzene (Table 5.1). At room temperature, the $S_w$ of the para isomer, which is a solid, is about half that of the meta isomer, which is a liquid, but the supercooled-liquid para isomer and the liquid meta isomer exhibit about the same $S_w^{(l)}$ and hence about the same $K_{ow}$, as expected. The same is true for phenanthrene and anthracene at room temperature in that the two solid isomers differ significantly in melting point (101 and $216°C$, respectively) and in water solubility ($\log S_w^{(s)} = -5.14$ and $-6.60$, respectively) but they display comparable supercooled-liquid $S_w^{(l)}$ and thus $K_{ow}$, as shown in Table 5.1.

The observed $\log K_{ow}$–$\log S_w$ correlation [Eq. (5.3)] together with the octanol–water ideal line [Eq. (5.2)] provides an effective means to account for the change in solute solubility in the octanol phase with increasing solute $\log K_{ow}$ or decreasing solute $\log S_w$. By treating octanol as a lipidlike substance, as substantiated later, one can see how the lipophilicity of a group of solutes varies with $S_w$. The lipophilicity of a solute should in a strict sense be related to the inverse of $\gamma_o^*$. The $\log \gamma_o^*$ values of most solutes, except highly insoluble ones such as DDT and HCB, are simply equal to the vertical distances between the ideal line and the experimental line in Figure 5.1. As noted, this vertical distance increases with decreasing $S_w$. This implies that in a homologous series of solutes, the higher-molecular-weight, less water-soluble compounds (i.e., the ones with larger $K_{ow}$ values) are not more lipophilic than the more water-soluble compounds. According to the $\gamma_o^*$ data, the solute affinity with octanol decreases with increasing $K_{ow}$ (or decreasing $S_w$), indicating that there is actually an increase in solute–octanol incompatibility as the solute molecular-weight increases. In essence, the higher $K_{ow}$ values, or lipid–water partition coefficients, for the latter solutes result from their much lower $S_w$ values rather than from their enhanced solubilities in octanol (or a lipid). To avoid the confusion of the term *lipophilicity* or *lipophilic* being used to refer to a compound, one must keep in mind that it only implies that the compound has a high lipid–water partition coefficient (i.e., its solubility in lipids is significantly higher than that in water). Thus, although all compounds with low $S_w$ values tend to be lipophilic, their solubilities in lipids usually bear no direct relation to the order of their solvent–water or lipid–water partition coefficients.

The correlation presented in Eq. (5.3) has also been found to give a reasonable account of the partition coefficients for many other classes of organic compounds, including moderately soluble alcohols, ketones, and ethers and sparingly soluble esters, alkyl halides, alkanes, and alkenes (Chiou et al., 1982b). This wide correlation for solutes of many classes presumably results from the unique molecular structure of $n$-octanol, which possesses a nonpolar chain of moderate size and a hydrophilic OH group; these impart to the molecule a unique polar-to-nonpolar balance with a weak-to-moderate polarity. This unique molecular property empowers octanol to accommodate a wide range of organic compounds with comparable solvencies through its hydrophilic OH and/or its nonpolar alkyl chain. As a consequence, the $\gamma_o^*$ values of a wide range of sparingly water-soluble solutes fall into a small range. We shall see later that the partition effects of solutes in some other solvent–water systems share some common features with those in the octanol–water system, while other systems manifest important differences in solute partition behavior as a result of large discrepancies in solvent composition and polarity.

## 5.3   HEPTANE–WATER SYSTEMS

The $n$-heptane/water mixture offers an extreme but instructive system for examining important differences in the partitioning of polar and nonpolar compounds into a highly nonpolar organic phase. As with the octanol–water system, the molecular-size differences between most solutes and heptane are usually not too large to negate the use of Raoult's law for treating solute partition with heptane. Note here that the mutual solubility of heptane and water is very small at room temperature, the solubility of heptane in water being about $9.5 \times 10^{-5}\,M$ and that of water in heptane being $5.3 \times 10^{-3}\,M$. Thus, Eq. (3.11) can be simplified by treating the molar volumes of water-saturated heptane and heptane-saturated water to be essentially the same as the molar volumes of the respective pure solvents. A further approximation can be made by assuming that the small amount of heptane in water has no significant effect on the solubility of solutes in the water phase. With these simplifications, Eq. (3.11) is reduced to

$$\log K_{hw} = -\log S_w - \log \overline{V}_h - \log \gamma_h \qquad (5.4)$$

where $K_{hw}$ is the heptane–water partition coefficient of the solute, $\overline{V}_h$ the molar volume of heptane (0.146 L/mol), and $\gamma_h$ the activity coefficient of the solute in heptane. The ideal line for the heptane–water system is therefore

$$\log K_{hw}^\circ = -\log S_w - \log \overline{V}_h \qquad (5.5)$$

where $\log K_{hw}^\circ - \log K_{hw} = \log \gamma_h$ applies for a solute at a particular $\log S_w$.

The $K_{hw}$ values for a series of organic solutes and the corresponding $K_{ow}$ values at room temperature are given in Table 5.2 for comparison. If polar and nonpolar solutes should exhibit comparable compatibilities with heptane as with octanol, an equally good linear correlation would exist between $\log K_{hw}$ and $\log S_w$, as noted for the octanol–water system. However, as shown in

**TABLE 5.2. Heptane–Water Partition Coefficients $(K_{hw})^a$ and Octanol–Water Partition Coefficients $(K_{ow})^b$ of Some Substituted Aromatic Compounds**

| Compound | $\log K_{hw}$ | $\log K_{ow}$ |
|---|---|---|
| Aniline | 0.040 | 0.90 |
| o-Toluidine | 0.544 | 1.29 |
| Benzaldehyde | 1.05 | 1.48 |
| Acetophenone | 1.14 | 1.58 |
| Anisole | 2.10 | 2.11 |
| Benzene | 2.26 | 2.13 |
| Toluene | 2.85 | 2.69 |
| Styrene | 3.11 | 2.95 |
| Benzoic acid | −0.72 | 1.85 |
| Phenylacetic acid | −1.07 | 1.30 |
| Nitrobenzene | 1.49 | 1.85 |
| Ethylbenzene | 3.43 | 3.15 |
| n-Propylbenzene | 4.11 | 3.68 |
| 1,3,5-Trimethylbenzene | 4.05 | 3.42 |
| t-Butylbenzene | 4.41 | 4.11 |
| Fluorobenzene | 2.45 | 2.27 |
| Chlorobenzene | 2.95 | 2.84 |
| Bromobenzene | 3.10 | 2.99 |
| Iodobenzene | 3.33 | 3.25 |
| o-Dichlorobenzene | 3.45 | 3.38 |
| m-Dichlorobenzene | 3.55 | 3.38 |
| p-Dichlorobenzene | 3.50 | 3.39 |
| 1,2,4-Trichlorobenzene | 4.06 | 4.02 |
| Biphenyl | 4.21 | 4.09 |
| 2-PCB | 4.76 | 4.54 |
| 2,2′-PCB | 5.05 | 4.80 |
| 2,4′-PCB | 5.34 | 5.10 |
| 2,4,4′-PCB | 5.86 | 5.62 |
| Hexachlorobenzene | 5.96 | 5.50 |
| p,p′-DDT | 6.66 | 6.36 |

[a] Data from Chiou and Block (1986).
[b] Data from data cited in Leo et al. (1971) and Chiou et al. (1982b).

Figure 5.2, the correlation between $\log K_{hw}$ and $\log S_w$ deviates considerably from linearity when relatively polar solutes with large $\log S_w$ are lumped together with the nonpolar solutes. Here an extreme phenomenon arises. The $\log K_{hw}$ values for the nonpolar solutes are comparable in magnitude with the $\log K_{ow}$ values and not far away from the heptane–water ideal line. By contrast, the values for the polar solutes sit far below the ideal line, as exemplified by the $\log K_{hw}$ values of highly polar benzoic acid ($\log S_w = -0.905$ mol/L), phenylacetic acid ($\log S_w = -0.485$ mol/L), and nitrobenzene ($\log S_w =$

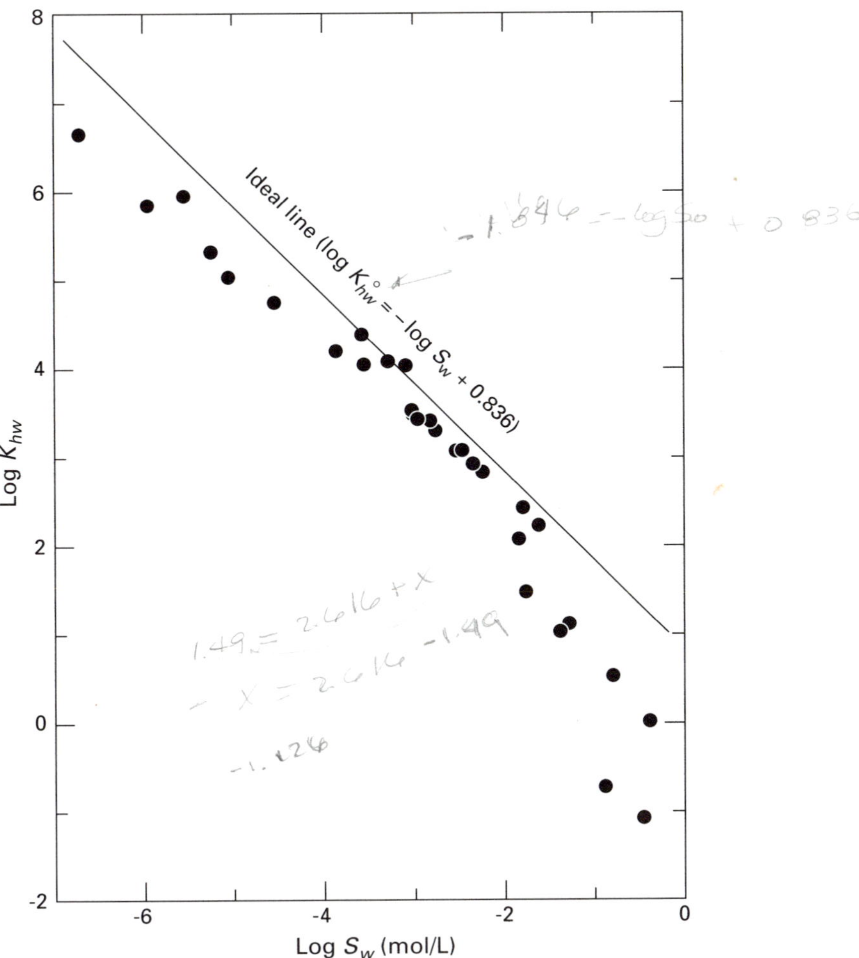



**Figure 5.2** Ideal line for solutes in heptane–water systems and the measured heptane–water partition coefficients ($K_{hw}$) of substituted aromatic compounds in Table 5.2 in relation to their water solubilities ($S_w$). [Data from Chiou and Block (1986). Reproduced with permission.]

–1.78 mol/L). For these three solutes, the $\log K_{ow}$ values deviate only moderately from the octanol–water ideal line. The observed sensitivity of a nonpolar solvent (heptane) in responding to solutes of vastly different polarities is consistent with the rule "like dissolves like." As a consequence, the $\gamma_h$ value varies sharply between solutes, much in reflection of the solute polarity. Thus, if a group of polar solutes are lumped with the nonpolar solutes, the correlation coefficient of $\log K_{hw}$ with $\log S_w$ will vary with the number and type of the polar solutes in the overall data set.

Because of the noted large variation in the response of nonpolar versus polar solutes to a nonpolar medium, the heptane–water system differs in one important respect from the octanol–water system: namely, that the solute water solubility ($S_w$) has a predominant effect on $K_{hw}$ only for relatively nonpolar solutes. This point is corroborated further in the later section by the effects of some polar and nonpolar substituents in benzene on the partition coefficients of substituted benzenes in heptane–water and octanol–water systems.

## 5.4   BUTANOL–WATER SYSTEMS

The $n$-butanol/water mixture represents an opposite extreme to the heptane–water system, in which the solvent phase is remarkably more polar than apolar heptane or weakly polar octanol. Note that butanol is the lowest-molecular-weight alcohol, whose polarity stays just below the level for it to be partially miscible with water; methanol, ethanol, and propanol are completely miscible with water. The solubility of butanol in water is $1.1\,M$ and the solubility of water in butanol is $9.4\,M$ at room temperature. This relatively high mutual solubility affects not only the molar volumes of the two solvent phases (water and butanol) but also, more critically, the solubility behavior of the solutes in the two equilibrium phases.

Although the experimental butanol–water partition coefficients ($K_{bw}$) are quite limited in number, the magnitude of $K_{bw}$ relative to $K_{ow}$ may be understood in terms of the solvent polarity and solvent–water mutual solubility. Dissolved water in octanol at $4.5 \times 10^{-3}\,M$ is known to reduce the solubility of DDT and hexachlorobenzene (two of the most water-insoluble compounds) in water-saturated octanol relative to that in pure octanol by about 20% (Chiou et al., 1982b). The related effect for more water-soluble solutes is expected to be less. Since the dissolved water in butanol at $9.4\,M$ is about 2000 times greater than in octanol and since butanol is much more polar a solvent than octanol, the solubility of nonpolar solutes in water-saturated butanol should be significantly lower than that in water-saturated octanol. Similarly, the greater (but not large) amount of butanol than octanol in water should enhance to a greater extent the solubility of nonpolar solutes in water than that exhibited by the dissolved octanol. Thus, for nonpolar solutes, $K_{bw} < K_{ow}$ is expected. On the other hand, for relatively polar or water-soluble solutes, one may expect $K_{bw} > K_{ow}$ to occur, although the difference may not be very substantial, mainly because the water-saturated butanol may act as a better partition phase than water-saturated octanol for solute partitioning; for such solutes, the solubility enhancement in water by either dissolved butanol or dissolved octanol would not be significant. Overall, the relatively high butanol–water mutual solubility would result in a large compression of the range of $K_{bw}$ relative to that of $K_{ow}$ for solutes that span a wide range of water solubility; the effect should be most noticeable for highly water-insoluble solutes because their water solubilities are sensitive to a dissolved organic solvent.

Westall (1983) determined the $K_{bw}$ values of benzene and chlorinated benzenes. For benzene ($K_{ow} = 135$), the $K_{bw}$ value is about one-half the $K_{ow}$ value. For trichlorobenzene ($K_{ow} \simeq 1050$) and tetrachlorobenzene ($K_{ow} \simeq 4900$) with a significantly reduced water solubility, the $K_{bw}$ values become less than one-fourth and one-fifth, respectively, of the corresponding $K_{ow}$ values. By contrast, for a series of highly water-soluble small organic acids, amines, and alcohols, Collander (1951) found that the $K_{bw}$ values are greater than the corresponding $K_{ow}$ values when $K_{ow} \leq 10$; the biggest deviation, by a factor of about 3.5, is with the most water-soluble solute ($K_{ow} \simeq 0.03$). These characteristics are consistent with the different solvent polarities of butanol and octanol and the related solvent–water mutual solubilities. For most sparingly water-soluble compounds, the $K_{bw}$ values would thus be small fractions of the $K_{ow}$ values; for a wide variety of solutes, the $K_{bw}$ values should fall into a shorter range than the $K_{ow}$ or $K_{hw}$ values. A similar result is expected for solutes in other organic phase–water mixtures where the organic solvent or medium is relatively polar in nature. With this consideration, the partition uptake of relatively water-insoluble solutes by such highly polar organic phases as proteins, cellulose, and carbohydrates should be very weak relative to the partition uptake by the oily substances such as petroleum hydrocarbons, waxes, and biological lipids. In a later section we will have an opportunity to look into the partition characteristics of solutes in the lipid–water system and see which solvent–water system examined so far best mimics solute partition behavior in the lipid–water system.

## 5.5 SUBSTITUENT CONTRIBUTIONS TO PARTITION COEFFICIENTS

The concept of substituent contribution to the partition coefficient of a substituted molecule with respect to a parent molecule was introduced by Fujita et al. (1964) in medicinal chemistry and pharmaceutical science for estimating the $K_{ow}$ values of some drugs and other chemicals in the absence of their experimental values. It has gained relatively good success when applied to small and structurally simple molecules but has had less success when extended to more complicated molecules. To understand the feasibility and limitation of the concept, one must unravel the physical basis associated with the contribution of a substituent to the partition coefficient of a reference (parent) molecule. Elucidation of the relevant factors on the substituent effect on $K_{ow}$ enables one to understand not only the observed effect and limitation in the octanol–water system but also the associated effect and limitation in other solvent–water systems.

According to the convention adopted by Fujita et al. (1964), when a substituent $X$ is incorporated into a parent molecule by replacing one of its H atoms, the impact on the partition coefficient of the substituted molecule is termed $\pi_X$, which is defined as

$$\pi_X = \log K_X - \log K_R \tag{5.6}$$

where $K_R$ is the partition coefficient of the parent solute (molecule) and $K_X$ is that of the substituted solute with a given solvent–water system. For substituted aromatic solutes, benzene is commonly chosen as the parent (or reference) compound.

Since $\pi_X$ is a derived quantity, the factors that affect it must be contained in the expression for partition coefficient, as illustrated with $K_{ow}$ and $K_{hw}$. In the absence of a specific solute interaction with a solvent or the dissociation of the solute, the $\pi_X$ in octanol–water and heptane–water systems is accounted for by Chiou et al. (1982a) by substituting Eqs. (5.1) and (5.4) into (5.6) to give, respectively,

$$\pi_X(\text{oct-water}) \simeq \log[(S_w)_R/(S_w)_X] - \log[(\gamma_o^*)_X/(\gamma_o^*)_R] \tag{5.7}$$

and

$$\pi_X(\text{hep-water}) \simeq \log[(S_w)_R/(S_w)_X] - \log[(\gamma_h)_X/(\gamma_h)_R] \tag{5.8}$$

where the $\log(\gamma_w/\gamma_w^*)$ term has been neglected because its contribution is small for most solutes in these two systems. If one designates the first term on the right-hand sides of Eqs. (5.7) and (5.8) as

$$\log[(S_w)_R/(S_w)_X] = \log[(\gamma_w)_X/(\gamma_w)_R] = \Delta_X \tag{5.9}$$

in which $\Delta_X$ indicates the change in solute solubility or activity coefficient in water when substituent $X$ is incorporated into the parent molecule, one obtains

$$\pi_X(\text{oct-water}) \simeq \Delta_X - \log[(\gamma_o^*)_X/(\gamma_o^*)_R] \tag{5.10}$$

and

$$\pi_X(\text{hep-water}) \simeq \Delta_X - \log[(\gamma_h)_X/(\gamma_h)_R] \tag{5.11}$$

By Eqs. (5.10) and (5.11), the substituent contribution to solute partition coefficient can be estimated in terms of the water solubilities of parent and substituted molecules and their compatibilities with a specific solvent.

The $\pi_X(\text{oct-water})$ and $\pi_X(\text{hep-water})$ values for solutes calculated from their $K_{ow}$ and $K_{hw}$ data and the respective $\Delta_X$ values from their liquid $S_w$ or $\gamma_w$ values using benzene as the parent molecule are presented in Table 5.3. A plot of $\pi_X(\text{oct-water})$ versus $\Delta_X$ is given in Figure 5.3, and a similar plot of $\pi_X(\text{hep-water})$ versus $\Delta_X$ is given in Figure 5.4. As noted, the $\pi_X(\text{oct-water})$ values approach the $\Delta_X$ values for all nonpolar substituents (e.g., alkyl and

**TABLE 5.3. Calculated Values of $\Delta_X$, $\pi_X$ (octanol–water), and $\pi_X$ (heptane–water) for Substituents in Benzene**[a]

| Compound | Substituent, $X$ | $\Delta_X$ | $\pi_X$(oct-w) | $\pi_X$(hep-w) |
|---|---|---|---|---|
| Benzene | — | 0 | 0 | 0 |
| Toluene | $CH_3$ | 0.60 | 0.56 | 0.59 |
| Styrene | $C_2H_3$ | 0.83 | 0.82 | 0.85 |
| Ethylbenzene | $C_2H_5$ | 1.20 | 1.02 | 1.17 |
| $o$-Xylene | 1-$CH_3$-2-$CH_3$ | 1.08 | 0.99 | 1.13 |
| $m$-Xylene | 1-$CH_3$-3-$CH_3$ | 1.09 | 1.07 | 1.28 |
| $n$-Propylbenzene | $n$-$C_3H_7$ | 1.66 | 1.55 | 1.85 |
| 1,3,5-Trimethylbenzene | 1,3,5-$(CH_3)_3$ | 1.46 | 1.29 | 1.79 |
| $t$-Butylbenzene | $t$-$C_4H_9$ | 1.96 | 1.98 | 2.15 |
| Fluorobenzene | F | 0.16 | 0.14 | 0.19 |
| Chlorobenzene | Cl | 0.72 | 0.71 | 0.69 |
| Bromobenzene | Br | 0.91 | 0.86 | 0.84 |
| Iodobenzene | I | 1.14 | 1.12 | 1.07 |
| $o$-Dichlorobenzene | 1-Cl-2-Cl | 1.34 | 1.22 | 1.19 |
| $m$-Dichlorobenzene | 1-Cl-3-Cl | 1.40 | 1.25 | 1.28 |
| 1,2,4-Trichlorobenzene | 1,2,4-$(Cl)_3$ | 1.93 | 1.89 | 1.80 |
| $\alpha,\alpha,\alpha$-Trifluorotoluene | $CF_3$ | 0.88 | 0.88 | 1.05 |
| Aniline | $NH_2$ | −1.24 | −1.23 | −2.22 |
| m-Chloroaniline | 1-$NH_2$-3-Cl | −0.27 | −0.25 | −1.55 |
| $o$-Toluidine | 1-$NH_2$-2-$CH_3$ | −0.83 | −0.84 | −1.72 |
| $m$-Toluidine | 1-$NH_2$-3-$CH_3$ | −0.79 | −0.73 | −1.72 |
| Phenol | OH | −1.70 | −0.67 | −3.18 |
| Benzoic acid | COOH | −0.73 | −0.28 | −2.98 |
| Phenylacetic acid | $CH_2COOH$ | −1.15 | −0.83 | −3.33 |
| Anisole | $OCH_3$ | 0.21 | −0.02 | −0.16 |
| Acetophenone | $COCH_3$ | −0.33 | −0.40 | −1.12 |
| Benzaldehyde | CHO | −0.23 | −0.65 | −1.21 |
| Nitrobenzene | $NO_2$ | 0.14 | −0.28 | −0.77 |
| Benzonitrile | CN | −0.26 | −0.57 | −1.36 |

*Source*: Data from Chiou et al. (1982a).

[a] The $\Delta$ values are calculated from Eq. (5.9) using the $\log \gamma_w$ values of substituted benzenes and $\log \gamma_w = 3.38$ for benzene as the parent solute at ~25°C. The $\pi_X$(oct-w) values and $\pi_X$(hep-w) values of the substituents are derived from the $\log K_{ow}$ and $\log K_{hw}$ values of the substituted benzenes with $\log K_{ow} = 2.13$ and $\log K_{hw} = 2.26$ for benzene.

halogen groups) and for many polar substituents (e.g., —$OCH_3$, —CHO, —$NO_2$, —CN, —$NH_2$), with the exception of —OH (phenol) and —COOH (benzoic acid). This observation is in accord with the earlier finding that $S_w$ is the principal determinant of $K_{ow}$ for most solutes, because the structure of octanol enables it to exhibit about the same solvency for solutes with a range of polarities. The data with phenol and benzoic acid, where $\pi_X$(oct-water) is much greater than $\Delta_X$, indicate that these highly polar solutes exhibit excep- tional compatibilities with (water-saturated) octanol, probably due to their

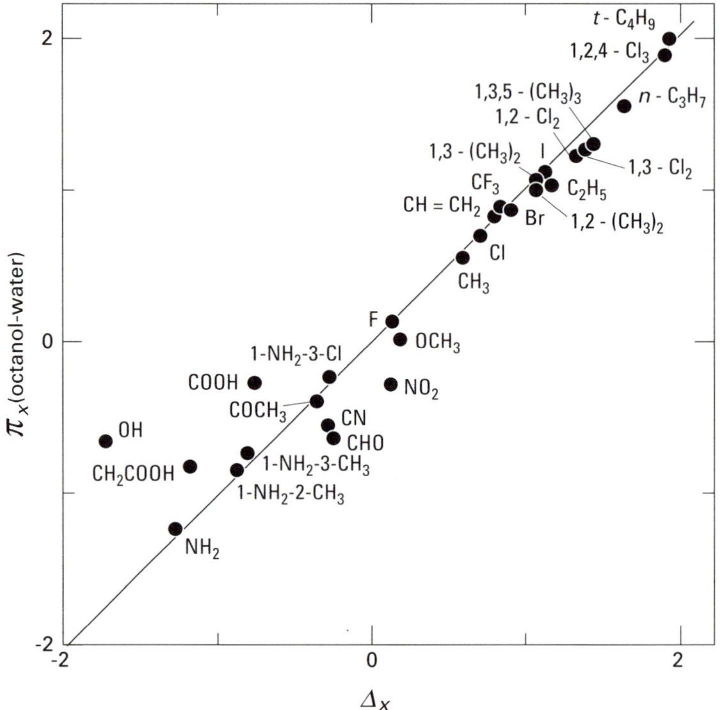



**Figure 5.3** Plot of $\pi_X$(octanol–water) versus $\Delta_X$ for common substituents with benzene as the reference standard. The line represents $\pi_X$ (octanal–water) = $\Delta_X$. [Data from Chiou et al. (1982a). Reproduced with permission.]

strong polar interactions and H-bonding effects with the solvent, making $\log\left[(\gamma_o^*)_X/(\gamma_o^*)_R\right]$ a large negative value.

In the heptane–water system, $\pi_X$(hep-water) approximates $\Delta_X$ only for relatively nonpolar substituted benzenes, in which the $\pi_X$(hep-water) values for alkylbenzenes are somewhat higher than the respective $\Delta_X$ value, which is expected because of the increased aliphatic–hydrocarbon contents in these solutes, which improve their compatibilities with aliphatic heptane. On the other hand, weakly and strongly polar groups, such as —OCH$_3$, —CHO, —NO$_2$, —CN, —NH$_2$, —COOH, and —OH, reduce greatly the affinity of the corresponding substituted benzenes with extremely nonpolar heptane, making $\pi_X$(hep-water) considerably lower than the respective $\Delta_X$ for the substituents.

In the octanol–water system, the results indicate that when $\pi_X$ approximates $\Delta_X$ for substitutent $X$ and $\pi_Y$ approximates $\Delta_Y$ for substitutent $Y$ in monosubstituted benzenes, the values of $\pi_{XX}$ (or $\pi_{YY}$) and $\pi_{XY}$ also approximate the corresponding values of $\Delta_{XX}$ (or $\Delta_{YY}$) and $\Delta_{XY}$ for disubstitutents $X,X$ and $X,Y$ attached to benzene. Supporting data are demonstrated with xylenes and dichlorobenzenes ($\pi_{XX} \simeq \Delta_{XX}$) and with toluidines and $m$-chloroaniline ($\pi_{XY}$

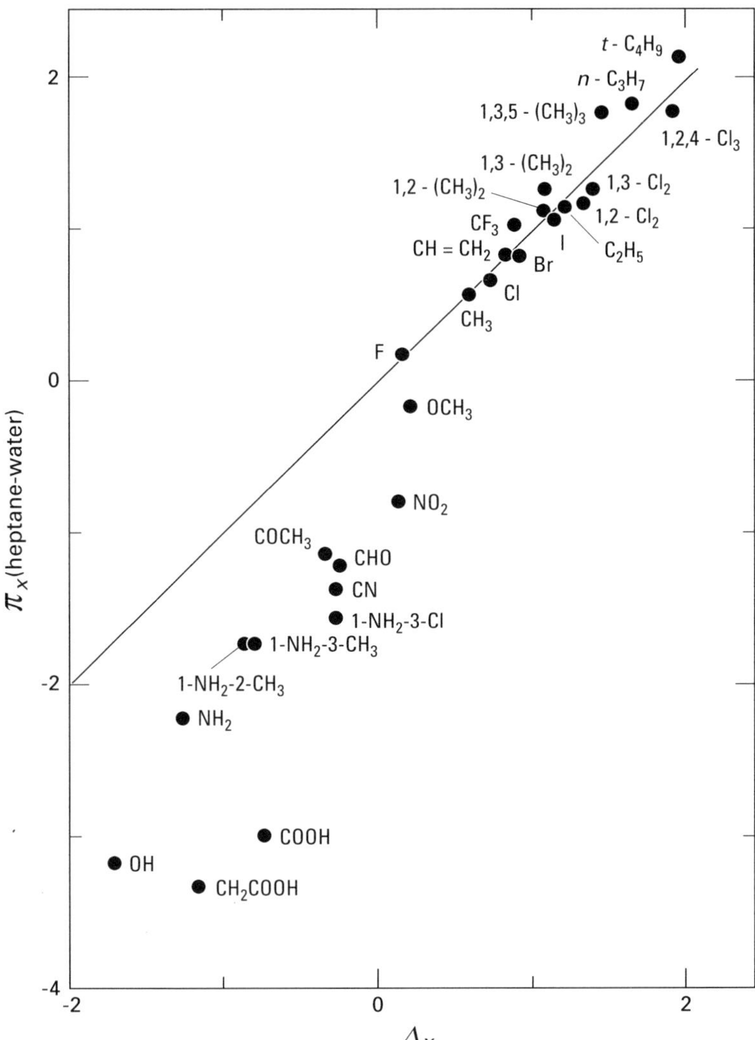


**Figure 5.4**  Plot of $\pi_X$(heptane–water) versus $\Delta_X$ for common substituents with benzene as the reference standard. The line represents $\pi_X$ (heptane–water) = $\Delta_X$. [Data from Chiou et al. (1982a). Reproduced with permission.]

$\simeq \Delta_{XY}$). It is recognized, however, that the magnitudes of $\pi_{XX}$ and $\pi_{XY}$ (or $\Delta_{XX}$ and $\Delta_{XY}$) in disubstitution are not necessarily additive of $\pi_X$ and $\pi_Y$ (or $\Delta_X$ and $\Delta_Y$) in monosubstitution. This is because the increment in the solute activity coefficient with addition of a substituent may vary from compound to compound (i.e., from benzene to a substituted benzene) and from solvent to solvent (e.g., from water to octanol). Since the *additivity rule* may not be strictly obeyed when more than one substituent is involved, it is better to consider the set of substituents as a whole rather than to treat them as a sum of independent components, although the latter approach offers a quick rough estimate of the $K_{ow}$ value of a solute with disubstituted or multisubstituted substituents with respect to that of the parent solute (in this case, benzene).

The preceding analysis of the relationship between $\pi_X$(oct-water) and $\Delta_X$ with benzene as a reference also applies reasonably well to many systems with other compounds as reference standards. For instance, when aniline ($\log K_{ow}$ = 0.90) is used as the reference, the values of $\pi_X$ and $\Delta_X$ are 0.98 and 0.98 for $X$ = Cl (meta); 0.39 and 0.42 for $X$ = —$CH_3$ (ortho); and 0.50 and 0.46 for $X$ = —$CH_3$ (meta). If toluene ($\log K_{ow}$ = 2.69) is used as the reference, $\pi_X$ and $\Delta_X$ are 0.43 and 0.48 for $X$ = —$CH_3$ (ortho) and 0.51 and 0.49 for —$CH_3$ (meta). These results agree with the earlier findings that the group contribution to $K_{ow}$ derives essentially from the variation of solute incompatibility with water, although $\pi_X$ may vary from one reference standard to another. Because of the sensitivity of apolar heptane (or another highly nonpolar solvent) to the polarity of the solute, a close relationship between $\pi_X$(hep-water) and $\Delta_X$ exists only for nonpolar substituents (e.g., alkyl and halogen groups) with benzene and other nonpolar compounds as the parent (reference) solutes.

As we have seen with the octanol–water and heptane–water systems, $\pi_X$ for a substituent would be numerically close to the respective $\Delta_X$ when the parent solute and its derivatives exhibit comparable solubilities in the (water-saturated) solvent and when the amount of the solvent dissolved in water is not large enough to affect significantly the solute solubility in water. In those solvent–water mixtures where the solvent–water mutual solubility is considerable, the resulting $\pi_X$ values might thus deviate more significantly from the respective $\Delta_X$ values calculated from solute solubilities (or activity coefficients) in water, although they might be correlated in some fashion with the corresponding $\pi_X$(oct-water) values for a series of substituted solutes. This is because the high solvent–water mutual saturation can affect unequally the solubilities of parent and substituted solutes in both water- and solvent-rich phases [i.e., the $\log(\gamma_w/\gamma_w^*)$ and $\log\gamma_o^*$ terms in Eq. (5.1) can differ significantly between the solutes].

## 5.6   LIPID–WATER SYSTEMS

### 5.6.1   Solubility of Solutes in Lipids

Knowledge of the partition behavior of compounds in lipid–water mixtures forms a crucial link to the potential for bioconcentration of contaminants into aquatic biotic species, such as fish, which constitutes an important part of our biological resources. The lipid–water mixture is also a system of special interest from the standpoint of solution theory because the molecular weights of most biological lipids are considerably greater than those of ordinary solutes and solvents but are substantially smaller than those of typical polymers. The molecular-size ratios of triglycerides (biological lipids) to common contaminants lie in a range of about 3 to 10. Such size disparity offers a rather unique system relative to the solute–polymer system, in which a huge molecular-size difference exists, and to the common solute–solvent system, in which the size difference is relatively small. As such, we have an opportunity to test critically the relative merits of Raoult's law and Flory–Huggins theory for the solute solubility or the solute partition coefficient with a lipid phase. In addition, it gives us a chance to find out which solvent (e.g., octanol, heptane, or others) best mimics the partition effects of organic compounds with a biological lipid.

Prior to our discussion of the solute partition in a lipid–water system, it is instructive to examine first the solubility data of common solutes in a lipid solvent. This will give us a clear picture on the merits of the Flory–Huggins model versus Raoult's law for handling solute solubility in lipids. Triglycerides are considered to be the lipids of most interest because they are an essential part of the lipids in animals and plants and because they have very unique molecular sizes, as mentioned earlier. Triolen (short for glyceryl trioleate, $C_{57}H_{104}O_6$; MW = 885.4) is selected as a model lipid because of its abundance and structural similarity to other triglycerides in organisms. It is selected also because it is a liquid at room temperature that greatly facilitates solubility measurements for solid compounds (note that most nonpolar liquids are completely miscible with triolein).

By combining Eqs. (2.5) and (3.25) for the activity of a solid compound, one obtains on the basis of Raoult's law the solubility of a solid compound in a solvent as

$$\ln x_{id}^{\circ} = \ln x^{s}\gamma^{\circ} = \frac{-\Delta \overline{H}_{fus}}{R}\frac{T_m - T}{TT_m} \tag{5.12}$$

where $x_{id}^{\circ}$ is the ideal mole-fraction solubility of a solid solute, $x^s$ the solid mole-fraction solubility if the solution is nonideal, and $\gamma^{\circ}$ is Raoult's activity coefficient to correct for the solution nonideality at saturation. The other terms are the same as defined earlier. For solutes exhibiting positive deviations from Raoult's law (i.e., $\gamma^{\circ} > 1$), the solid solubility at the point of saturation ($x^s$) cannot exceed $x_{id}^{\circ}$ if Raoult's law holds. Thus, if there is no specific solute–solvent interaction or solute–solute molecular association, the ideal solubility of a solid compound on a weight-fraction or molar-concentration basis is expected to decrease with increasing solvent molecular weight (or solvent molar volume) according to Raoult's law. This expectation follows from the reasoning that when the solvent molecular weight increases, the mass of the dissolved solid solute will have to decrease to maintain a constant solute mole fraction in solution.

Based on the experimental data shown later, the solid solubility observed in triolein often exceeds the Raoult's law ideal solubility limit as defined by Eq. (5.12), even when the molecular-size disparity between solute and triolein is only moderately large. To account satisfactorily for the solubility observed in triolein, Chiou and Manes (1986) modified the conventional Raoult's law by incorporating the Flory–Huggins model [Eq. (2.15) with $\chi = 0$] into Eq. (3.25), which leads to

$$\ln\phi_{at}^{\circ} - \phi_{at}^{\circ}\left(1 - \frac{\overline{V}}{\overline{V_t}}\right) = \frac{-\Delta\overline{H}_{\text{fus}}}{R}\frac{T_m - T}{TT_m} - \left(1 - \frac{\overline{V}}{\overline{V_t}}\right) \tag{5.13}$$

where $\phi_{at}^{\circ}$ is the volume-fraction athermal solubility of a solid solute and $\overline{V}_t$ is the molar volume of triolein (0.966 L/mol). Here the term *athermal solubility* is adopted to replace the conventional ideal solubility by Raoult's law, since the latter becomes invalid for a macromolecular system.

The suitability of Raoult's law [Eq. (5.12)] versus the Flory–Huggins model [Eq. (5.13)] for ordinary solutes with a lipid solvent is here examined against the measured solubilities of some relatively nonpolar solids in triolein, as shown in Table 5.4. The size disparity between triolein and the solutes based on their molar volumes falls into the range $\overline{V}_t/\overline{V} = 3.9$ to 8.5. Solubility data for solids having high melting points ($T_m$) and high heats of fusion ($\Delta\overline{H}_{\text{fus}}$) are excluded from consideration because the solid activity calculated is sensitive to uncertainties in $T_m$ and $\Delta\overline{H}_{\text{fus}}$. Since the solids and triolein selected have similar compositions and polarities, their solutions are not expected to deviate greatly from being ideal or athermal.

As shown in Table 5.4, the observed (mole-fraction) solubilities of the solids in triolein are higher than $x_{\text{id}}^{\circ}$ given by Eq. (5.12) by as much as 100%. On the other hand, the observed solid solubilities on a volume-fraction basis are either close to or lower than the respective athermal volume-fraction solubilities according to Eq. (5.13). The results are therefore in much better agreement with the Flory–Huggins model than with Raoult's law. Of particular significance are the data with lindane, fluoranthene, and DDT, which exhibit only moderate size disparities with triolein ($\overline{V}_t/\overline{V} = 4$ to 5). The magnitude of the negative deviation from Raoult's law is beyond the uncertainty of observed and calculated solubilities. Since the experimental data are well reconciled with the Flory–Huggins model (with $\chi = 0$) and since there is no convincing evidence for the occurrence of any strong specific interaction of these nonpolar solutes with triolein, the negative deviation observed with Raoult's law (i.e., $\gamma^{\circ} < 1$) is clearly an artifact of the model for which there is no physical justification.

A contrary finding in favor of Raoult's law over the Flory–Huggins model was reported by Shinoda and Hildebrand (1957, 1958) for some binary mixtures with molar–volume ratios as high as 9:1. However, these results are for rare mixtures of globular and compact molecules that do not conform to the Flory–Huggins postulate for chainlike molecules. As pointed out by Flory (1970), these rare mixtures do not fulfill the condition of equal accessibility of the total volume to molecular segments of the solute and solvent. For lipid triolein, the segments of the hydrocarbon chains are apparently relatively free

**TABLE 5.4. Solubilities of Solid Organic Compounds in Triolein (g per 100 g of triolein) and Related Physical Properties**[a]

| Compound | $T_m$ (K) | $\Delta \overline{H}_{\text{fus}}$ | MW | $\overline{V}$ | Solubility[b] | $T$ (K) | $x^{\circ}_{\text{ob}}$ | $x^{\circ}_{\text{id}}$ | $\phi^{\circ}_{\text{ob}}$ | $\phi^{\circ}_{at}$ |
|---|---|---|---|---|---|---|---|---|---|---|
| Naphthalene | 353 | 19.3 | 128.19 | 130 | $18.41 \pm 3.32$ | 296 | $0.556 \pm 0.045$ | 0.282 | $0.146 \pm 0.023$ | 0.133 |
| p-Dichlorobenzene | 326 | 18.2 | 147.01 | 114 | $56.70 \pm 7.92$ | 296 | $0.771 \pm 0.024$ | 0.506 | $0.286 \pm 0.028$ | 0.265 |
|  |  |  |  |  | 63.3 | 310 | 0.792 | 0.707 | 0.310 | 0.427 |
| Acenaphthene | 369 | 21.0 | 154.21 | 171 | $10.30 \pm 1.81$ | 296 | $0.369 \pm 0.041$ | 0.185 | $0.0948 \pm 0.0152$ | 0.0871 |
| Biphenyl | 344 | 18.5 | 154.21 | 155 | $19.93 \pm 6.47$ | 296 | $0.519 \pm 0.083$ | 0.350 | $0.153 \pm 0.043$ | 0.175 |
|  |  |  |  |  | 41.5 | 310 | 0.705 | 0.491 | 0.277 | 0.265 |
| 2,6-Dimethylnaphthalene | 383 | 24.2 | 156.23 | 155 | $5.47 \pm 1.64$ | 296 | $0.234 \pm 0.054$ | 0.107 | $0.0473 \pm 0.0136$ | 0.0928 |
| 2,3-Dimethylnaphthalene | 378 | 19.8 | 156.23 | 155 | $7.91 \pm 2.31$ | 296 | $0.304 \pm 0.063$ | 0.175 | $0.0671 \pm 0.0185$ | 0.152 |
| Fluorene | 389 | 18.9 | 166.23 | 165 | $9.56 \pm 1.01$ | 296 | $0.337 \pm 0.024$ | 0.160 | $0.0801 \pm 0.0080$ | 0.0742 |
| Phenanthrene | 374 | 18.6 | 178.24 | 170 | $10.22 \pm 2.37$ | 296 | $0.333 \pm 0.052$ | 0.206 | $0.0818 \pm 0.0175$ | 0.0980 |
| Fluoranthene | 384 | 19.0 | 202.26 | 200 | $7.82 \pm 1.75$ | 296 | $0.253 \pm 0.043$ | 0.171 | $0.0662 \pm 0.0140$ | 0.0828 |
| Lindane | 386 | 24.3 | 290.8 | 186 | $9.28 \pm 0.22$ | 298 | $0.224 \pm 0.004$ | 0.107 | $0.0526 \pm 0.0011$ | 0.0498 |
|  |  |  |  |  | 15.3 | 310 | 0.318 | 0.157 | 0.0824 | 0.0742 |
| p,p'-DDT | 382 | 25.4 | 354.49 | 250 | $8.00 \pm 3.10$ | 296 | $0.163 \pm 0.054$ | 0.0979 | $0.0487 \pm 0.0183$ | 0.0485 |
|  |  |  |  |  | $9.54 \pm 0.57$ | 298 | $0.192 \pm 0.008$ | 0.105 | $0.0579 \pm 0.032$ | 0.0520 |
|  |  |  |  |  | 10.05 | 310 | 0.200 | 0.156 | 0.0608 | 0.0789 |

*Source*: Data from Chiou and Manes (1986).

[a] $T_m$ = solute melting point; $T$ = system temperature; $\Delta \overline{H}_{\text{fus}}$ = solute molar heat of fusion (kJ/mol); MW = molecular weight; $\overline{V}$ = solute molar volume (mL/mol); $x^{\circ}_{\text{ob}}$ = observed solute mole-fraction solubility; $x^{\circ}_{\text{ib}}$ = ideal solute mole-fraction solubility by Raoult's law; $\phi^{\circ}_{\text{ob}}$ = observed solute volume-fraction solubility; and $\phi^{\circ}_{\text{at}}$ = athermal solute volume-fraction solubility by Eq. (5.13).
[b] Solubility data at $T$ = 296 K from Patton et al. (1984); data at $T$ = 310 K from Dobbs and Williams (1983); data at $T$ = 298 K from Chiou and Manes (1986).

to interact individually with other segments and with solute molecules, despite the fact that the chains are connected to one end. More generally, if the solvent has a considerably higher molecular weight than the solute and possesses many flexible segments, Raoults' law tends to overestimate the solute activity and therefore underestimate the solute solubility. This is because Raoult's law takes no account of the molecular size disparity between solute and solvent on the entropy of mixing. Whereas the same effect could occur in other systems with similar solute–solvent size disparities, the effect may well escape recognition in those systems in which there is significant solute–solvent incompatibility. Here the molecular incompatibility and size-disparity effects may offset each other, and the experimental data could then be interpreted erroneously as a confirmation of Raoult's law.

### 5.6.2   Lipid–Water Partition Coefficient

Information on the solute partition behavior in lipid–water mixtures is essential to an understanding of contaminant bioconcentration potentials in natural aquatic environments. Meanwhile, it offers a direct account of a chemical's lipophilicity as well as an important reference to the fish bioconcentration factor (BCF) observed. As before, we select triolein as the model lipid in our analysis of the lipid–water partition coefficient.

The preceding section showed evidence that Raoult's law is inappropriate for describing the solute solubility in triolein. We would expect the Raoult's law–based partition equation [Eq. (3.11)] to suffer the same drawback. This is despite the fact that it proved to be a reasonable model for octanol–water and heptane–water systems, in which the solute and solvent have comparable molecular sizes. The anticipated problem for common solutes in triolein–water mixtures may be appreciated more directly by considering the solvent molar volume term in Eq. (3.11), which, when substituted for triolein, gives

$$\log K_{tw} = -\log S_w - \log \overline{V}_t^* - \log \gamma_t^* - \log\left(\gamma_w \big/ \gamma_w^*\right) \qquad (5.14)$$

in which the small $\log(\gamma_w/\gamma_w^*)$ term may be dropped for most solutes, as rationalized in the earlier discussion on $K_{ow}$. The dependence of $K_{tw}$ on $\overline{V}_t^*$ in Eq. (5.14) implies that if one were to measure the partition coefficients of a solute with a series of solvents having similar compositions but very different molecular weights, the partition coefficient should decrease sharply as the solvent molar volume becomes very large. Thus, by Eq. (5.14), the $K_{tw}$ values measured should become considerably smaller than, say, the corresponding $K_{ow}$ values, since the molar volume of triolein is about eight times that of octanol and since triolein and octanol have quite similar molecular properties. However, the $K_{tw}$ data measured do not conform to this expectation. Alternatively put, analysis of $K_{tw}$ by Eq. (5.14) would force one to assume a fractional $\gamma_t^*$ value, as illustrated below, which could not be well justified.

Determined $K_{tw}$ values, corresponding $K_{ow}$ values, solubilities in water ($S_w$), and molar volumes ($\overline{V}$) of 38 organic solutes at room temperature (20 to 25°C) are listed in Table 5.5. The $S_w$ values for solid solutes are the values of their supercooled liquids, calculated from solid solubilities, heats of fusion ($\Delta\overline{H}_{\text{fus}}$), and melting points ($T_m$) according to Eqs. (3.9) and (3.25). For 1,2,3-trichlorobenzene, 1,3,5-trichlorobenzene, 1,2,3,4-tetrachlorobenzene, 1,2,3,5-tetrachlorobenzene, and pentachlorobenzene, which have low melting points ($T_m < 370\,\text{K}$), calculations were made with the assumption of $\Delta\overline{H}_{\text{fus}} = 56.5\,T_m$ (J/mol), along with the solid solubilities of 16.3, 10.6, 7.18, 3.23, and 0.385 mg/L. The molar volumes are those for solutes in the liquid state; densities of 1,2,3-trichlorobenzene, 1,2,3,4-tetrachlorobenzene, pentachlorobenzene, hexachlorobenzene, and DDT at their melting points were determined and used to calculate their $\overline{V}$ values. Liquid molar volumes of 1,3,5-trichlorobenzene and 1,2,3,5-tetrachlorobenzene were assumed to be the same as those of 1,2,3-trichlorobenzene and 1,2,3,4-tetrachlorobenzene. Liquid molar volumes of PCBs were approximated by using the densities of liquid Arochlor PCB mixtures that have approximately the same chlorine numbers as the individual PCBs.

We now show more explicitly the calculated $\gamma_t^*$ values by Raoult's law and their dependence on solute molecular size for the solutes in Table 5.5. For small solutes with $\overline{V}_t/\overline{V} > 6$, Eq. (5.14) leads to $\gamma_t^* = 0.27$ to $0.42$. This implied serious negative deviation from Raoult's law is not justified by the lack of specific solute–solvent interactions between these solutes and triolein, but rather, is an artifact of the model calculation (Chiou and Manes, 1986). As expected, the assumed molecular-size effect on $\gamma_t^*$ by Raoult's law becomes progressively reduced (i.e., the $\gamma_t^*$ increases toward 1) as the solute molecular size increases. Although the resulting $\gamma_t^*$ values for large solutes, such as hexachlorobenzene (HCB), DDT, and some PCBs, are greater than 1, they are not physically rigorous because the observed solubility of DDT and others cannot be well accounted for by Raoult's law, as shown earlier.

With the noted limitation of Raoult's law, Chiou (1985) treated the solute partition coefficient in a triolein–water mixture by application of the Flory–Huggins model [Eq. (3.13)], which gives

$$\log K_{tw} = -\log S_w\overline{V} - \left[\left(1 - \overline{V}/\overline{V}_t^*\right) + \chi\right]/2.303 - \log\left(\gamma_w/\gamma_w^*\right) \qquad (5.15)$$

where $\overline{V}$ is the molar volume of the solute. Other terms remain as defined earlier. The water content in triolein at 25°C is 0.11% by weight (or $5.6 \times 10^{-2}\,M$), which is significantly less than that in octanol ($2.3\,M$). This gives $\overline{V}_t^* = 0.919\,\text{L/mol}$, or $\log \overline{V}_t^* = -0.037$, on the assumption of volume additivity for triolein and water. To simplify the analysis further, again the term $\log(\gamma_w/\gamma_w^*)$ accounting for the solute solubility enhancement in water by dissolved triolein is assumed to be zero.

Since Eq. (5.15) accommodates effectively the measured $\log K_{tw}$ values for all the solutes, it is used as the basis for interpreting the solute behavior in

**TABLE 5.5. Water Solubilities and Partition Coefficients of Organic Compounds in Triolein–Water and Octanol–Water Systems**[a]

| Compound | $\overline{V}$ | $\log S_w^b$ | $\log S_w \overline{V}$ | $\log K_{ow}$ | $\log K_{tw}$ |
|---|---|---|---|---|---|
| Aniline | 0.0911 | −0.405 | −1.45 | 0.90 | 0.91 |
| o-Toluidine | 0.107 | −0.817 | −1.79 | 1.29 | 1.24 |
| Benzaldehyde | 0.102 | −1.51 | −2.50 | 1.48 | 1.58 |
| Acetophenone | 0.117 | −1.31 | −2.24 | 1.58 | 1.61 |
| Anisole | 0.109 | −1.85 | −2.82 | 2.11 | 2.31 |
| Benzene | 0.0894 | −1.64 | −2.69 | 2.13 | 2.25 |
| Toluene | 0.106 | −2.25 | −3.22 | 2.69 | 2.77 |
| Nitrobenzene | 0.102 | −1.78 | −2.77 | 1.85 | 2.15 |
| Ethylbenzene | 0.123 | −2.84 | −3.75 | 3.15 | 3.27 |
| n-Propylbenzene | 0.139 | −3.30 | −4.16 | 3.68 | 3.77 |
| 1,3,5-Trimethylbenzene | 0.139 | −3.09 | −3.95 | 3.42 | 3.56 |
| Fluorobenzene | 0.0938 | −1.80 | −2.83 | 2.27 | 2.33 |
| Chlorobenzene | 0.102 | −2.36 | −3.35 | 2.84 | 2.97 |
| Bromobenzene | 0.105 | −2.55 | −3.53 | 2.99 | 3.12 |
| Iodobenzene | 0.112 | −2.78 | −3.73 | 3.25 | 3.42 |
| o-Dichlorobenzene | 0.113 | −2.98 | −3.98 | 3.38 | 3.51 |
| m-Dichlorobenzene | 0.114 | −3.04 | −3.98 | 3.38 | 3.63 |
| p-Dichlorobenzene | 0.118 | (−3.03) | −3.96 | 3.39 | 3.55 |
| Hexachloroethane | | | | 4.14 | 4.21 |
| 1,2,3-Trichlorobenzene | 0.125 | (−3.74) | −4.64 | 4.14 | 4.19 |
| 1,3,5-Trichlorobenzene | 0.125 | (−3.82) | −4.72 | 4.31 | 4.36 |
| 1,2,3,4-Tetrachlorobenzene | 0.142 | (−4.24) | −5.09 | 4.60 | 4.68 |
| 1,2,3,5-Tetrachlorobenzene | 0.142 | (−4.53) | −5.38 | 4.59 | 4.69 |
| 1,2,4,5-Tetrachlorobenzene | 0.142 | | | 4.70 | 4.70 |
| Hexachlorobutadiene | 0.158 | −5.01 | −5.81 | 4.90 | 5.04 |
| Pentachlorobenzene | 0.166 | (−5.18) | −5.96 | 5.20 | 5.27 |
| Hexachlorobenzene | 0.186 | (−5.57) | −6.30 | 5.50 | 5.50 |
| Biphenyl | 0.155 | (−3.88) | −4.69 | 4.09 | 4.37 |
| 2-PCB | 0.174 | (−4.57) | −5.33 | 4.51 | 4.77 |
| 2,2′-PCB | 0.189 | (−5.08) | −5.57 | 4.80 | 5.05 |
| 2,4′-PCB | 0.189 | (−5.28) | −5.97 | 5.10 | 5.30 |
| 4,4′-PCB | | | | 5.58 | 5.48 |
| 2,4,4′-PCB | 0.204 | (−5.98) | −6.67 | 5.62 | 5.52 |
| 2,5,2′,5′-PCB | | | | 5.81 | 5.62 |
| 2,4,5,2′,5′-PCB | | | | 6.11 | 5.81 |
| 2,4,5,2′,4′,5′-PCB | | | | 6.72 | 6.23 |
| p,p′-DDT | 0.250 | (−6.74) | −7.34 | 6.36 | 5.90 |

*Source*: Data from Chiou (1985).

[a] $S_w$ = solute water solubility (mol/L); $\overline{V}$ = solute molar volume (L/mol); $K_{ow}$ = octanol–water partition coefficient; $K_{tw}$ = triolein–water partition coefficient.

[b] Values in parentheses are for the supercooled liquids.

triolein. By Eq. (5.15), since $\overline{V}/\overline{V}_i^*$ is neither constant nor approaching zero, no single ideal line relating $\log K_{tw}$ versus $\log S_w\overline{V}$ or $\log K_{tw}$ versus $\log S_w$ can be established to describe the solute incompatibility with triolein ($\chi/2.303$). The magnitude of $\chi/2.303$ can only be determined individually for each solute by reference to Eq. (5.15), in which $\log(\gamma_w/\gamma_w^*)$ is neglected for approximation. Calculated $\chi/2.303$ values are generally quite small yet positive ($<0.25$) for benzene derivatives with relatively large $\log S_w$; respective values for larger, less-soluble solutes, such as 2,4′-PCB (0.32), 2,4,4′-PCB (0.81), hexachlorobenzene (0.45), and DDT (1.1), are somewhat greater. True $\chi/2.303$ values for latter nonpolar solutes are probably smaller because of the neglect of $\log(\gamma_w/\gamma_w^*)$, which may be significantly greater than zero because of the sensitivity of their very small water solubilities to a small amount of triolein dissolved in water. However, since the sum of $\chi/2.303$ and $\log(\gamma_w/\gamma_w^*)$ is nevertheless small compared to $-\log S_w$ for all solutes and since the variation of $\overline{V}$ should also be relatively small, $\log S_w$ is evidently the principal determinant of $\log K_{tw}$. In this sense, the triolein–water system is quite similar to the octanol–water system as far as the solute partition is concerned. Therefore, the $K_{tw}$ values exhibit similar magnitudes as the corresponding $K_{ow}$ values for all solutes.

The noted small differences between $K_{tw}$ and $K_{ow}$ values appear to be related to the solute size. The $\log K_{tw}$ values for all simple nonpolar benzene derivatives tend to be slightly greater than $\log K_{ow}$ values, by 0.1 to 0.2. By contrast, $\log K_{tw}$ values of larger nonpolar solutes (hexachlorobenzene, some PCBs, and DDT) are either about the same as or smaller than the respective $\log K_{ow}$ values. The first result manifests that triolein is somewhat less polar than octanol, so that the solution of low-polarity solutes with water-saturated triolein is closer to being ideal or athermal than that with water-saturated octanol. The second result is not well understood. In addition to the effect of $\chi$ on $K_{tw}$, the transition in order between $K_{tw}$ and $K_{ow}$ could be related to the solute water-solubility enhancement, in which the $\log(\gamma_w/\gamma_w^*)$ term is much greater with dissolved triolein in water (at low mg/L levels) than with dissolved octanol in water (585 mg/L). This could potentially happen because triolein is eight times larger in size and is somewhat less polar than octanol (which would make triolein a more effective solubility enhancer per unit weight), although the dissolved triolein mass in water is far less than the dissolved octanol mass. The problem remains to be resolved. The influence of the polarity and molecular size of a dissolved organic matter on solute water solubility is studied in more detail in Chapter 7, where a series of natural macromolecules and synthetic organic materials are employed as the dissolved organic matter.

Because $\log S_w$ is the major factor for both $\log K_{tw}$ and $\log K_{ow}$, $\log K_{tw}$ can be estimated in terms of $\log S_w$ (or $\log S_w\overline{V}$) and $\log K_{ow}$. Plots of $\log K_{tw}$ versus $\log S_w\overline{V}$ and $\log K_{tw}$ versus $\log K_{ow}$ are given in Figures 5.5 and 5.6, respectively. The results show that although $\log K_{tw}$ is closely related to $\log S_w\overline{V}$ and to $\log K_{ow}$, the correlations show a noticeable curvilinearity when $\log K_{tw}$ or

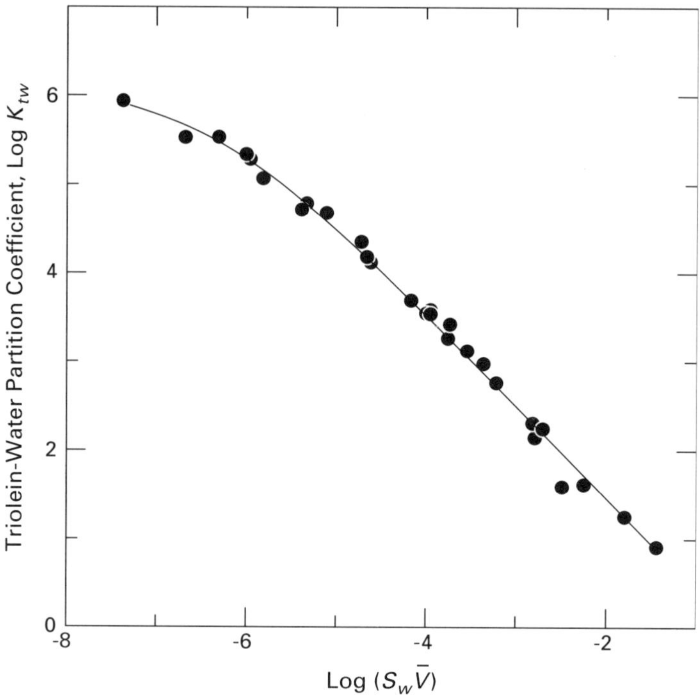

**Figure 5.5**  Plot of $\log K_{tw}$ versus $\log S_w \overline{V}$ for compounds in Table 5.5. [Data from Chiou (1985).]

$\log K_{ow}$ exceeds about 5.0, about the point that separates most substituted benzenes from comparatively larger molecules, such as PCBs and DDT. This nonlinearity appears to originate from the differences in $\chi/2.303$ and $\log(\gamma_w/\gamma_w^*)$ for small and large organic solutes.

Despite unresolved causes that contribute to the upper curvature in Figures 5.5 and 5.6, the $\log K_{tw}$ values for solutes of small to moderate sizes (where $\log K_{ow} < 5.0$) can be satisfactorily correlated in a linear form with respective $\log S_w \overline{V}$ and $\log K_{ow}$ values. The linear regression between $\log K_{tw}$ and $\log S_w \overline{V}$ gives

$$\log K_{tw} = -1.05 \log S_w \overline{V} - 0.646 \tag{5.16}$$

with $n = 23$ and $r^2 = 0.987$. The linear regression between $\log K_{tw}$ and $\log K_{ow}$ gives

$$\log K_{tw} = 1.00 \log K_{ow} + 0.105 \tag{5.17}$$

with $n = 25$ and $r^2 = 0.995$. Since the variation of $\overline{V}$ is relatively small, $\log S_w$ may be used to replace $\log S_w \overline{V}$ in Eq. (5.16), giving

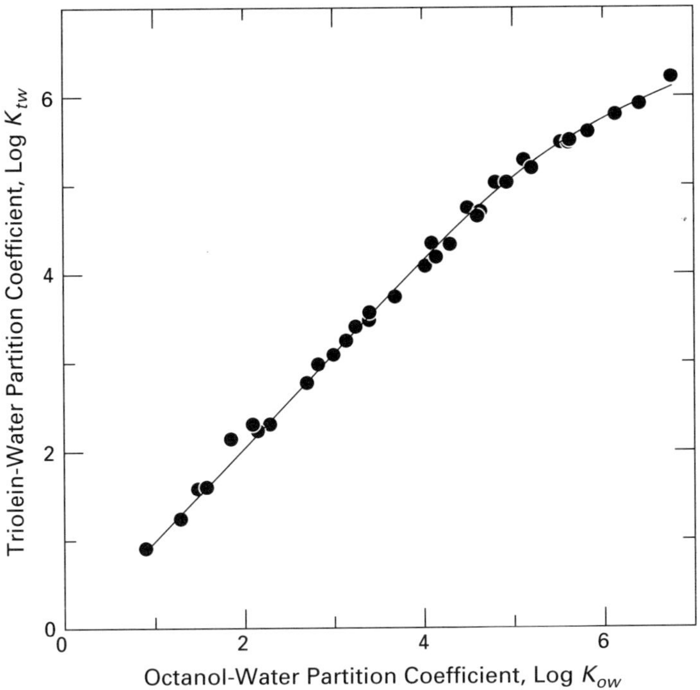



**Figure 5.6**   Plot of $\log K_{tw}$ versus $\log K_{ow}$ for compounds in Table 5.5. [Data from Chiou (1985).]

$$\log K_{tw} = -0.960 \log S_w + 0.537 \tag{5.18}$$

with $n = 23$ and $r^2 = 0.959$.

Because of the close proximity of $K_{tw}$ to $K_{ow}$ for most solutes studied, one would expect the relation between $\pi_X$ and $\Delta_X$ as noted for $K_{ow}$ to also be applicable for $K_{tw}$. In short, the close correspondence between $K_{tw}$ and $K_{ow}$ indicates that the overall hydrophilic/lipophilic character of triolein or a lipid is quite similar to that of octanol, and therefore octanol is reasonably representative of a biological lipid insofar as it behaves as a partition phase.


## 5.7   CORRELATIONS OF PARTITION COEFFICIENTS

It has long been recognized that the partition coefficients of a set of solutes with one solvent–water mixture (say, $K_{sw,1}$) may often be well correlated with their partition coefficients with a different solvent–water mixture (say, $K_{sw,2}$). The correlation is usually presented in logarithmic forms of $K_{sw,1}$ and $K_{sw,2}$ as

$$\log K_{sw,2} = a \log K_{sw,1} + b \tag{5.19}$$

In those systems where a good linear correlation between $\log K_{sw,2}$ and $\log K_{sw,1}$ is observed, the correlation is frequently called a *linear free-energy relationship* (LFER), since the logarithmic term of a partition coefficient (which is an equilibrium constant, $K$) is related to the molar free-energy change of the solute at some chosen standard state (i.e., $\Delta\overline{G}^\circ = -RT\ln K$). In this particular case, it refers to the free-energy change when 1 mole of the solute at unit concentration in one solvent (e.g., water) is transferred to a water-saturated organic solvent where the solute is kept at a unit concentration (note that the two solute solutions at the standard state specified are not in equilibrium so that $\Delta\overline{G}^\circ \neq 0$). LFERs developed for a specific set of solutes with some model solvents are useful for assessing the partition behavior of similar organic compounds with the same or compositionally similar solvents, biological components, and natural organic phases. The LFER correlation in the form of Eq. (5.19) was developed by Collander (1951) for systems where the two organic solvents with $K_{sw,1}$ and $K_{sw,2}$ are similar in composition or contain similar functional groups, such as *i*-butanol versus *i*-pentanol, or *n*-octanol versus oleyl alcohol. Earlier we have seen that *n*-octanol and triolein provide another case for such a linear correlation.

From the illustrated partition characteristics of solutes with different organic solvents (including lipids), it is recognized that the magnitude of $K_{sw}$ depends critically on the solute solubility in water and on the composition and polarity of the solvent. Consequently, the numerical values of $a$ and $b$ in Eq. (5.19) are expected to vary with respect to solute and solvent properties. From Eqs. (3.6) and (5.19), the coefficient $a$ is simply

$$a = \frac{d\log K_{sw,2}}{d\log K_{sw,1}} = \frac{d\log\left(\gamma_w^*/\gamma_o^*\right)_2}{d\log\left(\gamma_w^*/\gamma_o^*\right)_1} \tag{5.20}$$

and the constant $b$ is simply the value of $\log K_{sw,2}$ for a hypothetical (or extrapolated) solute in the series with $\log K_{sw,1} = 0$ (i.e., at $K_{sw,1} = 1$). The second expression in Eq. (5.20) suffices when the solute–solvent solution obeys Raoult's law to a good approximation. Thus, $a$ expresses the *rate* of change in $\log K_{sw,2}$ with respect to the *rate* of change in $\log K_{sw,1}$ for a selected set of solutes that spans a specific range of $\log K_{sw,1}$ and $\log K_{sw,2}$. Since the $b$ constant is in most cases an extrapolated value, often outside the range of actual $\log K_{sw,1}$, it defies a rigorous interpretation, especially when the selected set of solutes come from a diversity of classes. The coefficient $a$ itself is by no means indicative of the relative solvency of the two organic solvents involved, which is manifested instead by the magnitudes of $K_{sw,1}$ and $K_{sw,2}$ for any solute of interest.

The simplest case in which $a$ and $b$ in Eq. (5.19) can be well rationalized is when the two organic solvents in $K_{sw,1}$ and $K_{sw,2}$ have closely similar structures and polarities, such that they exhibit similar compatibilities with any of the solutes, and exhibit comparable solvent–water mutual saturation effects on solute solubility. The solvent pairs expected to comply with these requirements include, for example, hexane versus heptane (or other higher alkanes), $n$-butanol versus $i$-butanol, $n$-pentanol versus $s$-pentanol, and $n$-octanol versus triolein. In those systems, $a$ should be very close to 1, largely independent of the types of solutes included in the set, and $b$ should be small (i.e., close to zero). For other systems, the degree of the linear fit between $K_{sw,1}$ and $K_{sw,2}$ would vary to different extents with the polarities of individual solutes and with the compositions of the solvents, and therefore $a$ and $b$ derived from the regression of $\log K_{sw,2}$ against $\log K_{sw,1}$ could vary widely.

As we have seen from the $\log K_{ow}$ and $\log K_{hw}$ data presented earlier, a reasonably good linear relationship exists between $\log K_{ow}$ and $\log K_{hw}$ if we restrict our analysis to a group of relatively nonpolar solutes (i.e., the ones with relatively large $\log K_{ow}$ values). The relationship becomes meager when polar solutes are included in the group because they respond very differently to apolar hexane compared to weakly polar octanol. Therefore, if a statistical analysis of $\log K_{hw}$ against $\log K_{ow}$ is attempted for a mixed set of nonpolar and polar solutes, the results will not yield a good linear fit, and the resulting $a$ and $b$ values will be ambiguous. A good way to rectify this problem is to divide the mixed set of solutes into two or more subsets according to their polarities (or specific modes of molecular actions) (Leo et al., 1971). This treatment improves the correlation fit for each subset and allows for a better interpretation of the resulting $a$ and $b$ values for each solute type. Therefore, if one is to predict the partition coefficient of a test solute with an organic phase of interest from its partition coefficient with a reference solvent, it is essential that the test solute belong to the same or similar class of solutes for which a previous correlation has been established.

If the two organic phases with $\log K_{sw,1}$ and $\log K_{sw,2}$ contain similar polar group(s) but differ significantly in their overall polarities, linear correlations may then be observed to encompass many solute classes. Leo et al. (1971) showed, for example, that the $\log K_{sw,2}$ for a diversity of solutes with weakly and moderately polar solvents, such as oleyl alcohol, methylisobutyl ketone, ethyl acetate, $n$-, $s$-, and $t$-pentanol, cyclohanone, and $n$-butanol, could all be reasonably correlated with $\log K_{ow}$ as the reference. Let us consider $\log K_{bw}$ (but-water) versus $\log K_{ow}$(oct-water) in some detail. It is shown earlier that the $K_{bw}$ is "shrunk" progressively with increasing $K_{ow}$ for a series of low-polarity chlorinated benzenes, or alternatively that the difference between $(\gamma_w^*/\gamma_o^*)_{\text{oct-water}}$ and $(\gamma_w^*/\gamma_o^*)_{\text{but-water}}$ increases with increasing $K_{bw}$. The correlation of $\log K_{bw}$ (as $\log K_{sw,2}$) with $\log K_{ow}$ (as $\log K_{sw,1}$) for the series of solutes yields $a < 1$ (about 0.7) (Leo et al., 1971). This finding is expected because the $\log K_{bw}$ values for a group of solutes would fall into a shorter range than the corresponding $\log K_{ow}$ values, owing to the higher polarity of butanol (over octanol) and the greater butanol–water mutual saturation. In other words, as the solute water solubility decreases, the solute solubility in butanol-saturated water decreases (or the $\gamma_w^*$ increases) less rapidly than that in octanol-saturated water because of the high butanol content; concomitantly, the solute solubility in water-saturated butanol decreases (or the $\gamma_o^*$ increases) more rapidly than that in water-saturated octanol because of the higher butanol polarity and high water content in butanol phase.

It is recognized that using $\log K_{ow}$ as the reference partition constant tends to reduce the level of data scattering in the correlation analysis for a group of solutes. To understand this effect, one recalls that the observed correlation of $\log K_{ow}$ with $\log S_w$ [Eq. (5.3)] applies satisfactorily to a wide variety of solutes (except for highly polar ones), which is attributed to the unique polar versus nonpolar balance of the octanol molecule. As a consequence, the use of octanol–water as the reference system for a mixed class of compounds tends to enhance the goodness of correlation over the use of other solvent–water systems as the reference. In essence, $K_{ow}$ is much like an inverse of $S_w$, which accounts to a large extent for a solute's partition magnitude with a particular organic phase. Thus, the correlation analysis using $\log K_{ow}$ as the reference minimizes differences in the behavior of a diverse group of solutes with the reference system that could otherwise contribute to the overall data scattering.

## 5.8   BIOCONCENTRATION OF ORGANIC CONTAMINANTS

A major concern for environmental contamination is the extent to which pollutants concentrate from water into aquatic organisms such as fish. The extent of such concentration, termed the *bioconcentration factor* (BCF), is given by the ratio of the pollutant concentration in fish to that in water. For nonionic organic contaminants, there is good reason to believe that their bioconcentration would occur by partition into certain biological components. It has been observed that the concentrations of several refractory chlorinated contaminants (e.g., DDT and PCBs) in different fish species or in different tissues of a fish correlate well with the lipid contents in whole fish or in its tissues (Reinert, 1970; Roberts et al., 1977; Sugiura et al., 1979). This finding suggests that the polar biological components, such as protein and carbohydrate, have a relatively poor affinity for nonionic (especially, nonpolar) contaminants. The BCFs of some chlorinated benzenes with guppies, rainbow trout, and bluegill in laboratory systems were measured by Könemann and van Leeuwen (1980), Oliver and Nimii (1983), and Banerjee et al. (1984). Chiou (1985) measured the $K_{tw}$ (triolein–water) of a large set of organic solutes, including many chlorinated benzenes, and found a high correlation between $K_{tw}$ and reported BCFs when BCFs are normalized to the fish lipid content.

In most laboratory BCF studies, the test fish and contaminant are brought to *equilibrium* either in a static-water system with a fixed initial load of contaminant or in a flow-through system in which the contaminant level in water is kept constant during equilibration. Contaminant concentrations in fish and water (the latter being frequently fixed) are monitored until equilibrium is reached. A one-dimensional pharmacokinetic model has also been conceived

**TABLE 5.6. Comparison of Lipid-Based Bioconcentration Factors (BCF$_{lipid}$) and Triolein–Water Partition Coefficients ($K_{tw}$) of Some Organic Compounds**

| Compound | $\log K_{tw}$ | log (BCF)$_{lipid}$ | |
| --- | --- | --- | --- |
| | | Guppies[a] | Rainbow Trout[b] |
| $o$-Dichlorobenzene | 3.51 | | 3.51–3.80 |
| $m$-Dichlorobenzene | 3.63 | | 3.70–4.02 |
| $p$-Dichlorobenzene | 3.55 | 3.26 | 3.64–3.96 |
| Hexachloroethane | 4.21 | | 3.79–4.13 |
| 1,2,3-Trichlorobenzene | 4.19 | 4.11 | 4.15–4.47 |
| 1,2,4-Trichlorobenzene | 4.12 | | 4.19–4.56 |
| 1,3,5-Trichlorobenzene | 4.36 | 4.15 | 4.34–4.67 |
| 1,2,3,4-Tetrachlorobenzene | 4.68 | | 4.80–5.13 |
| 1,2,3,5-Tetrachlorobenzene | 4.69 | 4.86 | |
| 1,2,4,5-Tetrachlorobenzene | 4.70 | | 4.80–5.17 |
| Hexachlorobutadiene | 5.04 | | 4.84–5.29 |
| Pentachlorobenzene | 5.27 | 5.42 | 5.19–5.36 |
| Hexachlorobenzene | 5.50 | 5.46 | 5.16–5.37 |

*Source*: Data from Chiou (1985).

[a] Data of Könemann and van Leeuwen (1980).
[b] Data of Oliver and Nimii (1983).

by Banerjee et al. (1984) for estimating the BCF from the rate of contaminant disappearance from water ($k_1$) and the rate of contaminant depuration in fish ($k_2$), in which $k_1/k_2$ = BCF for the whole fish. In all cases, the laboratory system for attainment of equilibrium with fish is relatively well controlled.

The measured $\log K_{tw}$ values from Chiou (1985) and respective $\log$ (BCF)$_{lipid}$ values for some chlorinated benzenes with two fish species from the studies of Könemann and van Leeuvan and of Oliver and Nimii are given in Table 5.6. A plot of $\log$ (BCF)$_{lipid}$ versus $\log K_{tw}$ is presented in Figure 5.7. For the 13 chlorinated compounds studied, there is good consistency between $K_{tw}$ and (BCF)$_{lipid}$. In most cases the agreement is within a factor of 2, which is about as good as can be expected, since the combined error with $K_{tw}$ and BCF can often be equally large. The $\log$ (BCF)$_{lipid}$ values exhibit virtually no systematic differences between the two fish species despite that the BCFs were measured at different contaminant concentrations. This suggests that BCFs are largely concentration independent, as would be expected for contaminant partition at low concentrations.

The close agreement between $\log K_{tw}$ and $\log$ (BCF)$_{lipid}$ manifests that a physical partition between fish lipids and external water rather than uptake by feeding is mainly responsible for fish bioconcentration of relatively non-polar organic compounds (Smith et al., 1988). A practically linear correlation is found by plotting $\log$ (BCF)$_{lipid}$ versus $\log K_{tw}$ using the combined BCFs with guppies and rainbow trout,

$$\log(\text{BCF})_{lipid} = 0.957 \log K_{tw} + 0.245 \qquad (5.21)$$

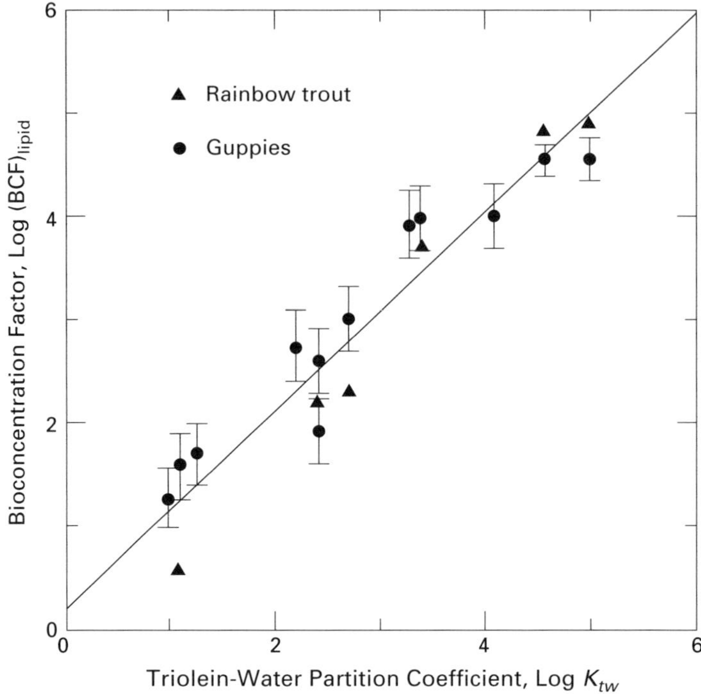


**Figure 5.7**   Correlation between $\log(\text{BCF})_{lipid}$ and $\log K_{tw}$ for compounds in Table 5.6 with guppies and rainbow trout. [Data compiled by Chiou (1985).]

with $n = 18$ and $r^2 = 0.915$. The correlation above is not statistically different from $\log(\text{BCF})_{lipid} = \log K_{tw}$ at the 95% confidence level. When the log $(\text{BCF})_{lipid}$ values of compounds are correlated with the corresponding $\log K_{ow}$ values, the result gives

$$\log(\text{BCF})_{lipid} = 0.893 \log K_{ow} + 0.607 \qquad (5.22)$$

with $n = 18$ and $r^2 = 0.904$. Although Eq. (5.22) is statistically different from $\log(\text{BCF})_{lipid} = \log K_{ow}$ at the 95% confidence level for the range of log $(\text{BCF})_{lipid}$ in Table 5.6, the difference between Eqs. (5.21) and (5.22) is relatively small, and one cannot be sure that the correlation with $K_{ow}$ will be statistically different from that with $K_{tw}$ for a wider range of the data set. Because of the correspondence between $K_{tw}$ and $K_{ow}$, octanol is therefore a good surrogate for biological lipids and thus $K_{ow}$ gives a reasonable estimate of $(\text{BCF})_{lipid}$ for nonpolar organic contaminants.

Several potential factors can contribute to discrepancies between $(\text{BCF})_{lipid}$ and $K_{tw}$ (or $K_{ow}$). Compounds that are unstable in water or that are readily metabolized by organisms, to the extent that the degradation rate is greater than the rate of equilibration, will give anomalous BCF values because of the inability of the system to reach true equilibrium state. Similarly, the BCF values obtained with short exposure times before steady-state concentrations in both biotic and water phases are reached could differ significantly from equilibrium values. For compounds with very large $K_{ow}$ or very small $S_w$ (e.g., DDT and some PCBs), the times for establishment of equilibrium would be very extended because of their large BCF values with fish, which require a much greater amount of water solution to be transported through fish gills than for compounds with a considerably smaller $K_{ow}$ or larger $S_w$. The more limited diffusion rates for larger molecules might also prolong the time for equilibrium. Whereas experimental BCF values for polar solutes are scarce, the $(BCF)_{lipid}$ of these compounds would probably be significantly higher than their respective $K_{tw}$ (or $K_{ow}$) values because of their additional partition or specific interactions with polar biological components (e.g., protein). As shown later, certain dissolved macromolecular materials (e.g., humic substances) present at low concentrations in natural water can significantly enhance the water solubility of some extremely insoluble compounds (e.g., DDT and some PCBs), thus decreasing their apparent BCF values.

The field BCF data are expectedly more complicated because the contaminant concentration may vary significantly with time and with location and because many biotic species (e.g., fish) are not confined to a fixed local environment. Contaminant concentrations and apparent BCFs determined with fish and water samples collected at a given time from a specific site are therefore the integrated results of these variables. As such, there would be large uncertainties concerning the achievement of equilibrium of contaminants between fish and water in natural systems. Pereira et al. (1988) studied field $(BCF)_{lipid}$ data for a number of chlorinated compounds on four fish species (Atlantic croaker, blue crab, spotted sea trout, and blue catfish) sampled from selected sites of the Calcasieu River estuary in Louisiana. Swackhammer and Hites (1988) conducted a similar field BCF investigation for chlorinated compounds on lake trout and white fish from the Siskiwit Lake, Isle Royale, Lake Superior. The $(BCF)_{lipid}$ data of Pereira et al. are given in Table 5.7 and a corresponding plot of $\log(BCF)_{lipid}$ versus $\log K_{tw}$ is presented in Figure 5.8.

As noted, although the field $(BCF)_{lipid}$ values exhibit significant scattering between fish species in response to the dynamic nature of the ecosystem and other variables, most data points are within one order of magnitude of the equilibrium correlation line, $(BCF)_{lipid} = K_{tw}$. The data scattering is virtually random, showing no obvious pattern with a specific fish species. The overall trend of field-based $(BCF)_{lipid}$ is surprisingly consistent with that found in well-controlled laboratory studies. Although true equilibrium is probably rarely achieved between water and fish in a dynamic estuarine system, the results are clearly supportive of the lipid model for bioconcentration of relatively nonpolar organic compounds.

From the information presented, a good approximation for the BCF of a relatively water-insoluble compound at equilibrium with a biotic species can

**TABLE 5.7.  Lipid-Based Bioconcentration Factors (BCF$_{lipid}$) of Chlorinated Organic Compounds in Four Biota Species in the Calcasieu River Estuary, Louisiana**

|  | log (BCF)$_{lipid}$ | | | |
| --- | --- | --- | --- | --- |
| Compound | Atlantic Croakers | Blue Crabs | Spotted Sea Trout | Blue Catfish |
| *o*-Dichlorobenzene | 3.94 | 4.46 | 3.79 | 3.82 |
| *m*-Dichlorobenzene | 3.60 | 3.86 | 3.25 | 3.40 |
| *p*-Dichlorobenzene | 3.91 | 4.53 | 4.09 | 3.51 |
| 1,3,5-Trichlorobenzene | 4.40 | 4.45 | 3.51 | 4.22 |
| 1,2,4-Trichlorobenzene | 4.76 | 4.90 | 3.54 | 4.68 |
| 1,2,3-Trichlorobenzene | 4.54 | 4.77 | 3.13 | 4.49 |
| 1,2,3,5-Tetrachlorobenzene | 5.05 | 5.20 | 4.27 | 4.90 |
| 1,2,3,4-Tetrachlorobenzene | 5.46 | 5.70 | 4.68 | 5.30 |
| Pentachlorobenzene | 5.93 | 6.12 | 4.96 | 5.57 |
| Hexachlorobenzene | 6.42 | 6.71 | 5.96 | 5.98 |
| Hexachloro-1,3-butadiene | 4.50 | 3.97 | 4.06 | 4.55 |

*Source*: Data from Pereira et al. (1988).

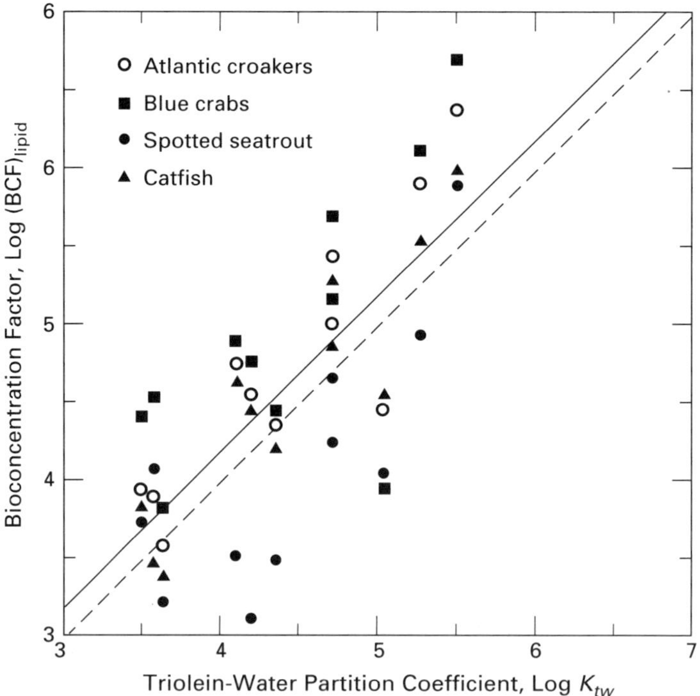


**Figure 5.8**  Correlation between log (BCF)$_{lipid}$ and log $K_{tw}$ for compounds in Table 5.7 with four different fish species. The dashed line represents log (BCF)$_{lipid}$ = log $K_{tw}$. [Data from Pereira et al. (1988).]

be obtained from the $K_{tw}$ of the compound and the lipid content of the biotic species as

$$\text{BCF} = \phi K_{tw} \tag{5.23}$$

where $\phi$ is the lipid fraction of a biotic species (on the wet-weight basis) and $K_{tw}$ is the triolein–water partition coefficient of the compound. For practical purposes, $K_{tw}$ may be replaced by $K_{ow}$, as rationalized before. By Eq. (5.23), one estimates the concentration of a relatively insoluble contaminant in a biotic species as

$$[C_i]_{\text{bio}} = [C_i]_w \phi K_{tw,i} \tag{5.24}$$

where $[C_i]_{\text{bio}}$ is the equilibrium concentration of contaminant $i$ with whole biotic species, $[C_i]_w$ the corresponding concentration in water, and $K_{tw,i}$ (or $K_{ow,i}$) the $K_{tw}$ (or $K_{ow}$) for contaminant $i$. Similar calculations can be extended to all other contaminants by use of Eq. (5.24) to obtain individual contaminant concentrations or their sum with a given biotic species.

# 6 Adsorption of Vapors on Minerals and Other Solids

## 6.1 INTRODUCTION

Before investigating the sorptive characteristics of organic contaminants either as vapors or as solutes onto composite soil or sediment samples, it is instructive that one first examine the adsorptive behavior of some vapors on basic minerals (and other natural solids) that occur in common soils. In Chapter 7 we will see that the principal fraction of the soil to adsorb a nonionic compound is the mineral matter and, in some special cases, some carbonaceous material (e.g., charcoal-like substance) admitted into the soil. Since the process of adsorption is competitive in nature, as pointed out earlier, the amount of a contaminant adsorbed by a mineral matter would depend critically on the competitive power of other species coexisting in the system. In natural environments, the most important competitive species is probably water, because of its ubiquity and abundance.

In this chapter we examine first the adsorption data of $N_2$ vapor on a few representative minerals and solids and then use the $N_2$ adsorption data to calculate the surface areas and micropore volumes of the samples. We then consider how the surface areas of some natural solids achieved by improper analytical methods lead to serious discrepancies from the results by the standard BET-$N_2$ method. Subsequently, we compare the vapor uptakes of a model organic compound (benzene) and water onto these samples to give one a good sense of the relative adsorptive powers of these samples for a nonionic organic compound and water. The comparative adsorption of an organic and water vapors by minerals renders a crucial link to organic–solute sorption to composite soils or sediments from water solution, where the solvent (water) competes forcefully for mineral adsorption, thus strongly depressing the solute adsorption. Similarly, the contrasting feature for organic vapor and water adsorption on activated carbon sets the ground for the weak competition of water against the organic–solute adsorption onto a carbonaceous substance, such as natural charcoal.

## 6.2 NITROGEN ISOTHERM AND SOLID SURFACE AREA

The surface phenomena of solids are frequently related to their subdivision or porosity, which in turn is evaluated in terms of their surface areas. The universally accepted standard method for measurement of surface area is the method of Brunauer, Emmett, and Teller (BET) (Brunauer et al., 1938; Adamson, 1967; Gregg and Sing, 1982), in which one determines the adsorption isotherm of any of a number of suitable vapor adsorbates (e.g., $N_2$) on the solid (adsorbent) of interest. Suitable adsorbates must be chemically inert to the solid, not subject to molecular sieving, and confined to the exterior of the solid (i.e., they must exhibit no significant penetration or site specific interaction with the solid). The BET model [Eq. (4.7)] calculates the monolayer capacity ($Q_m$) from the adsorption isotherm; the surface area per adsorbate molecule ($a_m$) is estimated from the liquid density as

$$a_m = 1.09(M/d_l N)^{2/3} \tag{6.1}$$

where $M$ is the molecular weight of the adsorbate, $d_l$ is the adsorbate liquid density, and $N$ is the Avogadro number. Using $d_l = 0.808\,\text{g/cm}^3$ for liquid $N_2$ at 77 K gives $a_m = 16.2 \times 10^{-20}\,\text{m}^2$ per $N_2$ molecule. The solid samples prior to adsorption studies are usually outgassed under vacuum (or under an inert gas) at some selected temperature.

Although $N_2$ vapor at the liquid $N_2$ boiling point of 77 K is the adsorbate most frequently employed for surface-area determination, the method is by no means limited to $N_2$, and a wide variety of other suitable adsorbates (e.g., krypton) produce reasonably consistent results on the same solid. The surface area of a solid, considered to be the solid–gas or solid–vacuum interfacial area, which is external to the material, is assumed both to preexist and to be unaffected by the measurement. Any vapor that may potentially react with or penetrate into the solid is not appropriate for surface-area determination. The surface area is therefore an intrinsic property of the solid; that is, within the precision of the method, it should be independent of the choice of acceptable adsorbates (Brunauer et al., 1938). There has been a serious confusion in the earlier soil-science literature about the surface areas of soils and clays to which we shall attend later.

We now consider the adsorption data of $N_2$ vapor at liquid $N_2$ temperature (77 K) on selected natural solids given in Table 6.1. The solid samples examined were outgassed under a stream of helium at temperature ranging from 100°C, as for activated carbon, to 150 to 200°C for the rest. Plots of the $N_2$-vapor uptake ($Q$) versus $N_2$ relative pressure ($P/P°$) for relatively pure silica (Alfa Aesar Co.), alumina (calcined, Alcoa Co.), goethite (Ward's Natural Science), Georgia kaolinite (KGa-2), K-saturated Arizona montmorillonite (K-SAz-1), Ca-saturated Arizona montmorillonite (Ca-SAz-1), and activated carbon (CAL grade, Calgon Co.) are shown in Figures 6.1 to 6.3. For the three mineral oxides and the two clays, the $N_2$ isotherms are type II, whereas for activated carbon the isotherm is more like type IV, of the Brunauer classification. Among the mineral oxides and clays, certain differences are visible with respect to their (concave-downward) curvatures at low $P/P°$ and the related

**TABLE 6.1. $N_2$ Monolayer Capacities, Total Surface Areas, Micropore Volumes by**
**$t$-Plot, and Nonporous Surface Areas of Selected Solids**

| Solid | $Q_m(N_2)$ (mg/g) | Total Surface Area ($m^2$/g) | Micropore Volume (mL/g) | Nonporous Area ($m^2$/g) |
|---|---|---|---|---|
| Silica (Alfa Aesar Co.) | 2.43 | 8.2 | $7 \times 10^{-4}$ | 6.9 |
| Alumina (Alcoa Co.) | 2.74 | 9.3 | $3 \times 10^{-4}$ | 9.2 |
| Goethite (Ward's Nat. Sci.) | 0.77 | 2.7 | $2 \times 10^{-4}$ | 2.6 |
| Kaolinite (KGa-2) | 6.03 | 21.0 | 0 | 21.1 |
| Ca-SAz-1 (homoionic) | 20.5 | 71.5 | $3.0 \times 10^{-2}$ | 17.2 |
| K-SAz-1 (homoinoic) | 27.9 | 97.1 | $3.9 \times 10^{-2}$ | 26.4 |
| Activated carbon (Calgon Co.) | 300 | 986 | 0.48 | 110 |

*Source*: Data on kaolinite, Ca-montmorillonite (Ca-SAz-1), and K-montmorillonite (K-SAz-1) from Rutherford et al. (1997) and the rest from C. T. Chiou (unpublished research).

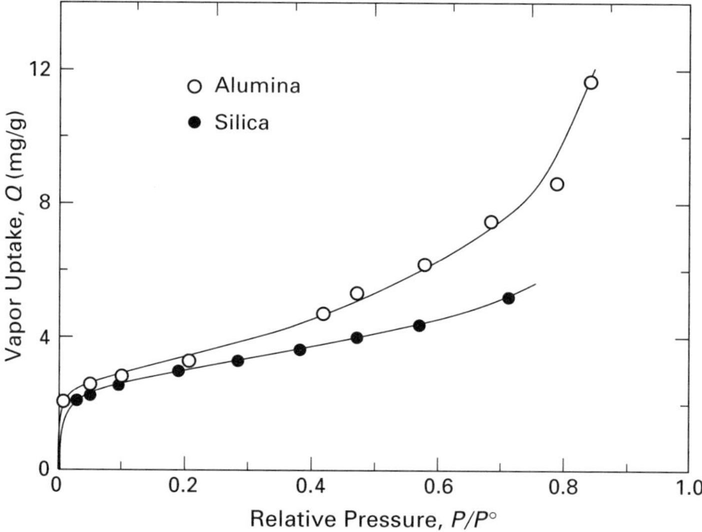


**Figure 6.1**   $N_2$-vapor adsorption at 77 K on alumina and silica. The solids are identified in Table 6.1.

shapes at higher $P/P°$, in reflection of the pore-size distribution and pore geometry and thereby of the energetic heterogeneity of the solid (adsorbent). The curvature at low $P/P°$ is related to the heat of adsorption for the first-layer adsorbate [i.e., the $C$ value in BET the model, Eq. (4.6)], in which the energetic heterogeneity is most pronounced. The relatively flat adsorption isotherm along with exceptionally high $N_2$ uptake for activated carbon is indicative of a very large quantity of highly energetic fine pores (in molecular dimensions) in the solid. Most natural solids tend to display a large range of

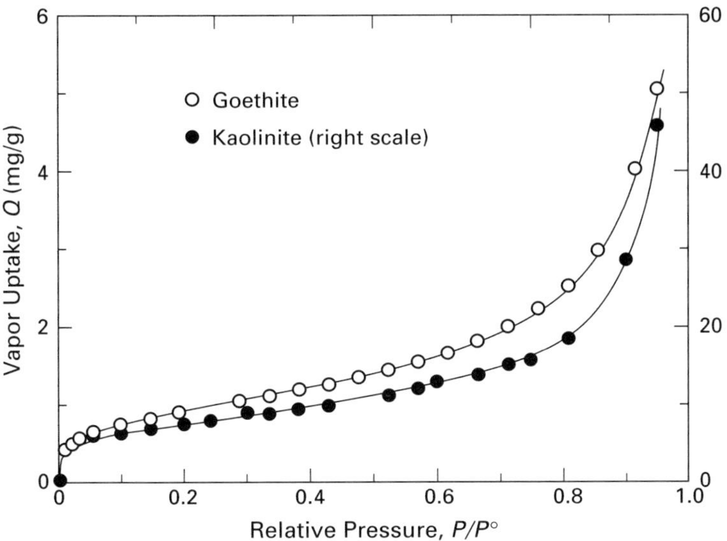


**Figure 6.2**  $N_2$-vapor adsorption at 77 K on goethite and kaolinite. The solids are identified in Table 6.1. [Data on kaolinite from Rutherford et al. (1997).]

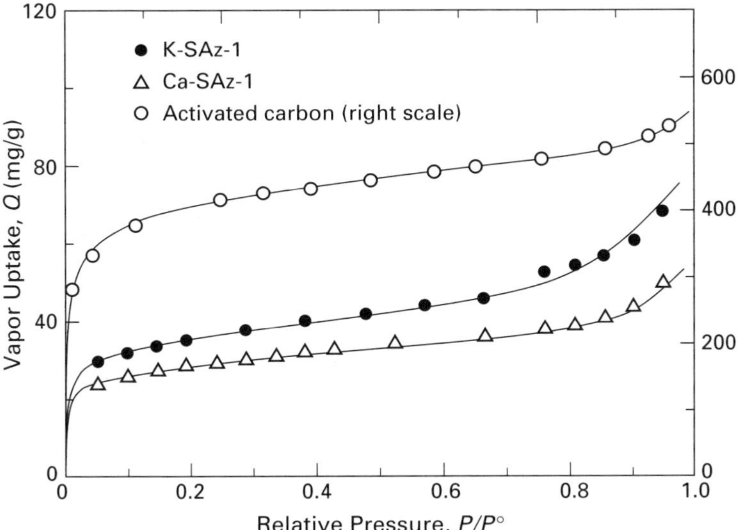


**Figure 6.3**  $N_2$-vapor adsorption at 77 K on K-SAz-1, Ca-SAz-1, and activated carbon. The solids are identified in Table 6.1. [Data on K-SAz-1 and Ca-SAz-1 from Rutherford et al. (1997).]

their preexisting pore sizes, as depicted here for several mineral oxides in accordance with their $N_2$-isotherm shapes.

The $N_2$ monolayer capacities [i.e., $Q_m(N_2)$] and BET surface areas for the solids, as determined by Eq. (4.7), are presented in Table 6.1. For a solid with a relatively smooth surface, on which the BET model is based, the completion of $N_2$ monolayer occurs usually at $P/P° = 0.05$ to $0.30$. The data for such mineral samples as silica, alumina, goethite, and kaolinite fall closely into this range. By contrast, the $P/P°$ at $Q_m(N_2)$ for Ca-SAz-1 (0.015), K-SAz-1 (0.018), and activated carbon (0.020) are noticeably less than the common lower limit of 0.05. Results for the latter solids, especially the activated carbon with a huge $Q_m(N_2)$, signify the massive $N_2$ vapor condensation into highly adsorbing preexisting micropores of the solids, on which the exactness of the measured surface areas become somewhat meager (because the BET model was formulated for relatively open surfaces). For the former solids, there exists presumably no significant amount of preexisting micropores.

## 6.3  MICROPORE VOLUME

The noted differences in BET surface for the solids reflect their subdivision, or more exactly their porosities, because the apparent sizes of the particles make a relatively minor contribution to the surface area. For example, a nonporous solid with a geometric particle size of $1\,\mu m$ has a surface area of about $1\,m^2/g$. The pores contributing most to the measured surface area are those called *micropores*, which have a diameter or a slit width less than about 20 angstroms (Å), where $1\,Å = 0.1\,nm$. Pores with dimensions between 20 and $500\,Å$ are called *mesopores*, and those larger than $500\,Å$, *macropores* (Gregg and Sing, 1982). The micropore volume associated with the micropores may be estimated by some developed analytical methods. The $t$-method of de Boer et al. (1966) and the $\alpha_s$-method of Sing (1970) are frequently used in this respect to determine the micropore volume of a solid by use of a set of suitable inert-vapor adsorption data. The method consists of plotting the adsorption data of a (nonpolar) vapor on a test solid against that of the same vapor on a nonporous standard solid, as detailed below.

In the $t$-plot of de Boer et al., the adsorbed liquid $N_2$ volume ($V$) on a test solid is plotted against the statistical thickness ($t$) of the adsorbed $N_2$ layer on a nonporous standard solid to yield the micropore volume and nonporous (open) surface area of the test solid. The relation between $V$ and $t$ is given by Remy and Poncelet (1995) as

$$V = V_m + 10^{-4} S_0 t \tag{6.2}$$

where $V$ is the adsorbed volume of the condensed liquid $N_2$ on the test solid (mL/g), $V_m$ the volume of $N_2$ adsorbed onto the solid's micropores (mL/g), $S_0$ the nonporous surface area (i.e., the area associated with the nonporous structure) of the solid ($m^2$/g), $t$ the statistical thickness of the adsorbed $N_2$ layer in angstroms on a reference nonporous solid, and $10^{-4}$ is a conversion factor. To convert the mass of $N_2$ adsorbed into the liquid volume at 77 K, a liquid $N_2$ density of 0.808 g/mL is generally assumed. The value of $t$ as a function of $P/P°$ is calculated from the adsorption data of $N_2$ on the reference solid. A universal $t$-curve of $N_2$ on nonporous solids has been developed (de Boer et al., 1966), which gives

$$t = \{13.99/[0.034 - \log{(P/P°)}]\}^{0.5} \tag{6.3}$$

where $t$ is in angstroms. If the plot of $V$ versus $t$ gives a straight line passing through the origin, the test solid is considered to be free of micropores. For a microporous test solid, the $t$-plot yields a straight line at high $t$ and a concave-down curve at low $t$; the extrapolation of the upper linear line to $t = 0$ gives a slope of $S_0$ and an intercept of $V_m$.

The $\alpha_s$ plot of Sing (1970) is an alternative method of the $t$-plot. In this method, the amount of a (nonpolar) vapor adsorbed at some fixed $P/P°$ on a reference solid is first normalized to the amount at $P/P° = 0.4$ (i.e., $\alpha_s = Q/Q_{0.4}$) to produce a standard $\alpha_s$-curve rather than a $t$-curve. The $\alpha_s$-curve is then used to construct an $\alpha_s$-plot from the isotherm of the vapor on a test solid, just as is the $t$-curve used to construct a $t$-plot. The reference solid is chosen to be a nonporous solid having a chemical composition similar to that of the test solid. Similar to $t$-plot, the $\alpha_s$-plot gives a straight line passing through the origin if the test solid is free of micropores; for a microporous solid, the $\alpha_s$-plot yields a straight line at high $\alpha_s$ and a curve at low $\alpha_s$, and the extrapolated intercept from the upper linear line gives the micropore volume. If the test sample contains a large number of mesopores, an upward deviation from a straight line will occur at relatively high $t$ and $\alpha_s$. In general, if the $N_2$ data are used together with an appropriate reference solid, the micropore volume of the test solid determined by the $\alpha_s$-plot should be the same as that obtained by the $t$-plot.

Unlike the $t$-plot, the applicability of the $\alpha_s$-plot is not restricted to the $N_2$ adsorption data. This enables separate $\alpha_s$-plots to be constructed from the data of $N_2$ and other suitable vapors for the test solid. The $\alpha_s$-plot offers an advantage in elucidating the sizes of various fine pores of a solid of interest by adopting appropriate reference nonpolar vapors of specified molecular sizes (Rutherford et al., 1997). When molecular sieving is observed by use of a large reference adsorbate but not with a small adsorbate, the microporosity detected based on the former should then be lower.

The calculated micropore volumes and the open surface areas of the solids by $t$-plot with $N_2$ data are shown in Table 6.1. As noted, the three mineral oxides and kaolinite (KGa-2) are virtually free of microporosity, since the total BET surface areas are practically the same as the open surface areas derived from the $t$-plot. In the absence of micropores, the surface areas of solids should generally be relatively small in magnitude. On the other hand, certain solids, such as K-SAz-1, Ca-SAz-1, and especially activated carbon, display very large micropore volumes, and thus very high BET surface areas, in reflection of their microporous structures. As expected, there is good positive correlation between the measured BET areas and the micropore volumes of the solids.

## 6.4  IMPROPER SURFACE-AREA MEASUREMENT

We now turn to a subject of special relevance to the surface-area determination by use of vapor data for natural solids. As stated earlier, a suitable adsorbate must be chemically inert, not subject to molecular sieving, and confined only to the exterior of the solid. Whereas the molecular-sieving effect can be realized more intuitively, the effect of adsorbate reaction and penetration into a bulk solid often escapes our immediate detection. Thus, if the vapor-uptake isotherm of an adsorbate, other than an inert vapor (e.g., $N_2$ or Kr), is to be used, particularly at room temperature, the *monolayer capacity* obtained with either the BET equation or other methods could deviate greatly from the BET monolayer capacity of, say, $N_2$, measured at the liquid $N_2$ temperature. Examples from the soil science literature are: (1) the monolayer capacities of water on clays based on the BET plot (Mooney et al., 1952); and (2) the monolayer capacities of polar liquids, such as ethylene glycol (EG) (Dyal and Hendricks, 1950; Bower and Gschwend, 1952) or ethylene glycol monoethyl ether (EGME) (Carter et al., 1965; Heilman et al., 1965; Eltantawy and Arnold, 1973), on clays and organic matter achieved under some evacuated conditions. Although recently, the BET method with inert gases (e.g., $N_2$) has also frequently been employed for surface-area determination in soil science, the terminology that has emerged to distinguish the different surface-area values by these different methods has become increasingly confusing, if not chaotic.

In soil science it has been assumed empirically that the amount of a polar solvent retained by a clay mineral or even organic matter when the solvent–solid mixture is evacuated to below a certain pressure over a certain length of time represents an approximate monolayer-adsorption capacity on the solid. By this estimated monolayer capacity and the molecular area of the solvent, the *total surface* of the solid is then calculated. For distinction, the surface area obtained by standard BET-$N_2$ method is considered to be the *external surface*. The difference between the two is portrayed to be the *internal surface* of the solid. Since the internal surface as derived is clearly impervious to inert gases (e.g., $N_2$), it deserves critical scrutiny. Although the same term has long been used in allied surface science, the implied internal surface there is nonetheless accessible to inert gases. As rationalized below, the internal surface adopted in soil science is largely an artifact of the calculation rather than a true surface in that the extensive solvent penetration into bulk solids is taken uncritically as formation of internal monolayers. In his classic book published in 1945, Brunauer made the following insightful and pertinent comments on the *internal surface* and *external surface* of a solid:

*Most adsorbents are highly porous bodies with tremendously large internal surfaces. The external surface, even that visible under the best microscope, only constitutes a small fraction of the total surface.*

*However, as long as the gas does not penetrate into the field of force that exists between the atoms, ions, or molecules inside the solid, it is considered to be on the outside, even it is adsorbed on the internal surface of the adsorbent.*

*If the gas enters inside of a solid, two things may happen: either the gas merely dissolves in it, forming a solid solution, or it reacts with the solid and forms a new compound.*

The comments above clearly convey the point that the large *internal surfaces* of the porous solids (e.g., activated carbon) are internal only when viewed with respect to the outer granule boundary, but are nevertheless *external* to the bulk solid and accessible to an inert gas, as later emphasized by Chiou et al. (1992). Therefore, if the vapor penetrates significantly into a solid and if such vapor data are used for surface-area calculations, a mistakenly large total surface and hence an erroneous internal surface will result. As a *reductio ad absurdem*, if one were to measure the uptake of, say, butane vapor by bulk hexane liquid, the resulting data could easily be misinterpreted as resulting from a presumed very large "internal surface" of the hexane.

A comparison of the surface areas obtained by the BET method by use of an inert vapor ($N_2$) and a polar vapor (e.g., EGME) for solids of different makeups helps to pinpoint the effect of solvent penetration on the determined surface areas. To meet this objective, let us consider the vapor uptakes of $N_2$ at liquid $N_2$ temperature (77 K) and EGME at room temperature on a series of natural and synthetic solids (Chiou et al., 1993), as shown in Table 6.2. The isotherm data on kaolinite (KGa-2), alumina (Quanta Chrome Co.), Ca-montmorillonite (Ca-SAz-1, unpurified), illite (Fifthian, Illinois), a mineral (Woodburn) soil, and a peat soil (Everglades, Florida) are shown, respectively, in Figures 6.4 to 6.6. The $N_2$ isotherms on all the solids are notably nonlinear with a type II shape. At liquid $N_2$ temperature, the $N_2$ vapor uptake by any solid takes place virtually all by surface adsorption, as there is little possibility for $N_2$ penetration into the solid. The respective EGME isotherms on all solids at room temperature are similarly nonlinear, with a generally sharper curvature except for the peat, the latter exhibiting instead a practically linear isotherm.

As noted in Figure 6.4, the EGME uptake capacities on alumina and kaolinite are only moderately higher than the corresponding $N_2$ capacities. The same applies to the $N_2$ and EGME data with sand, hematite, and synthetic hydrous iron oxide (Chiou et al., 1993). This result is indicative of the surface coverage, since EGME has a greater molecular weight (MW = 88) and a somewhat higher density ($d_l$ = 0.93 g/mL) than $N_2$ ($d_l$ = 0.808 g/mL) and since, according to Eq. (6.1), the adsorbate mass per unit area is proportional to $d_l^{2/3} M^{1/3}$. On the other hand, the EGME uptake capacities on highly expanding Ca-SAz-1, nonexpanding illite (0.4% organic matter), Woodburn soil (21%

**TABLE 6.2.** N$_2$ Monolayer Capacities, $Q_m(N_2)$; BET-N$_2$ Surface Areas (SA); EGME Equivalent Monolayer Capacities, $Q_m(EGME)_{eq}$; EGME Apparent Monolayer Capacities, $Q_m(EGME)_{ap}$; N$_2$ Relative Pressures at $Q_m(N_2)$, $(P/P°)_m^°$; EGME Relative Pressures at $Q_m(EGME)_{ap}$, $(P/P°)_m^*$, with Selected Minerals and Soils[a]

| Sample | $Q_m(N_2)$ | $(P/P°)_m^°$ | SA(m$^2$/g) | $Q_m(EGME)_{eq}$ | $Q_m(EGME)_{ap}$ | $(P/P°)_m^*$ | $\dfrac{Q_m(EGME)_{ap}}{Q_m(EGME)_{eq}}$ |
|---|---|---|---|---|---|---|---|
| Ottawa sand (Fisher Sci.) | 0.032 | 0.20 | 0.11 | 0.042 | 0.035 | 0.09 | 0.83 |
| Hematite (Ward's Nat. Sci.) | 0.14 | 0.14 | 0.50 | 0.18 | 0.18 | 0.10 | 1.00 |
| Alumina (Quanta Chrome Co.) | 0.87 | 0.08 | 3.03 | 1.13 | 1.02 | 0.03 | 0.90 |
| Kaolinite (KGa-2) | 6.03 | 0.11 | 21.0 | 7.84 | 7.90 | 0.11 | 1.01 |
| Illite (Fithian, IL) | 19.3 | 0.06 | 67.2 | 25.1 | 38.0 | 0.10 | 1.51 |
| Ca-SAz-1 (natural) | 21.8 | 0.08 | 75.9 | 28.3 | 215 | 0.02 | 7.60 |
| NHIO[b](9.1% Fe) | 3.31 | 0.11 | 11.5 | 4.30 | 7.98 | 0.03 | 1.86 |
| SHIO[c](62.4% Fe) | 53.4 | 0.12 | 186 | 69.4 | 61.2 | 0.03 | 0.88 |
| Peat (Everglades, FL) | 0.36 | 0.12 | 1.26 | 0.47 | NA[d] | NA[d] | NA[d] |
| Woodburn soil (Corvallis, OR) | 3.22 | <0.01 | 11.2 | 4.19 | 13.4 | 0.02 | 3.20 |

*Source*: Data from Chiou et al. (1993).

[a] All $Q_m$ values are in mg/g.

[b] A natural hydrous iron oxide from Redding, California.

[c] A synthetic hydrous iron oxide.

[d] Not available, because the EGME isotherm is linear, not subject to analysis by the BET model.

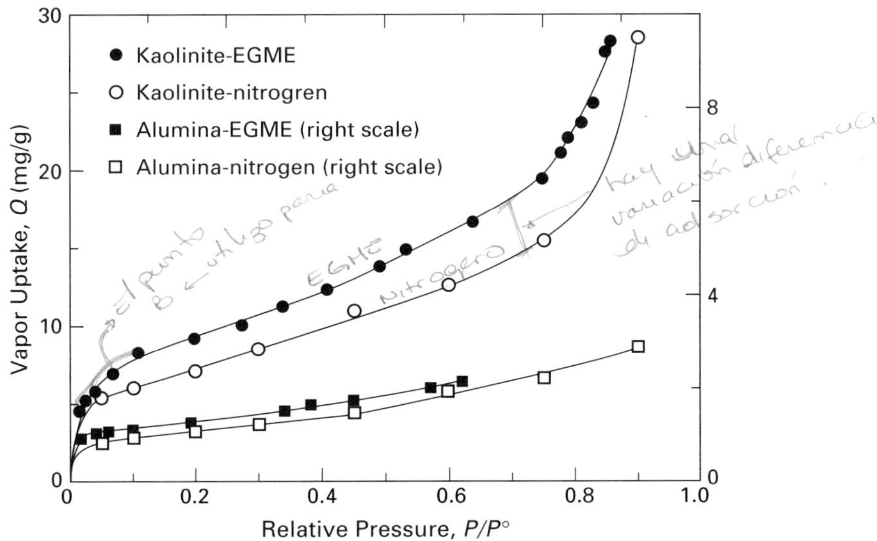



**Figure 6.4**  Uptake of EGME vapor at room temperature and $N_2$ vapor at 77 K by kaolinite and alumina. The solids are identified in Table 6.2. [Data from Chiou et al. (1993). Reproduced with permission.]

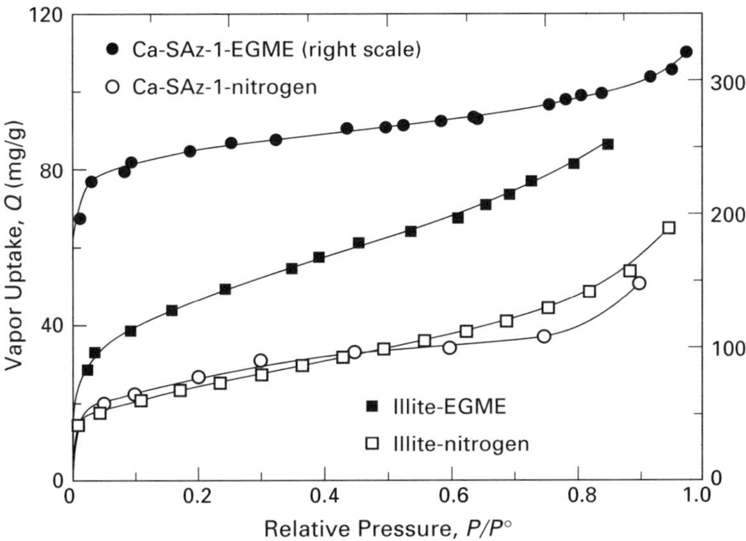



**Figure 6.5**  Uptake of EGME vapor at room temperature and $N_2$ vapor at 77 K by Ca-SAz-1 and illite. The solids are identified in Table 6.2. [Data from Chiou et al. (1993). Reproduced with permission.]

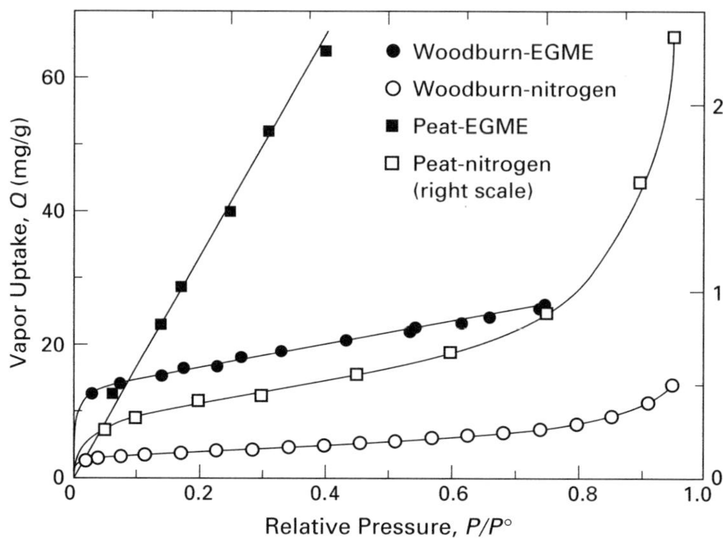



**Figure 6.6**   Uptake of EGME vapor at room temperature and $N_2$ vapor at 77 K by Woodburn soil and peat. The solids are identified in Table 6.2. [Data from Chiou et al. (1993). Reproduced with permission.]

clay and 1.9% organic matter), and peat (86% organic matter) are considerably higher than the corresponding $N_2$ capacities.

In view of the fact that the EGME isotherms are visually type II in shape for all solids, except for peat, one could apply the BET equation (4.7) to determine the apparent EGME monolayer capacities with the solid samples and then compare them with the values equivalent to the respective $N_2$ monolayer capacities as a means to validate the results. The calculated surface areas by standard BET-$N_2$ method, the $N_2$ monolayer capacities, $Q_m(N_2)$, the apparent EGME monolayer capacities, $Q_m(\text{EGME})_{\text{ap}}$, and the EGME monolayer capacities equivalent to the $N_2$ monolayer capacities, $Q_m(\text{EGME})_{\text{eq}}$, are presented in Table 6.2. In calculating the $Q_m(\text{EGME})_{\text{eq}}$ from the corresponding $Q_m(N_2)$, a constraint is applied to maintain the constancy of the surface area, such that

$$\frac{Q_m(\text{EGME})_{\text{eq}}}{M(\text{EGME})} a_m(\text{EGME}) = \frac{Q_m(N_2)}{M(N_2)} a_m(N_2) \tag{6.4}$$

where $M(\text{EGME})$ is the molecular weight of EGME, $M(N_2)$ the molecular weight of $N_2$, $a_m(\text{EGME})$ the molecular area of EGME, and $a_m(N_2)$ the molecular area of $N_2$ (16.2 Å$^2$). Whereas the approximate $a_m(\text{EGME})$ value can be obtained from Eq. (6.1) using the liquid density of EGME (0.93 g/mL), which gives $a_m(\text{EGME}) = 32.2$ Å$^2$, a better value of $a_m(\text{EGME}) = 40.0$ Å$^2$ is found from the calculated BET monolayer capacity of EGME on reference solids of known surface areas (Chiou et al., 1993); the latter is then used for converting $Q_m(N_2)$ to $Q_m(EGME)_{eq}$ by Eq. (6.4), which then gives $Q_m(EGME)_{eq} = 1.30Q_m(N_2)$. Whereas a comparison of $Q_m(EGME)_{ap}$ with $Q_m(EGME)_{eq}$ provides an interesting test on the extension of the BET method with a polar vapor for systems in which the solvent exhibits no significant penetration into the bulk solid, it must be understood that the $Q_m$ monolayer values so determined with different vapors (including $N_2$) do not necessarily occur at the same $P/P°$.

Inspection of the data in Table 6.2 shows that the $Q_m(EGME)_{ap}$ values for such solids as sand, hematite, alumina, kaolinite, and synthetic hydrous iron oxide (SHIO) determined by use of EGME data with the BET equation are virtually the same as the $Q_m(EGME)_{eq}$ values based on the respective $Q_m(N_2)$ values by the standard BET-$N_2$ method. In other words, if one were to determine the surface areas of these solids using $Q_m(EGME)_{ap}$ together with $a_m(EGME)$, the results would be practically the same as obtained by the standard BET-$N_2$ method. This finding manifests the fact that uptake of EGME by these solids is confined essentially to external solid surfaces without any significant solvent penetration into the solid (and without any significant molecular sieving by the solid). For these solids, the results validate and extend the utility of the BET model for surface-area determination with EGME (or with similar polar vapors) as long as there is no significant bulk–solid penetration, a basic assumption and requirement of the BET model. It is noted, however, that while the observed $(P/P°)_m$ values for $N_2$ at $Q = Q_m(N_2)$ fall into the common range (0.05 to 0.30) for inert gases, the observed $(P/P°)_m$ values for EGME at $Q = Q_m(EGME)_{ap}$ are generally lower. This effect suggests that polar EGME adsorbs more efficiently than does $N_2$ (or other nonpolar vapors) onto the surfaces of these solids through additional polar and/or possibly H-bonding forces.

For the remaining solids, one sees that the ratios of $Q_m(EGME)_{ap}$ to $Q_m(EGME)_{eq}$ are larger than 1, ranging from 1.5 for illite, 3.2 for Woodburn soil, and 7.6 for Ca-SAz-1. In this case, if one were to determine the surface areas of these solids by use of the respective $Q_m(EGME)_{ap}$ values, the calculated surface areas would be higher than those by the standard BET-$N_2$ method to varying extents. Such discrepancies are caused by EGME penetration into solids, which is not a surface phenomenon and thus violates the basic assumption of the BET model for surface-area determination.

For a strongly expanding clay (e.g., Ca-montmorillonite), the excessive uptake of a polar solvent (e.g., water, EG, and EGME) actually occurs by cation solvation (McNeal, 1964; Dowdy and Mortland, 1967; Tiller and Smith, 1990) rather than by forming internal monolayers that are intercalated between silicate layers of the clay as assumed previously (Dyal and Hendricks, 1950; Quirk, 1955; Carter et al., 1965). Here the ion-dipole complex formed initially as a result of the $Ca^{2+}$ solvation with EGME may be considered as "forming a new compound," according to the view of Brunauer (1945), since the $Ca^{2+}$ solvation force is very potent. In a more recent study by Chiou and

Rutherford (1997), it has been shown, however, that after the initial cation solvation by EGME or water, the expanded clay interlayer spacings create more voids, on which polar molecules may adsorb at high $P/P°$. Evidently, the very large initial uptake of EGME (or other polar solvents) by a strongly solvating clay is driven by a powerful cation solvation force.

In a similar study by Mooney et al. (1952) on the water vapor uptake by montmorillonites in various cationic forms, the calculated BET surface areas with water data are likewise orders of magnitude larger than the values based on $N_2$ data, as would be anticipated. Since the uptake of EGME (or other polar vapors) by an expanding clay depends sensitively on the specific clay cation (Chiou and Rutherford, 1997), the $Q_m(EGME)_{ap}$ value or the resulting surface area would vary sensitively with the cation in the clay (e.g., $Ca^{2+}$ versus $K^+$), even when the surface areas by $N_2$ data of the clay in different cationic forms are fairly comparable. Thus, although K-SAz-1 montmorillonite has a somewhat higher BET-$N_2$ surface area than Ca-SAz-1 montmorillonite, as shown earlier, the latter clay exhibits a much greater water-vapor uptake than does the former, to be illustrated later, due to the fact that $Ca^{2+}$ is a far more powerful hydrating cation than $K^+$ (Cotton and Wilkinson, 1966).

The virtually linear EGME isotherm on peat, together with the peat's very small BET-$N_2$ surface area ($1.26\,m^2/g$), is good evidence for the presumption that the EGME uptake by peat at room temperature occurs primarily by partition (i.e., penetration) into the peat's organic matter matrix (Chiou et al., 1993). Further elaboration of the organic compound partition into soil organic matter is presented in Chapter 7. Since the EGME isotherm is practically linear (rather than type II), the BET model does not apply, and there is no theoretical $Q_m(EGME)_{ap}$. The partition effect, as reflected by the relatively linear uptake of EGME on peat, is analogous to the solubilization of organic substances into amorphous polymers (Flory, 1941; Huggins, 1942). The extrapolated limiting capacity of EGME with peat at $P/P° = 1$, normalized to the organic content of the peat, is $250\,mg/g$, which is comparable in magnitude with the finding of Bower and Gschwend (1952) that 170 to $250\,mg$ of EG is retained by $1\,g$ of soil organic matter under some evacuation condition. Bower and Gschwend thus obtained a value of 560 to $800\,m^2/g$ as the apparent surface area of soil organic matter by calculations in which the observed uptake of EG on soil organic matter was ascribed to surface adsorption rather than to bulk solubility.

The data for illite and natural hydrous iron oxide, which show moderate ratios of $Q_m(EGME)_{ap}$ to $Q_m(EGME)_{eq}$, reflect a moderate amount of EGME penetration, probably due to some combined effect of dissolution into organic matter and cation solvation with the clay component in the sample. On illite, the small number of exchangeable cations and a small amount of organic impurity (ca. 1.5% organic matter) could lead to the discrepancy observed. For the natural hydrous iron oxide, which is a relatively impure material, the nonsurface uptake by small amounts of expanding clay and organic matter could easily give a $Q_m(EGME)_{ap}$ to $Q_m(EGME)_{eq}$ ratio of 1.86. For Woodburn soil, which has a ratio of 3.2, the sample is known to contain 1.9% organic matter and 21% clay of unspecified form; the calculation shows that the effect of partition into soil organic matter is less significant than the effect of cation solvation (Chiou et al., 1993).

On the premise that nonpolar organic vapors partition far less efficiently than polar vapors into soil organic matter, shown in Chapter 7, the analysis above would suggest that use of the isotherms of nonpolar organic vapors along with the BET model should give a reasonable estimate of the surface areas of minerals and low-organic-content soils. This is because these vapors will not engage in cation solvation and they partition (dissolve) poorly into relatively polar soil organic matter (Chiou et al., 1993). This point is well manifested by the relatively consistent BET surface areas given in Table 6.3 for many low-organic-content soils and minerals by using the isotherms of $N_2$ at 77 K and various nonpolar organic vapors at room temperature.

An illustrative example of the earlier confusion in surface-area determination in soil science is underscored as follows: The specific surface area of a soil clay sample (i.e., the surface area per unit solid mass) is an operational concept in which the numerical value found for a given soil clay depends on a specific experimental method used (Sposito, 1984). Contrary to this point of view, it is recognized that as long as the adsorbates used in surface-area determination conform to Brunauer's requirement that they do not penetrate the bulk solid, the surface-area determination is clear, unambiguous, and largely independent of the adsorbate employed. As seen, elimination or minimization of the adsorbate penetration into soil organic matter and/or expanding clay minerals leads to essentially the same surface area for a natural solid by use of different adsorbates, since the surface area is an intrinsic property of the solid.

**TABLE 6.3. Comparison of Surface Areas of Minerals and Low-Organic-Carbon Soils Determined by the BET Equation Using $N_2$ and Low-Polarity Vapors as Adsorbates**

| | | Surface Area ($m^2/g$) | |
| --- | --- | --- | --- |
| Sample | % OC | BET-$N_2$ | BET(vapor) |
| Woodburn soil | 1.10 | 11.2 | $13.2^a$; $13.8^b$ $10.9^c$; $12.1^d$ |
| Ashurst field soil | 2.41 | 1.9 | $3.3^e$ |
| Ashurst garden soil | 4.55 | 6.3 | $4.6^e$ |
| Whittlesey Black Fen soil | 0.29 | 71.0 | $50.5^e$ |
| Boston silt | 2.66 | 28.6 | $23.2^e$ |
| Bentonite (Wyoming) | — | 65.0 | $61.0^e$ |
| Webster soil | 3.02 | 4.2 | $5.0^f$ |
| Kaolinite | 0.07 | 13.6 | $9.0^f$; $10^g$ |

*Source*: Data from Chiou et al. (1993).

[a] With benzene; [b] With chlorobenzene; [c] With *m*-dichlorobenzene; [d] With 1,2,4-trichlorobenzene; [e] With ethylene dibromide; [f] With toluene; [g] With *p*-xylene.

## 6.5    ADSORPTION OF WATER AND ORGANIC VAPORS

In the $N_2$ uptake by solids, the vapor is adsorbed mainly via the induced dipole (London) forces between the vapor and the solid, since $N_2$ has zero dipole moment, although its small quadrupole moment may contribute to a weak polar interaction with the solid. By contrast, water has a large dipole moment and an exceptional H-bonding power; thus the adsorption of water vapor on a solid is expected to depend critically on the surface property of the solid. If water vapor penetrates a solid as well as adsorbing on the solid surface, the result will be more complicated. In general, a polar vapor should adsorb much more efficiently onto a polar solid surface than onto a nonpolar surface because of powerful polar and H-bonding interactions between adsorbate and adsorbent (solid). The adsorption of a relatively nonpolar vapor onto various solids should be largely independent of the surface polarity, since the adsorption derives primarily from the dispersion forces between adsorbate and solid surfaces. The first situation is especially merited for the adsorption of water vapor onto various natural solids. In sharp contrast to water, benzene is largely apolar, having only a very weak H-bonding character (Barton, 1975). Thus, the adsorption of benzene vapor on a solid should be insensitive to the solid polarity, and the adsorption data should be largely representative of the adsorptive behavior of other nonpolar or weakly polar organic vapors. The purpose for comparing the adsorption data of water with the data of benzene on organic-matter-free solids is to give the reader a simple and clear picture of the relative adsorptive powers of water and uncharged organic compounds on the various minerals and natural solids. This information helps explain the competitive adsorption of water against nonionic organic solutes onto the polar soil mineral component.

The adsorption isotherms of water and benzene vapors on a range of solid samples, where the adsorbed mass per unit weight of the solid, $Q$ (mg/g), is plotted against the relative pressure $(P/P°)$, are illustrated in Figures 6.7 to 6.13. In comparing the adsorbed masses of different vapors on a given solid, it is necessary to take into account the difference in density of the condensed adsorbates, since an adsorbate with a higher density will invariably lead to a greater adsorbed mass when a given solid surface area or a given pore space is occupied. For vapors condensed onto a smooth (open) surface area, the mass of a condensed liquid (adsorbate) is related approximately to its $d_l^{2/3}M^{1/3}$ value, as mentioned earlier. When vapors condensed into a pore space, as with activated carbon or charcoal, the adsorbed mass per unit filled volume is approximately proportional to $d_l$ of the adsorbate. Thus, if the adhesive force between adsorbate and solid is greater than the adsorbate's cohesive force and if the adhesive force involved is primarily of London force (Manes, 1998), one would expect benzene vapor to exhibit a somewhat greater mass uptake than water vapor on a smooth surface; the reverse would apply for vapor adsorption onto a pore space. We will subsequently examine the adsorptive behavior of water with some minerals and other solids.

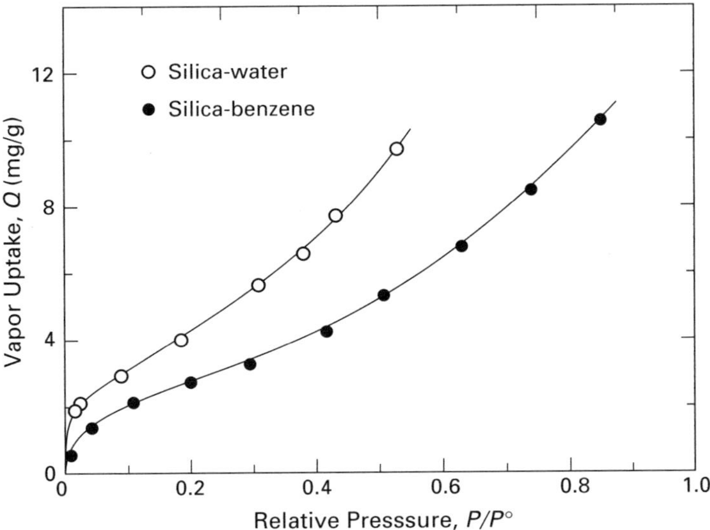


**Figure 6.7**  Uptake of water and benzene vapors at room temperature by silica. The solid is identified in Table 6.1. [Data from C. T. Chiou (unpublished research).]

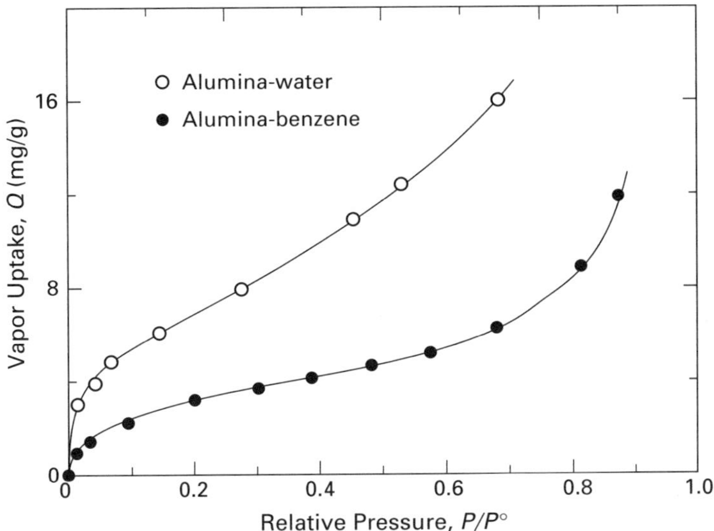


**Figure 6.8**  Uptake of water and benzene vapors at room temperature by alumina. The solid is identified in Table 6.1. [Data from C. T. Chiou (unpublished research).]

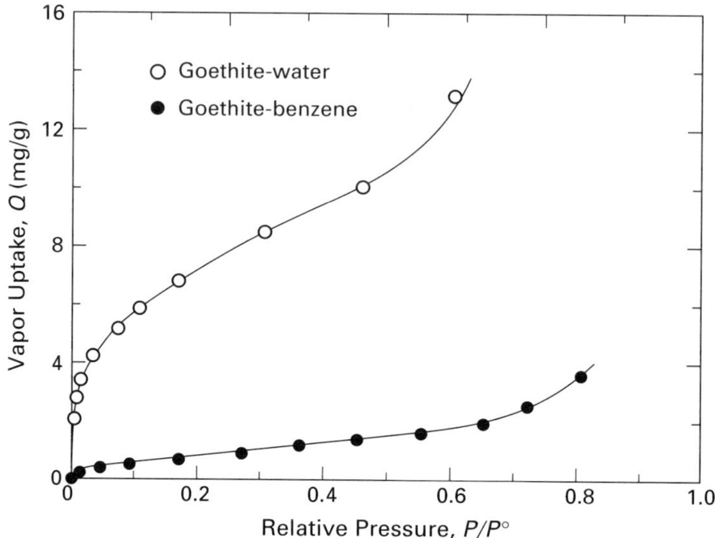

**Figure 6.9**  Uptake of water and benzene vapors at room temperature by goethite. The solid is identified in Table 6.1. [Data from C. T. Chiou (unpublished research).]

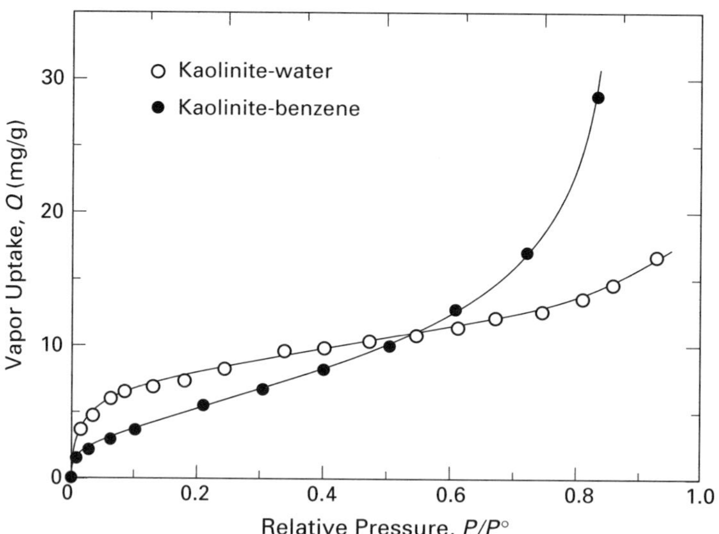

**Figure 6.10**  Uptake of water and benzene vapors at room temperature by kaolinite. The solid is identified in Table 6.1. [Water data from Chiou and Rutherford (1997) and benzene data from C. T. Chiou (unpublished research).]

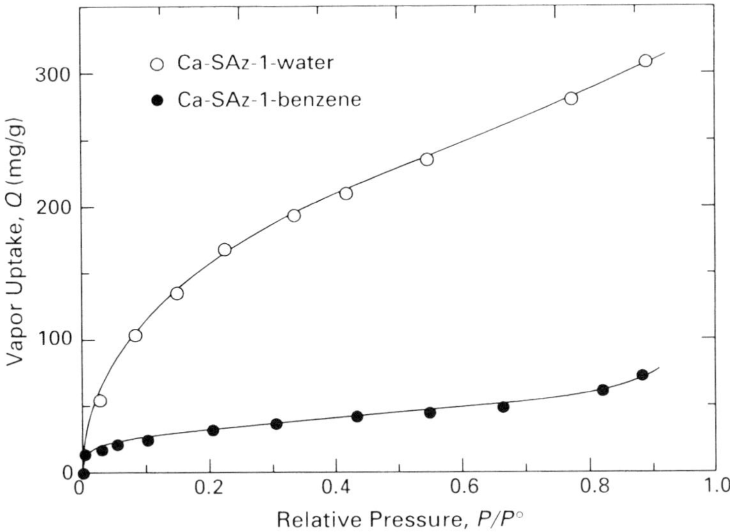

**Figure 6.11** Uptake of water and benzene vapors at room temperature by Ca-SAz-1. The solid is identified in Table 6.1. [Water data from Chiou and Rutherford (1997).]

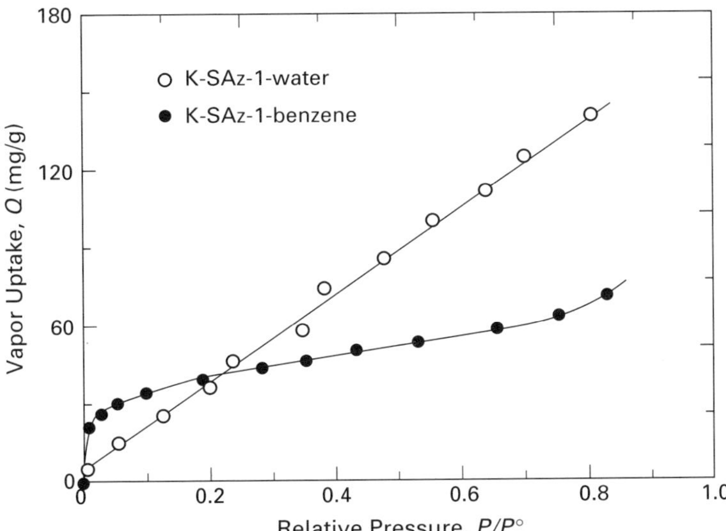

**Figure 6.12** Uptake of water and benzene vapors at room temperature by K-SAz-1. The solid is identified in Table 6.1. [Water data from Chiou and Rutherford (1997).]

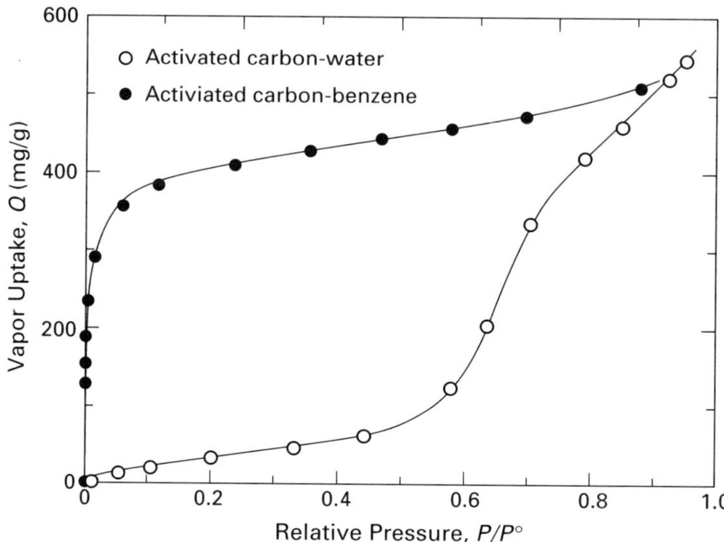


**Figure 6.13**   Uptake of water and benzene vapors at room temperature by activated carbon. The solid is identified in Table 6.1. [Data from C.T. Chiou (unpublished research).]

For all mineral oxides investigated, except K-exchanged montmorillonite (K-SAz-1), the water-vapor isotherms display markedly greater adsorption capacities than the corresponding benzene-vapor isotherms. This observation demonstrates that the enhanced water uptake is promoted by forces other than London forces. However, an extreme opposite behavior occurs with activated carbon, on which benzene vapor exhibits a remarkably greater adsorption than does water vapor. For nonexpanding minerals having no solvating cations, such as silica, alumina, and iron oxide (goethite), the observed higher water versus benzene uptake may be reasonably ascribed to the enhanced polar and H-bonding interactions of water with polar solid surfaces.

On montmorillonite, a 2:1 clay with siloxane plane surfaces, the water uptake varies greatly with the clay cation; Ca-SAz-1 shows considerably higher water than benzene uptake over the entire $P/P°$ range, whereas K-SAz-1 shows lower water than benzene uptake at the low $P/P°$ but higher water than benzene uptake at high $P/P°$. A similar result has been found for water and $N_2$ on these two clays (Chiou et al., 1997). In the earlier discussion, it is recognized that the uptake of polar vapors (including water) by montmorillonite depends sharply on the solvating power of the cation, which readily explains the difference observed here on water uptakes by Ca-SAz-1 and K-SAz-1. The water-uptake data suggest that siloxane surfaces are not sufficiently polar to effectively attract water, so that the water uptake by montmorillonite is governed mainly by the extent of cation hydration. (Note that extensive hydration of cations on surfaces may also reduce the surface sites otherwise available for adsorption.) For K-SAz-1, the relatively weak and linear water uptake appears to reflect a gradual increase in $K^+$ hydration with increasing humidity ($P/P°$); for Ca-SAz-1, the strong $Ca^{2+}$ hydration leads to a sharp and nonlinear water uptake (Chiou et al., 1997). On kaolinite, the water uptake is only moderately higher than the benzene uptake, as the solid has a siloxane plane on one side and a presumably more hydrophilic gibbsite plane on the other. The water uptake by kaolinite may also involve a small hydration with a residual amount of cations situated on clay edges. Thus, except possibly for montmorillonite with weak solvating cations, water is adsorbed generally more strongly than benzene.

On activated carbon, the very weak water uptake relative to an organic vapor uptake has been ascribed to the small polarizability per unit (liquid) volume of water and to the exceptionally large water cohesive energy density (Manes, 1998), that is, the cohesive energy of water is much greater than the adhesive energy between water and nonpolar activated-carbon surfaces. The observed small water uptake at low relative humidity ($P/P°$) probably results from adsorption onto the ash (mineral) of activated carbon and from the polar and H-bonding interactions of water with trace oxygenated impurities on the carbon surface. At high $P/P°$, the condensation of water vapor into the carbon pore space is promoted by the lower energy needed to concentrate the vapor to saturation for adsorptive condensation. The similarity in maximum adsorbed (liquid) volume between benzene and water (ca. 0.64 mL/g for benzene and 0.57 mL/g for water) is characteristic of the adsorbate pore-filling process with a highly porous adsorbent.

The strong water adsorption on most minerals has a direct consequence on the efficiency of mineral adsorption of organic solutes (contaminants) from water solution. The adsorption of nonpolar and weakly polar solutes from water on most minerals is expected to be greatly suppressed by water (because of their weak adsorptive competition against water). As to montmorillonite clays, or such clays in soils or sediments, it is also reasonable to expect that their adsorption of nonpolar solutes from water will probably not be significant. In rare cases where soils have a very high montmorillonite content that is saturated primarily with poor solvating cations, their adsorption of highly polar solutes from water may then become significant. The adsorption of organic solutes from water by activated carbon, or possibly by charcoal-like natural substance, should not be strongly suppressed by water. In the next chapter, we will be concerned with the uptake of organic compounds from water and other media to soils (or sediments), the latter comprising both organic and mineral matter as their basic constituents. The preceding account on adsorption of water vapor against an organic vapor (benzene) provides a good background for elucidating the roles of mineral and organic matter in the soil uptake.

# 7 Contaminant Sorption to Soils and Natural Solids

## 7.1 INTRODUCTION

Soil may be defined as a collection of natural bodies synthesized in profile form from a variable mixture of broken and weathered minerals and decaying organic matter that covers the earth surface (Brady, 1974). Discounting the water content in soil, which varies geographically with the soil and with the depth in a vertical soil profile, the principal soil components are mineral matter and organic matter. Minerals are composed of aluminosilicates and oxides in various crystalline and amorphous forms and vary considerably in individual quantity and physical size, the latter ranging from very small colloidal clay particles (<2 μm) to relatively large sand particles (>50 to 60 μm). The soil organic matter, which originates primarily from biologically degraded plant tissue and becomes part of the underlying horizons by infiltration or physical incorporation, consists of a heterogeneous makeup of organic constituents, such as lignins, carbohydrates, protein, fats, and waxes. It contains a large fraction of operationally defined *humic substance*, which is yellow to brown in color and moderately refractory to biological degradation (Stevenson, 1985). Although the molecular structures of humic substances have not been well characterized, they are known to be high-molecular-weight amorphous materials (frequently referred to as *humic polymers*) with significant polar-group contents. In ordinary soils, more than 90% of the dry organic matter is made up of carbon, hydrogen, and oxygen, with minor amounts of nitrogen, sulfur, and phosphorus. Except for relatively rare organic-rich soils, which are termed *organic soils*, ordinary soils are rich in minerals and are referred to as *mineral soils*. The organic matter content for most of these mineral soils falls between 0.5 and 3.0% by weight.

Studies of the sorption of organic compounds to soil began with the advent of pesticides for pest and weed controls in the 1940s, which called for an understanding of their interaction with and persistence in agricultural soils. In the 1970s, research in this field became more active and intense after wide varieties and large quantities of pesticides and industrial organic wastes were found throughout the environment, raising public concerns as to their long-term environmental impacts. Over the years, research in this field has shed important light on the characteristics of chemical–soil interactions in terms of soil organic and mineral components and of the properties of organic compounds. Understanding of the mechanistic roles of organic and mineral matter in soil and the effect of water on their individual functions enables us to make reasonable predictions of the sorption behavior of a wide variety of organic compounds under different system conditions.

Sorption data are most often analyzed at equilibrium (or near-equilibrium) conditions. Although concentrations of contaminants in soil, water, and other phases in natural systems frequently deviate from those at equilibrium, the equilibrium data serve as an essential guide to the direction of contaminant movement at a particular point in time and to the likely consequence of an earlier contamination event. A comparison of the field data with equilibrium values also enables one to elucidate whether a compartment (such as soil or sediment) functions as a sink (to receive a given contaminant) or as a source (to release a given contaminant) under specified conditions. Such information is often valued in the characterization of a contamination site.

In the description of the roles of soil or sediment organic matter (SOM) and minerals in uptake of organic compounds and pesticides, select terminology is often used to refer to the mechanism involved. The term *sorption* is used to denote the uptake of a contaminant (solute or vapor) by soil or sediment without reference to a specific mechanism (i.e., by adsorption and/or partition). The terms *partition* and *adsorption* are used to refer to specific processes involved. As one recalls, the term *partition* refers to a process in which the sorbed material penetrates into the entire network of an organic phase by forces common to solution, whereas the term *adsorption* refers to condensation of vapors or solutes on the surfaces or interior pores of a solid (adsorbent).

The term *organic matter* is used to refer to the bulk of the organic content in soil (or sediment). While variations in composition between humic and non-humic organic matter in soil can influence the overall behavior of the organic matter, it is generally not possible to separate their effects because practically all sorption studies have been carried out with intact soils. In general, except for top surficial soils where a significant amount of undecomposed (or poorly decomposed) plant litter may exist, the SOM normally contains mainly well-humified organic material.

## 7.2 BACKGROUND IN SORPTION STUDIES

### 7.2.1 Influences of Mineral Matter, Organic Matter, and Water

To better understand the sorption of an organic compound to a soil or a natural solid under a particular system condition, it is helpful to have a brief overview of important and unique sorption characteristics in relation to the soil or solid composition and the water content associated with it. The highly heterogeneous nature of soil samples from different geographic sources greatly complicates the resulting sorption of organic contaminants. This made it a formidable challenge for scientists to interpret the wide range of soil sorption data. In earlier studies of pesticide–soil interactions, the soil was generally assumed to be a single adsorbent, or at best a mixed adsorbent of some kind, analogous to other well-defined conventional adsorbents. Although this view reconciled to a large extent the sorptive behavior of organic pesticides on relatively dry soils and minerals, it ran into serious technical difficulties in explaining the sorption data with water-saturated soils. We shall see later that the different sorptive characteristics with relatively dry and water-saturated soils are directly responsible for the change in chemical activity, bioavailability, and toxicity of contaminants sorbed to the soil.

The adsorptive character of soils and minerals has been illustrated unequivocally in earlier studies on the vapor uptake of chloropicrin (Stark, 1948), ethylene dibromide (Hanson and Nex, 1953; Wade, 1954), and methyl bromide (Chisholm and Koblitsky, 1943) by water-unsaturated soils and minerals, in which the vapor uptake is suppressed by soil moisture. The uptake of parathion and lindane by soils from hexane solution exhibits a similar suppression by soil moisture (Yaron and Saltzman, 1972; Chiou et al., 1985). These observations indicate that organic compounds and water compete for adsorption on initially water-unsaturated soil minerals, which comprise most of the available surface area of the soil solid. Moreover, the isotherms measured for the uptake of pesticides from either the vapor phase or from a nonpolar solvent (e.g., hexane) on water-unsaturated soils and minerals are commonly nonlinear, characteristic of an adsorption process.

In keen contrast to the findings above, the uptake of the same nonionic compounds from water by soils, or from vapor phase by water-saturated soils, displays uniquely different features. Most notably, the extent of soil uptake for given organic compounds shows a strong dependence on the SOM content (Kenaga and Goring, 1980; Means et al., 1980; Kile et al., 1995). The uptake of organic vapors by wet soils displays a similar effect (Wade, 1954; Leistra, 1970). The predominant effect of SOM content in this case is demonstrated by the relative invariance of the sorption coefficients of given organic compounds among soils, or size fractions of soil, when the coefficients are normalized to the SOM content (Karickhoff et al., 1979; Kenaga and Eoring, 1980; Kile et al., 1995). The sorption isotherms of nonionic compounds on water-saturated soils are all relatively linear (Chiou et al., 1979; Karickhoff et al., 1979; Schwarzenbach and Westall, 1981; Sun and Boyd, 1991; Rutherford et al., 1992) and are not strongly temperature dependent, exhibiting only small exothermic heats of sorption (Mills and Biggar, 1969; Spencer and Cliath, 1970; Yaron and Saltzman, 1972; Pierce et al., 1974; Chiou et al., 1979). Moreover, the soil uptake of binary nonpolar solutes from water occurs without a significant competition between the solutes (Schwarzenbach and Westall, 1981; Chiou et al., 1983, 1985) in contrast to the strong competitive effects found in sorption by dry soil from the vapor phase and from nonpolar organic solvents (Chisholm and Koblitsky, 1943; Wade, 1954; Spencer et al., 1969; Mills and Biggar, 1969; Yaron and Saltzman, 1972; Chiou and Shoup, 1985; Chiou et al., 1985; Pennell et al., 1992; Thibaud et al., 1993).

Despite the fact that the relationship observed between soil uptake and SOM content had greatly simplified assessments on the uptake of nonionic organic compounds from water by soils, there was no single widely accepted view on the sorptive mechanism with SOM in the literature before 1979. Prior to that time, one popular view considered SOM as a high-surface-area adsorbent (Bower and Gschwend, 1952; Bailey and White, 1964) capable of adsorbing nonionic organic compounds by hydrophobic interactions (Weed and Weber, 1974; Browman and Chesters, 1977; Mingelgrin and Gerstl, 1983). Such a hydrophobic adsorption concept, however, is not supported by common adsorption criteria and in particular by the observed soil sorption data in aqueous systems. Moreover, the earlier accepted surface area for SOM (550 to $800\,m^2/g$), reported by Bower and Gschwend (1952) based on the ethylene glycol (EG) retention method, was later shown to be largely an artifact of the high solubility of polar EG in relatively polar SOM (Chiou et al., 1990, 1993). Using some high-organic-content soils (peat and muck) as a model for SOM, the surface area of SOM as measured by the standard BET method (with $N_2$ vapor as the adsorbate) is actually only about $1\,m^2/g$ (Chiou et al., 1990; Pennell et al., 1995), which is nearly three orders of magnitude lower than the value assumed earlier.

In attempts to reconcile the inconsistency in reported SOM surface areas, Pennell and Rao (1992) consider that the large difference between the values obtained by the polar-solvent retention method and by the standard BET-$N_2$ method represents the internal surface of the SOM. Although the term *internal surface* has been used concurrently in surface science, it is well understood there that the internal surface, which extends inward the porous channels of a solid (e.g., activated carbon), is freely accessible to an inert gas such as $N_2$, as noted in Chapter 6. Therefore, if an assumed internal surface is impervious to an inert gas, it is more a reflection of solvent penetration into the SOM solid matrix, as pointed out earlier by Brunauer (1945). As we will see later, the excess uptake of an organic vapor (especially, a polar vapor) over that of $N_2$ gas by SOM is more properly interpreted in terms of the vapor partition.

Along with these unique features for soil uptake from water, we also recall that water vapor exhibits a generally much greater adsorption than an organic (benzene) vapor on various dry minerals, as elucidated in Chapter 6. Based on this disparity, one would then expect water, as a solvent, to strongly suppress the adsorption of an organic solute onto a soil mineral, because the adsorption process is competitive. Thus, the many outstanding features in the sorption of organic solutes from water solution, such as the virtual isotherm linearity and the dependence of the uptake on soil organic content, can only be reconciled readily and logically with the partition-dominated solute uptake by SOM.

### 7.2.2  Soils as a Dual Sorbent for Organic Compounds

A major advance in the description of the sorption process of organic compounds with soil (or sediment) started with the proposition by Chiou and co-workers (1979, 1981) that the SOM behaves primarily as a partition medium, rather than a conventional solid adsorbent. In addition to the recognized dependence of soil sorption on SOM content, they showed that the sorption of relatively nonpolar solutes from water is essentially linear from low to high relative concentrations (ratios of equilibrium concentrations to solute solubilities) and that the equilibrium heats of sorption for the solutes are less exothermic than their heats of condensation from water. In related studies (Chiou et al., 1983, 1985), they also showed that the soil sorption of binary solutes from water exhibits no significant solute competition. The inability of the soil mineral fraction to adsorb nonionic organic compounds from water significantly is attributed to strong dipole interactions of water with minerals, which suppress the adsorption of these compounds on minerals. In keeping with the idea of solute partitioning into SOM, it was observed that solutes (as liquids or supercooled liquids) with higher SOM-normalized sorption coefficients ($K_{om}$) or soil-organic-carbon-normalized sorption coefficient ($K_{oc}$) exhibit generally lower limiting sorption capacities on SOM. By application of the Flory–Huggins model to account for solute solubility in (amorphous) soil organic phase, Chiou et al. (1983) developed a partition equation to account for the magnitudes of the observed sorption coefficients. This analysis led to the recognition that the primary factor affecting the sorption coefficients of slightly water-soluble organic compounds is the solubility of the compounds (as liquids or supercooled liquids) in water. The frequently observed empirical correlation between the normalized sorption coefficient ($K_{om}$ or $K_{oc}$) and octanol–water partition coefficient ($K_{ow}$) of the solutes was recognized to be the consequence that the solute solubility in water is the major determinant of both $K_{om}$ and $K_{ow}$ values (Chiou et al., 1982b, 1983). The notion that the SOM acts essentially as a partition medium for the organic solute uptake is reinforced by the later finding that the SOM has a low surface area (about $1\,m^2/g$) (Chiou et al., 1990), which is far too small to account for solute uptake by SOM by adsorption.

The different characteristics in the sorption of nonionic organic compounds from aqueous and nonaqueous systems are reconciled with the postulate that the soil (or sediment) behaves as a dual sorbent: The mineral matter functions as a conventional adsorbent and the SOM as a partition medium (Chiou et al., 1979, 1981, 1983, 1985; Chiou and Shoup, 1985). The linear isotherms and other characteristics in aqueous systems are attributed to the solute partition into SOM and a concomitant suppression of adsorption on mineral matter by water. The nonlinear isotherms and higher sorption capacities on dry soils are ascribed to adsorption on minerals, which predominates over the simultaneous partition in SOM. The schematic plots in Figure 7.1 depict the relative sorptive effects of mineral matter and SOM of a mineral soil that contains a moderate amount of SOM (say, 1 to 2%). The scales for these two effects are not exact but are drawn to emphasize the dominant role of either the adsorption on minerals (for dry soil) or the partition into SOM (for water-saturated soil). For dry soil, as shown in Figure 7.1$a$, the much greater nonlinear adsorption with mineral matter than the linear partition with SOM gives rise to a

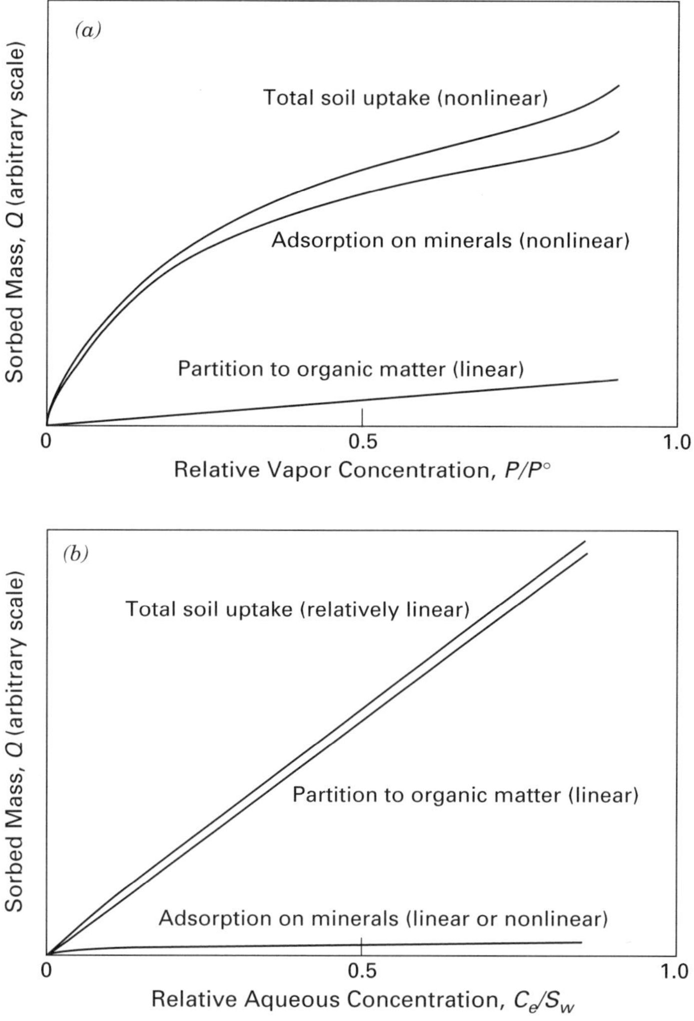



**Figure 7.1**  Schematic plots of contributions by adsorption on soil mineral matter and by partition to soil organic matter to the uptake of an organic compound by a mineral soil. (*a*) Uptake from vapor phase on dry soil; (*b*) Uptake from water solution on wet soil.

high and nonlinear overall soil sorption (the sum of the two contributions). For water-saturated soil, as shown in Figure 7.1*b*, the mineral adsorption is sharply suppressed by water and the partition in SOM predominates to produce a relatively low and linear total sorption.

The specific roles of SOM and mineral matter provide the point of departure for understanding the diverse and often seemingly contradictory sorption behaviors of organic contaminants with soils from water solution, from organic solvent solution, and from the vapor phase. These topics and related systems are treated in some detail later. Since there is a continuum of the organic matter content in soil, the analysis and discussion of the sorption data in aqueous systems is restricted to soils (or sediments) having more than 0.1 to 0.2% organic content so that the effect of SOM on contaminant uptake is significant enough to be reliably quantified.

## 7.3  SORPTION FROM WATER SOLUTION

### 7.3.1  General Equilibrium Characteristics

We begin by looking at the sorption data for relatively nonpolar organic compounds (solutes), and then look at the data for relatively polar organic compounds, because of some characteristic differences in their behaviors. Whereas the demarcation between polar and nonpolar compounds is not straight-forward, *polar compounds* are considered to be those that possess significant polar groups in their molecular structures, and *nonpolar compounds* those that contain little or no polar groups. Some common strong polar groups are —OH, —NH$_2$, —COOH, —CO—, and —NO$_2$, as illustrated in Table 5.3. Polar groups enhance molecular interactions of the compounds by polar forces and H-bonding with each other and with other polar compounds. In general, the effect of a polar group in a molecule is more significant for small molecules than for large molecules. Polar organic solutes generally exhibit low partition from water to a water-immiscible (or partially miscible) organic phase relative to nonpolar solutes because the former have a high affinity for water. For instance, as shown in Table 5.3, the $\log K_{ow}$ (octanol–water) values of highly polar organic solutes are generally <2.

In water solution the sorption isotherms for relatively nonpolar organic compounds on soils or sediments are usually virtually linear, as has been demonstrated in a number of studies (see, e.g., Yaron and Saltzman, 1972; Chiou et al., 1979, 1983; Karickhoff et al., 1979; Means et al., 1980; Schwarzenbach and Westall, 1981; Kile et al., 1995). Similar isotherm linearity has been reported for the soil uptake of volatile nonpolar pesticide vapors onto water-saturated soils (Leistra, 1970; Spencer and Cliath, 1970). In some studies where slight isotherm curvatures were shown (see, e.g., Mingelgrin and Gerstl, 1983), the extent of isotherm nonlinearity (either concave upward or downward) appears to be comparable with the normal range of data scatter and hence cannot be distinguished from a linear isotherm. This is especially true in sorption studies of low-organic-content soils when the solute uptake is computed by the difference in solute concentrations in water before and after equilibration; here the uncertainty in detecting small concentration changes is expected to be relatively large. Since the curvatures in allegedly nonlinear isotherms for some relatively nonpolar pesticides (e.g., ethylene dibromide and lindane) (Mingelgrin and Gerstl, 1983) are quite small and show no consistent shape, the results offer no clear evidence for strong solute adsorption over the con-centration range studied.

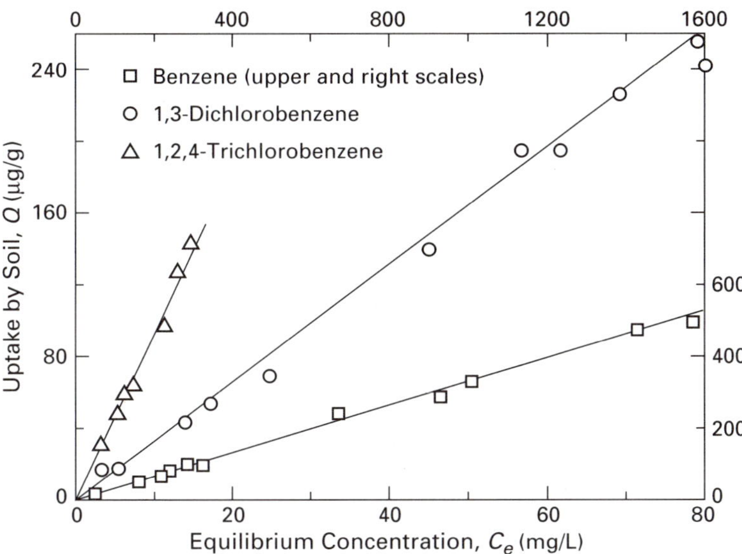



**Figure 7.2**   Sorption of benzene, 1,3-dichlorobenzene, and 1,2,4-trichlorobenzene from water on Woodburn soil ($f_{om}$ = 0.019) at 20°C. [Data from Chiou et al. (1983). Reproduced with permission.]

While in previous studies the linear sorption isotherms were observed for solutes in the low concentration range, such linear isotherms also extend to high relative concentrations ($C_e/S_w$) for sparingly water-soluble solutes, where $C_e$ is the equilibrium solute concentration and $S_w$ is the solute solubility in water. Figure 7.2 shows typical linear isotherms for the sorption of benzene, 1,3-dichlorobenzene, and 1,2,4-trichlorobenzene from water on a Woodburn soil which contains 1.9% SOM ($f_{om}$ = 0.019) (Chiou et al., 1983). The benzene isotherm is linear with $C_e/S_w$ up to about 0.90. Similar linear isotherms for many halogenated organic liquids on a Willamette silt loam ($f_{om}$ = 0.016) (Chiou et al., 1979) are shown in Figure 7.3, where, for example, 1,2-dichlorobenzene exhibits linearity with $C_e/S_w$ up to 0.95. This wide isotherm linearity together with the dependence of soil sorption on $f_{om}$ is illustrative of solute partition into an organic phase (in this case, SOM) as the dominant sorption pathway. Here the low soil uptake of the low-polarity solutes results from both the low SOM content and the low partition efficiency of the solutes with relatively polar SOM; the isotherms are thus essentially linear rather than concave upward in shape (see Chapter 3, section 3.5). As noted, the sorption capacities ($Q$) of many of the solutes normalized to the SOM content are <10% of the SOM weight. The weak adsorption of nonpolar solutes on soil minerals may be attributed to the strong competitive adsorption of water for polar mineral surfaces. The normalized $K_{om}$ values of these halogenated solutes (i.e., $K_{om} = K_d/f_{om}$, where $K_d$ is the soil–water distribution coefficient) and their water solubilities at 20°C are given in Table 7.1.

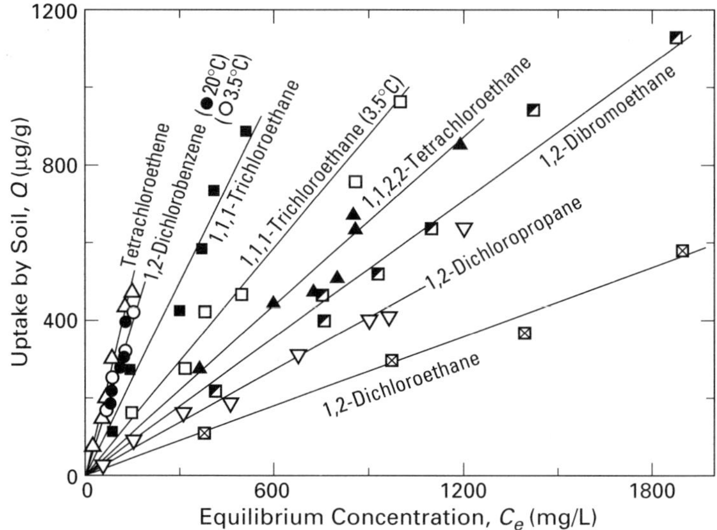


**Figure 7.3** Sorption of halogenated organic liquids on Willamette silt loam ($f_{om}$ = 0.016) at 20°C. [Data from Chiou et al. (1979). Reproduced with permission.]

**TABLE 7.1. Normalized Sorption Coefficients of Halogenated Organic Liquids from Water on Willamette Silt Loam ($K_{om}$) and Corresponding Liquid Solubilities in Water ($S_w$) at 20°C**

| Compound | $S_w$ (mg/L) | $K_{om}$ |
|---|---|---|
| 1,2-Dichloroethane | 8450 | 19 |
| 1,2-Dichloropropane | 3570 | 27 |
| 1,2-Dibromoethane | 3520 | 36 |
| 1,1,2,2-Tetrachloroethane | 3230 | 46 |
| 1,1,1-Trichloroethane | 1360 | 104 |
| 1,2-Dibromo-3-chloropropane | 1230 | 75 |
| 1,2-Dichlorobenzene | 148 | 180 |
| Tetrachloroethene | 200 | 210 |

*Source*: Data from Chiou et al. (1979).

Although the idea of solute partition to SOM was suggested earlier by Swoboda and Thomas (1968) as a possible mechanism for parathion uptake by soil from water, it did not gain widespread acceptance because of the lack of other supporting evidence. As a matter of fact, there had been serious misconception about the occurrence of linear sorption isotherms. As noted for sparingly water-soluble solutes and pesticides with soil, the isotherm linearity was thought by many to be a result of solutes' low concentrations in water that restricted the soil adsorption capacity in a low and linear range (Mingelgrin

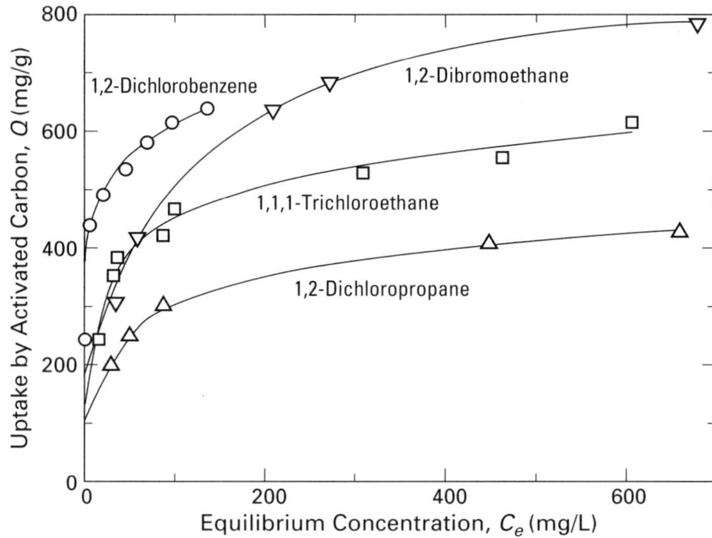


**Figure 7.4** Adsorption of selected halogenated organic liquids from water on Pittsburgh CAL (12 × 40) activated carbon at 20°C. [Data from Chiou (1981). Reproduced with permission.]

and Gerstl, 1983; MacIntyre and Smith, 1984). Although the adsorption of single solutes (and vapors) may be linear at very low relative concentrations ($C_e/S_w$) (i.e., in the Henry's law concentration region), the observed sorption linearity that extends over a wide range of $C_e/S_w$ (as with the soil uptake from water) should not be confused with the linear range of an overall nonlinear adsorption isotherm. To make this point evident, one may, for example, compare the adsorption isotherms of 1,2-dichlorobenzene, 1,1,1-trichloroethane, 1,2-dichloropropane, and 1,2-dibromoethane on activated carbon (Figure 7.4) with their sorption isotherms on a Willamette silt loam (Figure 7.3). The isotherms of the compounds on activated carbon are linear only at very low equilibrium concentrations relative to their water solubilities, whereas the sorption isotherms on soils show no obvious indication of a curvature even at concentrations approaching saturation.

An important feature associated with the linear sorption of organic compounds to soil is that the molar heat of sorption of the compound is constant, independent of its loading on soil (Chiou et al., 1979). This effect may readily be understood in terms of the calculated (equilibrium) heat of sorption of a compound using its linear isotherms obtained at two temperatures. A schematic plot of the linear isotherms of a compound at temperature $T_1$ and $T_2$ (in Kelvin) is presented in Figure 7.5, with $T_2 > T_1$. Let $Q$ be the mass of the compound sorbed by a unit mass of soil (or, more closely, by a unit mass of SOM) and $C_e$ be the concentration in water of the compound in equilibrium with a given $Q$ on soil. Each linear isotherm is assumed to cover a wide range

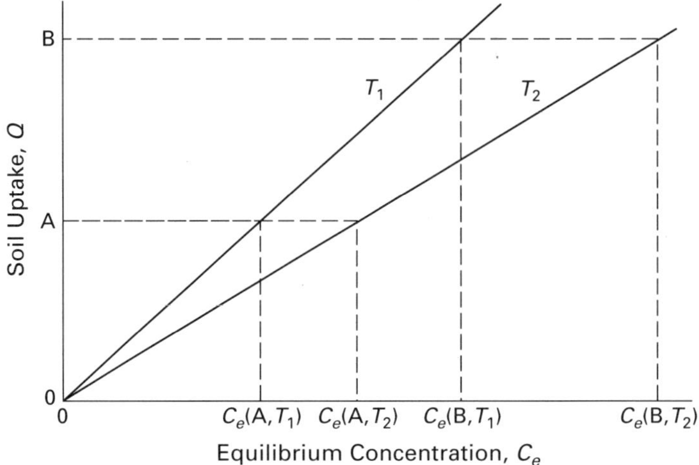


**Figure 7.5**  Schematic plot of the linear solute sorption from water by soil ($Q$) versus the equilibrium solute concentration ($C_e$) at temperatures $T_1$ and $T_2$, with $T_2 > T_1$.

of $C_e$ relative to the compound solubility in water at the system temperature. The isotherms are drawn such that the soil uptake at $T_1$ is greater than that at $T_2$, as is usually observed for most organic solutes; however, a reverse temperature dependence may take place if the compound shows abnormal (i.e., exothermic) heat of solution in water over a temperature range, as noted with 1,1,1-trichloroethane (Chiou et al., 1979).

The molar isosteric heat of solute sorption at a given uptake capacity can be obtained by use of the Clausius–Clapeyron equation (4.16). At the capacity $Q_A$, for example, the equation gives

$$\Delta\overline{H}(Q_A) = -R\frac{\ln[C_e(A,T_2)/C_e(A,T_1)]}{1/T_1 - 1/T_2} \tag{7.1}$$

where $R$ is the gas constant (8.31 J/mol·K) and $C_e(A,T_2)$ and $C_e(A,T_1)$ are the equilibrium concentrations corresponding to $Q_A$ at temperature $T_2$ and $T_1$, respectively. The molar heat of sorption at capacity $Q_B$ [i.e., $\Delta\overline{H}(Q_B)$] can be obtained similarly by substituting $C_e(B,T_2)$ for $C_e(A,T_2)$ and $C_e(B,T_1)$ for $C_e(A,T_1)$ in Eq. (7.1). Because the isotherms are linear, one finds that $C_e(A,T_2)/C_e(A,T_1) = C_e(B,T_2)/C_e(B,T_1)$ and hence that $\Delta\overline{H}(Q_A) = \Delta\overline{H}(Q_B)$. By repeating the same calculations at other loadings, one thus concludes that the molar heat of sorption is constant and independent of the loading capacity. Because of the linearity of the isotherms, the concentration ratio in Eq. (7.1) is equal to the ratio of the sorption coefficient (the slope of the isotherm) at $T_1$ to that at $T_2$.

The calculated molar heats of sorption for most organic compounds and pesticides on soil in water are generally less exothermic than $-\Delta\overline{H}_w$ and are largely independent of sorption capacities (Mills and Biggar, 1969; Yaron and Saltzman, 1972; Pierce et al., 1974; Chiou et al., 1979, 1985). The same is true of the sorption of organic compounds from the vapor phase by water-saturated soils, such as found for ethylene dibromide (Wade, 1954) and lindane (Spencer and Cliath, 1970), in which the heats of sorption are less exothermic than the heats of vapor condensation $(-\Delta\overline{H}_v)$. These results are inherently consistent with the conceived partition uptake of nonionic organic compounds by the soil organic matter of water-saturated soils.

The relation between $\Delta\overline{H}$ and $\Delta\overline{H}_w$ for soil sorption in aqueous systems can be explained readily by the temperature dependence of the normalized isotherms, as shown in Figure 7.6, using the relative solute concentration $(C_e/S_w)$ as the abscissa. By such a normalized plot, one usually finds a reverse temperature dependence for the sorption of organic solutes to soil (Yaron and Saltzman, 1972; Mills and Biggar, 1969); that is, at a given solute loading on soil,

$$\frac{C_e(T_1)}{S_w(T_1)} > \frac{C_e(T_2)}{S_w(T_2)} \tag{7.2}$$

or

$$\frac{S_w(T_2)}{S_w(T_1)} > \frac{C_e(T_2)}{C_e(T_1)} \tag{7.3}$$

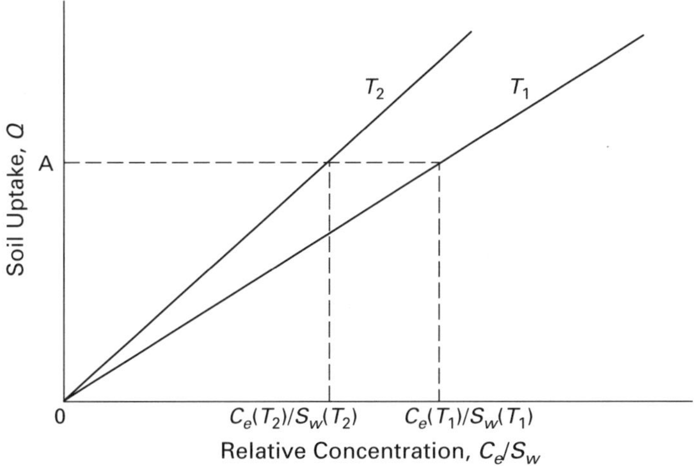


**Figure 7.6**  Schematic plot of the linear solute sorption from water by soil $(Q)$ versus the relative solute concentration $(C_e/S_w)$ at temperatures $T_1$ and $T_2$, with $T_2 > T_1$.

which can be expressed as

$$\frac{d \ln S_w}{dT} > \frac{d \ln C_e}{dT} \tag{7.4}$$

Since one finds that

$$\frac{d \ln S_w}{dT} = \frac{\Delta \overline{H}_w}{RT^2} \tag{7.5}$$

and that

$$\frac{d \ln C_e}{dT} = \frac{\Delta \overline{H}_d}{RT^2} = -\frac{\Delta \overline{H}}{RT^2} \tag{7.6}$$

with $\Delta \overline{H}_d$ denoting the molar heat of desorption (i.e., $\Delta \overline{H}_d = -\Delta \overline{H}$), one therefore finds that

$$\Delta \overline{H} > -\Delta \overline{H}_w \tag{7.7}$$

which means that the heat liberated when 1 mole of solute is sorbed to the soil is less exothermic than the solute's reverse heat of solution in water. Equation (7.7) explains the small exothermic heats of sorption of DDT (Pierce et al., 1974), lindane and $\beta$-BHC (Mills and Biggar, 1969), parathion (Yaron and Saltzman, 1972), and 1,1,1-trichloroethane (Chiou et al., 1979) on soil.

Equation (7.7) is actually a consequent form of Eq. (3.23) for the molar heat of partition of a solute between an organic solvent and water. Since the heats of solution of organic compounds in an organic phase ($\Delta \overline{H}_o$) are generally positive but small due to their improved compatibilities, $\Delta \overline{H}$ will be small for compounds with low positive $\Delta \overline{H}_w$ values and may even become positive (endothermic) for compounds with abnormal (negative) $\Delta \overline{H}_w$ values. For example, Chiou et al. (1979) showed that the $\Delta \overline{H}$ for 1,2-dichlorobenzene sorption by soil from water is nearly zero because of its low $\Delta \overline{H}_w$ and that the $\Delta \overline{H}$ for 1,1,1-trichloroethane is positive because of its negative $\Delta \overline{H}_w$ in the temperature range 3.5 to 20°C. One may conclude from these data that in systems where the $\Delta \overline{H}$ values are negative, such exothermic heats originate primarily from condensation of the solutes from water ($-\Delta \overline{H}_w$) and that with $\Delta \overline{H}_o$ being normally positive, interactions between SOM and solute ($\Delta \overline{H}_o$) are normally endothermic, as usually is the case for the heat of solution. One finds small and nearly constant exothermic heats ($\Delta \overline{H}$) (and hence small temperature coefficients) in solute partition equilibria as a result of the partial cancellation in heat between $\Delta \overline{H}_o$ and $\Delta \overline{H}_w$ according to Eq. (3.23).

The sorption data of $p,p'$-DDT are especially worth noting because DDT is a solid with a large heat of fusion ($\Delta \overline{H}_{fus}$), about 25 kJ/mol (Plato and

Glasgow, 1969), and is highly incompatible with water, which should make $\Delta \overline{H}_w$ much greater than 25 kJ/mol. The sorption coefficient of DDT with soil or sediment normalized to the SOM content ($K_{om}$) is approximately $1.5 \times 10^5$ (Pierce et al., 1974; Shin et al., 1970), while the heat of sorption ($\Delta \overline{H}$) at equilibrium is about $-8.4$ to $-16.8$ kJ/mol (or $-12.6 \pm 4.2$ kJ/mol) (Pierce et al., 1974). Thus the observed $\Delta \overline{H}$ is far less exothermic than $-\Delta \overline{H}_w$, as would be expected for a solid solute with large $\Delta \overline{H}_{fus}$ in partition equilibrium [see Eq. (3.23) and the discussion thereafter]. Based on these values, one can calculate the standard entropy change for the transfer of DDT from water to the SOM as

$$\Delta \overline{G}^\circ = -RT \ln K_{om} \tag{7.8}$$

and

$$\Delta \overline{S}^\circ = (\Delta \overline{H}^\circ - \Delta \overline{G}^\circ)/T \tag{7.9}$$

where $\Delta \overline{G}^\circ$ is the (molar) standard free energy change for the transfer of 1 mole (or a unit mass) of the solute from water at unit concentration to the SOM phase at unit concentration, and $\Delta \overline{H}^\circ$ and $\Delta \overline{S}^\circ$ are the corresponding enthalpic and entropic changes at the said standard state. Since the $\Delta \overline{H}$ for a solute in a partition process is largely independent of the solute concentration, the $\Delta \overline{H}^\circ$ value at the standard state is essentially equal to the $\Delta \overline{H}$ value at the point of equilibrium. Now, if one takes $K_{om} \simeq 1.5 \times 10^5$ and $\Delta \overline{H}^\circ \simeq \Delta \overline{H} = -12.6$ kJ/mol for DDT, one gets $\Delta \overline{S}^\circ \simeq 58$ J/mol·K at $T = 298$ K. Although the calculated $\Delta \overline{S}^\circ$ value for DDT is subject to some uncertainty because of the inaccuracy of the $\Delta \overline{H}$ value, it is nonetheless indicative of a relatively small change in molar entropy for the transfer of DDT from water into SOM at the standard state, as would be expected for a partition process. Such a small entropy change is in sharp contrast to a usually very large entropy decrease when a trace component adsorbs strongly from a solvent (water) onto an adsorbent. In analyzing the sorption process with entropy, it is important that the $\Delta \overline{S}^\circ$ at the standard state, rather than the $\Delta \overline{S}$ at equilibrium, be employed. This is because the $\Delta \overline{S}$ values for solutes at equilibrium between any two phases (where $\Delta \overline{G} = 0$) will always be negative whenever the process (adsorption or partition) proceeds exothermically. We shall consider later the heat effect associated with the soil sorption of organic compounds in nonaqueous systems.

Accountability of the solute partitioning into the SOM phase is further substantiated by the estimated magnitude of the solute solubility in SOM. Since the isotherm is practically linear, the solubility of a solute in SOM may be determined by

$$S_{om} = S_w \cdot K_{om} \tag{7.10}$$

where $S_{om}$ is the solute solubility in SOM and $S_w$ the solute solubility in water. For solid DDT with $S_w = 5.5\,\mu g/L$ (Weil et al., 1974) and $K_{om} \simeq 1.5 \times 10^5$ at 25°C, one therefore gets $S_{om} \simeq 830\,mg/kg$, or 0.83 g/kg. By comparison, the solubility of DDT in pure octanol is about 42 g/L (Chiou et al., 1982b), which is some 50 times its solubility in SOM. The low estimated solubility of DDT in SOM is much expected for a relatively nonpolar solid compound in a polar macromolecular amorphous material. It is evident from these data that the very high sorption coefficient of DDT results primarily from its extremely low water solubility, which gives rise to a high partition coefficient.

On the premise of solute partition, one expects organic compounds with high water solubility to also exhibit high $S_{om}$ values because these compounds are usually also more compatible with organic solvents. For example, as shown in Table 7.2, benzene, with $S_w = 1780\,mg/L$ and $K_{om} \simeq 18$, gives $S_{om} \simeq 32\,g/kg$ according to Eq. (7.10), which is about 40 times greater than the $S_{om}$ value of solid DDT (due partly to the fact that benzene is a liquid and DDT is a solid). This is consistent with the fact that benzene is completely miscible with octanol and most organic solvents. As shown in Table 7.2, the $S_{om}$ values for solid compounds are smaller because of the melting-point effect. Thus, although the $S_{om}$ values for a given solute vary somewhat among soils or sediments due to compositional differences in their organic matters, the magnitudes of the $S_{om}$ values fall largely into the range to be expected for low-polarity organic compounds in relatively polar organic polymers. To explain differences in soil uptake of organic compounds from water, Mingelgrin and Gerstl (1983) suggested that the less polar the compound, the more it will tend to adsorb on a hydrophobic surface (SOM) from a polar solvent (water), while removing solvent molecules from that surface. This *hydrophobic adsorption* concept is not consistent with the fact that the limiting uptake

**TABLE 7.2. Estimated Solubilities of Some Organic Liquids and Solids in Soil Organic Matter by Use of Eq. (7.10) and the Sorption Data on Woodburn Soil**

| Compound | $K_{om}$ | $S_w$ (mg/L) | $S_{om}$ (mg/g) |
|---|---|---|---|
| Liquids | | | |
| Benzene | 18.2 | 1,780 | 32.4 |
| Chlorobenzene | 47.9 | 491 | 23.5 |
| *o*-Dichlorobenzene | 186 | 148 | 27.5 |
| *m*-Dichlorobenzene | 170 | 134 | 22.8 |
| 1,2,4-Trichlorobenzene | 501 | 48.8 | 24.5 |
| Solids | | | |
| *p*-Dichlorobenzene | 159 | 72.0 | 11.5 |
| 2-PCB | 1,700 | 3.76 | 6.4 |
| 2,2′-PCB | 4,790 | 0.717 | 3.4 |
| 2,4′-PCB | 7,760 | 0.635 | 4.9 |
| 2,4,4′-PCB | 24,000 | 0.115 | 2.8 |
| Lindane | 360 | 7.8 | 2.8 |

*Source*: Data from Chiou et al. (1983).

capacity of DDT with SOM is far less than that of benzene. Further, the observation that polar organic liquids exhibit much higher uptake than nonpolar organic liquids on a high-organic-content peat soil (Chiou and Kile, 1994) is intrinsically consistent with solute partition rather than with hydrophobic adsorption.

The sorption characteristics of individual solutes from a binary or multisolute system provide another means for distinction between partition and adsorption. In their study of the simultaneous sorption of pyrene and phenanthrene from water by river sediments, Karickhoff et al. (1979) observed no discernible sorptive interference between the two compounds (although no explicit single-solute versus binary-solute data were presented). Similarly, Chiou et al. (1983, 1985) found no apparent sorptive competition between $m$-dichlorobenzene and 1,2,4-trichlorobenzene and between parathion and lindane as binary solutes from water on soils over the concentrations investigated. The isotherm data for parathion and lindane are presented in Figure 7.7. Thus, for these relatively nonpolar solutes and the soils studied, no apparent solute competition occurred, while a strong competitive effect would be expected if adsorption were the dominant process. The lack of significant competition between nonpolar solutes simplifies the determination of individual sorption coefficients in multisolute systems.

The apparent noncompetitive effect for relatively nonpolar solutes reflects the dominance of solute partition in SOM, as a result of the strong adsorptive competition of water for soil minerals. The partition (i.e., solubilization) of the

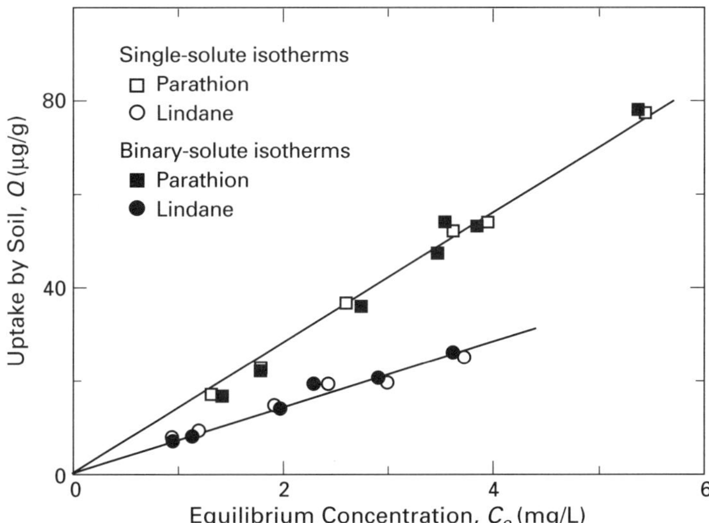



**Figure 7.7**  Sorption of parathion and lindane as single and binary solutes from water on Woodburn soil ($f_{om} = 0.019$) at 20°C. [Data from Chiou et al. (1985). Reproduced with permission.]

solutes in SOM here is promoted by the low solubilities (i.e., poor compatibilities) of the solutes in water, which in effect accounts for the large difference between the sorption coefficients of DDT and benzene, as discussed earlier. The weak adsorption of relatively nonpolar compounds on minerals from water has been documented, for instance, for lindane and dieldrin on sand (Boucher and Lee, 1972); lindane on Ca-bentonite (Chiou et al., 1984); DDT on montmorillonite (Pierce et al., 1974); and phenanthrene on alumina, kaolinite, silica, and silica gel (Huang et al., 1996). These results reflect the strong adsorptive competition of water for minerals, in keeping with the water-versus-benzene vapor adsorption data presented in Chapter 6. The weak adsorption of phenanthrene from water on minerals (Huang et al., 1996) results in an essentially linear isotherm, in which the heat of adsorption is found to be less exothermic than the heat of condensation from water $(-\Delta \overline{H}_w)$ (Huang and Weber, 1997). This suggests that a nonpolar solute in water could only concentrate to some extent near the mineral surface without being condensed to form a separate (condensed) phase.

Even for the sorption of more polar solutes, such as parathion on clay (Saltzman et al., 1972), phenol on goethite (Yost and Anderson, 1984), and 2,4-D on montmorillonite (Haque and Sexton, 1968), the sorption also tends to be rather weak. In most previous solute-sorption studies with clays, small amounts of organic matter in the samples were usually ignored. Since a trace amount of organic matter in unpurified clay minerals would have a significant impact on the solute uptake from water, neglecting this effect could seriously distort the data interpretation. Thus the relatively high uptake of 2,4-D by unpurified illite clay (Haque and Sexton, 1968) might partly be an artifact of the organic impurities in the clay.

Despite that the sorption data of low-polarity compounds in water are illustrative of the dominance of solute partition in soil/sediment organic matter, the situation may be more complicated for relatively polar organic contaminants under some conditions. For example, a close inspection of the isotherms of 2-chlorophenol and 2,4-dichlorophenol on soils, as reported by Boyd et al. (1989), reveals that the isotherms display a discernible concave-downward curvature at low relative concentrations $(C_e/S_w)$ but a good linearity at high $C_e/S_w$. In other words, the sorption coefficient is enhanced at low $C_e/S_w$ but remains nearly constant at high $C_e/S_w$. A similar effect was documented for several polar substituted ureas (herbicides) on soils by Spurlock and Biggar (1994), where the lowering of solute $C_e/S_w$ by about three orders of magnitude to levels of $10^{-4}$ to $10^{-5}$ resulted in an increase of the sorption coefficients by about a factor of 3.5. Boyd (1982) found that the sorption coefficients of some phenolic compounds at low $C_e/S_w$ in their single-solute systems were reduced by some 10 to 30% in binary- and ternary-solute systems.

To account for the enhanced sorption coefficients of polar ureas at low $C_e/S_w$ on soils, Spurlock and Biggar (1994) proposed a general nonlinear solute–SOM partition model that couples a linear partition to SOM matrix with a nonlinear specific interaction with active SOM groups. This specific interaction applies only to polar solutes and is considered to be most significant at low $C_e/S_w$, due to "site" availability. At high $C_e/S_w$, with the specific sites nearly saturated, the polar solutes then exhibit a linear partition into the entire SOM. Although the presumed polar solute–SOM specific interaction as a potential source of sorption nonlinearity seems sensible, it is debatable that it could be portrayed as a partition (solubility) process, since it operates only over a short range of solute concentrations. Much more work is needed to substantiate the assumed specific interaction between solute and SOM and related active SOM sites and active solute groups.

By considering the dual mechanistic functions of the soil, the unsuppressed adsorption of polar solutes on certain minerals could also lead to significant nonlinear sorption by soils at low $C_e/S_w$ (Chiou, 1995). On some special soil minerals (e.g., certain montmorillonite clays as discussed in Chapter 6), water may not sufficiently suppress the adsorption of polar organic solutes as it does the low-polarity solutes. In contrast, the residual mineral adsorption of low-polarity solutes should be relatively weak and linear over a large concentration range, as noted for phenanthrene (Huang et al., 1996). The consequence of this unsuppressed adsorption with soil minerals or of the specific interaction with SOM to the overall solute sorption to a mineral soil having a significant SOM content is illustrated schematically in Figure 7.1$b$. Since the presumed unsuppressed mineral adsorption or the solute–SOM specific interaction for polar solutes is significant largely at low $C_e/S_w$, the effect may overwhelm the linear partition with SOM at low $C_e/S_w$, while the partition contribution usually prevails at moderate to high $C_e/S_w$. As a result, the isotherm would be nonlinear at low $C_e/S_w$ but virtually linear at moderate to high $C_e/S_w$. The solute concentration at which the linear partition and the nonlinear contribution cross is expected to be a function of the SOM content, the magnitude of $K_{om}$ or $K_{oc}$, and the mineral type and content, which determine the relative magnitudes of partition and adsorption.

Because soils and sediments from different geographic locations differ considerably in their compositions, there is no simple a priori way to predict the magnitude of sorption nonlinearity and the specific cause for its occurrence. The possibilities of all mechanisms contributing to the concentration dependence of the polar-solute sorption coefficient have to be settled by experiments together with relevant soil and solute physical and chemical properties. The potential nonlinear sorption of polar solutes to soils and sediments at low concentrations is important because the concentrations of many polar solutes in natural water (usually at low µg/L to low mg/L) fall into a low $C_e/S_w$ range, owing to their high water solubilities. However, one should keep in mind that the magnitude of nonlinear sorption that one observes in single-solute systems is usually attenuated in multisolute systems because of the solute competition, as shown later. Thus the data derived from single-solute studies may not properly reflect the actual solute behavior in natural systems. The nonlinear effect for low-polarity solutes should be far less serious in consideration of their wider isotherm linearity and their generally lower water solubilities (which raise their relative concentrations). Later in this chapter, examples of non-linear sorption for both polar and nonpolar solutes on soils and the potential sources of such nonlinearity will be brought into a better perspective. As a logical sequence of our presentation, we shall first consider the sorptive behavior of relatively nonpolar solutes.

### 7.3.2    Effect of Soil Organic Matter versus Sediment Organic Matter

In the study of solute sorption coefficients ($K_{om}$ or $K_{oc}$), a subject of practical interest is how much the SOM medium property varies between soils and between sediments to affect the sorption coefficient ($K_{om}$ or $K_{oc}$) of an organic contaminant. This information is critical to whether soils or sediments from dispersed geographic locations need to be studied individually (if $K_{om}$ or $K_{oc}$ values vary widely) or can be treated rather indiscriminately (if $K_{om}$ or $K_{oc}$ values are relatively invariant). Before the more extensive investigation by Kile et al. (1995), the literature data on this subject were limited to only a few selected solutes and to a relatively small set of soil samples examined by different analytical techniques (Bailey and White, 1964; Goring, 1967). Based on $K_{oc}$ data from different reports for a few selected solutes, Kenaga and Goring (1980) observed that the $K_{oc}$ variation between soils is generally less than a factor of 3 to 4. Mingelgrin and Gerstl (1983) indicated that the $K_{oc}$ could vary by as large as a factor of 10 or greater, based on selected $K_{oc}$ values of some pesticides. Utilizing a correlation of $K_{oc}$ with (O + N)/C weight ratio of natural organic matter, Rutherford et al. (1992) estimated the $K_{oc}$ variation of non-polar solutes between soils to be less than a factor of 3, based on the range of (O + N)/C values for common soil organic matter.

Since the different analytical procedures employed led inevitably to $K_{om}$ or $K_{oc}$ variation, especially for soil or sediment samples with very low organic contents, a more accurate account of the $K_{om}$ or $K_{oc}$ variation between soils or sediments could only be achieved through the use of a large set of soils and sediments with significant SOM contents to be analyzed by consistent and rigorous analytical methods. With this consideration, Kile et al. (1995) measured the $K_{oc}$ values of two relatively nonpolar solutes, carbon tetrachloride (CT) and 1,2-dichlorobenzene (DCB), on 32 "normal" soils and 36 "normal" bed sediments collected from diverse geographic locations in both the United States and China. For all samples, solute concentrations in both water and soil/sediment were solvent-extracted and analyzed by gas chromatography. Partition data of low-polarity solutes with the SOM of soils and sediments should best detect differences in SOM polarity and composition, if any, because the solubility of nonpolar solutes is sensitive to the organic medium polarity (see Chapter 5) and because the adsorption of such solutes on soil/sediment minerals should be most effectively suppressed by water.

The sources of soils and bed sediments, the sample BET-N$_2$ surface areas (SA), the organic carbon contents (% OC), and the measured $K_{oc}$ values for CT and DCB from Kile et al. (1995) are shown in Table 7.3. Sorption isotherms

**TABLE 7.3. Sources of Soils, Bed Sediments, and Suspended Solids and Their Surface Areas (SA), Organic Carbon Contents (OC), and Measured Partition Coefficients ($K_{oc}$) of CT and DCB at Room Temperature**

| No. | Source | SA ($m^2/g$) | % OC | $K_{oc}$, CT | $K_{oc}$, DCB |
|---|---|---|---|---|---|
| | *Soil Samples* | | | | |
| 1. | Burleigh Co., North Dakota (U.S. EPA reference soil 2) | 7.85 | 2.40 | 52 | 263 |
| 2. | Oliver Co., North Dakota (U.S. EPA reference soil 3) | | 1.43 | 53 | 277 |
| 3. | Pierre, South Dakota (U.S. EPA reference soil 7) | 22.4 | 2.21 | 63 | 319 |
| 4. | West-central Iowa (U.S. EPA reference soil 10) | 8.84 | 2.04 | 58 | 248 |
| 5. | Manchester, Ohio (U.S. EPA reference soil 12) | 9.38 | 2.25 | 57 | 230 |
| 6. | Columbus, Kentucky (U.S. EPA reference soil 19) | 3.75 | 1.73 | 67 | 308 |
| 7. | Anoka, Minnesota | 1.07 | 1.08 | 61 | 261 |
| 8. | Piketon, Ohio | 7.77 | 1.49 | 53 | 263 |
| 9. | Marlette soil, East Lansing, Michigan | 3.99 | 1.80 | 45 | 223 |
| 10. | Spinks soil, East Lansing, Michigan | 1.51 | 1.03 | 65 | 318 |
| 11. | Elliot, Illinois (International Humic Substances Society reference soil) | 9.79 | 2.90 | 49 | 252 |
| 12. | Woodburn soil, Corvallis, Oregon | 11.2 | 1.26 | 65 | 296 |
| 13. | Renslow soil, Kittitas Co., Washington | 11.6 | 2.40 | 59 | 340 |
| 14. | Sanhedrin soil, Mendocino Co., California | 7.88 | 6.09 | 68 | 344 |
| 15. | Cathedral soil, Fremont Co., Colorado | 5.58 | 3.12 | 74 | 407 |
| 16. | Wellsboro soil, Otsego Co., New York | 5.73 | 3.47 | 68 | 383 |
| 17. | Fangshan District, Beijing, China | 4.96 | 5.61 | 54 | 262 |
| 18. | Anda, Heilongjiang, China | | 2.83 | 67 | 288 |
| 19. | Jinxian Co., Jiangxi, China | | 0.34 | 53 | 327 |
| 20. | Nanjing, Jiangsu, China | | 1.08 | 61 | 236 |
| 21. | Changshu, Jiangsu, China | | 1.77 | 55 | 253 |
| 22. | Xuyi Co., Jiangsu, China | 54.0 | 0.67 | 53 | 257 |
| 23. | Jinhu Co., Jiangsu, China | | 4.02 | 64 | 306 |
| 24. | Hongze Co., Jiangsu, China | 22.8 | 0.81 | 70 | 313 |
| 25. | Dushan Co., Guizhou, China | 8.20 | 2.54 | 61 | 315 |

(*Continued*)

**TABLE 7.3.** *Continued*

| No. | Source | SA ($m^2/g$) | % OC | $K_{oc}$, CT | $K_{oc}$, DCB |
|---|---|---|---|---|---|
| 26. | Gangcha Co., Qinghai, China | 4.21 | 1.12 | 62 | 295 |
| 27. | Xinghai Co., Qinghai, China | 2.86 | 0.16 | 59 | 264 |
| 28. | Luochuan Co., Shanxi, China | | 0.46 | 66 | 315 |
| 29. | Yishan Co., Guangxi, China | 40.2 | 0.66 | 66 | 275 |
| 30. | Yangchun Co., Guangdong, China | | 0.83 | 64 | 293 |
| 31. | Xuwen Co., Guangdong, China | | 0.64 | 55 | 257 |
| 32. | Qiongzhong Co., Hainan, China | 4.85 | 0.34 | 62 | 304 |

*Bed-Sediment Samples*

| No. | Source | SA ($m^2/g$) | % OC | $K_{oc}$, CT | $K_{oc}$, DCB |
|---|---|---|---|---|---|
| 1. | Isaacs Creek at Ohio River, near Ripley, Ohio (U.S. EPA reference sediment 11) | 20.2 | 1.50 | 66 | 301 |
| 2. | Mississippi River, near Columbus, Kentucky (U.S. EPA reference sediment 18) | 22.1 | 0.79 | 103 | 476 |
| 3. | Illinois River, near Lacon, Illinois (U.S. EPA reference sediment 22) | 3.39 | 2.20 | 116 | 572 |
| 4. | Kaskaskia River, Illinois (U.S. EPA reference sediment 25) | 7.60 | 0.99 | 90 | 444 |
| 5. | Mississippi River (Pool 2), St Paul, Minnesota | 5.90 | 1.50 | 94 | 441 |
| 6. | Mississippi River (Pool 11), Guttenburg, Iowa | 4.86 | 1.13 | 87 | 370 |
| 7. | Mississippi River (Pool 26), Alton, Illinois | 15.5 | 1.40 | 91 | 387 |
| 8. | Mississippi River, Helena, Arkansas | 12.8 | 1.60 | 109 | 534 |
| 9. | Yazoo River, Vicksburg, Mississippi | 19.7 | 0.58 | 119 | 499 |
| 10. | Mississippi River, St. Francisville, Louisiana | 8.09 | 0.40 | 119 | 549 |
| 11. | Lake Charles, adjacent to the Calcasieu River, Lake Charles, Louisiana | 13.3 | 1.97 | 112 | 536 |
| 12. | Marine sediment from Suisin Bay, northern San Francisco Bay, site 408.1 | 15.7 | 1.17 | 105 | 516 |
| 13. | Marine sediment from Suisin Bay, northern San Francisco Bay, site 416 | 21.6 | 1.48 | 107 | 532 |

**TABLE 7.3.** *Continued*

| No. | Source | SA $(m^2/g)$ | % OC | $K_{oc}$, CT | $K_{oc}$, DCB |
|---|---|---|---|---|---|
| 14. | Marine sediment from Suisin Bay, northern San Francisco Bay, site 433 | 21.3 | 1.78 | 106 | 532 |
| 15. | Tangwang River, Yichun, Heilongjiang, China | 12.8 | 4.73 | 106 | 551 |
| 16. | Songhuajiang River, Majiadukou, Jiling, China | | 1.12 | 117 | 553 |
| 17. | Tumen River, Helong Co., Jiling, China | 4.93 | 1.99 | 93 | 420 |
| 18. | Xuanwu Lake, Nanjing, Jiangsu, China | | 4.12 | 103 | 557 |
| 19. | Guchen Lake, Gaochun Co., Jiangsu, China | | 1.24 | 94 | 554 |
| 20. | Lake Hongze, Sihong Co., Jiangsu, China | 29.9 | 1.04 | 101 | 455 |
| 21. | Zhujiang River, Guangzhou, Guangdong, China | | 3.37 | 95 | 545 |
| 22. | Yellow River, Zhengzhou, Henan, China | | 0.11 | 112 | 589 |
| 23. | Yinghe River, Lushan Co., Henan, China | 1.85 | 0.11 | 101 | 477 |
| 24. | Ziya River, Ci Co., Hebei, China | 5.83 | 2.19 | 106 | 599 |
| 25. | Ganjiang River, Ruijin Co., Jiangxi, China | 5.32 | 0.70 | 112 | 542 |
| 26. | Zishui River, Chengbu Co., Hunan, China | 8.97 | 2.82 | 92 | 437 |
| 27. | Liuyanghe River, Liuyang Co., Hunan, China | | 0.29 | 116 | 555 |
| 28. | Youshui River, Xuanen Co., Hubei, China | 11.9 | 1.22 | 92 | 451 |
| 29. | Niqu River, Louhuo Co., Sichuan, China | 4.84 | 0.39 | 96 | 474 |
| 30. | Huaihe River, Bengbu, Anhui, China | 17.6 | 0.50 | 112 | 535 |
| 31. | Huaihe River, Huainan, Anhui, China | 8.21 | 0.45 | 91 | 466 |
| 32. | Jinghe River, Jingyuan Co., Ningxia, China | 12.1 | 0.73 | 108 | 584 |
| 33. | Sangonghe River, Fukang Co., Xinjiang, China | 4.00 | 0.38 | 103 | 499 |
| 34. | Yaluzangbu River, Lazi Co., Tibet, China | 4.94 | 0.45 | 107 | 526 |
| 35. | Lake Pumo, Langkazi Co., Tibet, China | 3.87 | 1.94 | 101 | 539 |

*(Continued)*

**TABLE 7.3.** *Continued*

| No. | Source | SA ($m^2$/g) | % OC | $K_{oc}$, CT | $K_{oc}$, DCB |
|---|---|---|---|---|---|
| 36. | Niyanghe River, Gongbujiangda Co., Tibet, China | 3.12 | 0.54 | 93 | 487 |

*Suspended Solids*

| No. | Source | SA ($m^2$/g) | % OC | $K_{oc}$, CT | $K_{oc}$, DCB |
|---|---|---|---|---|---|
| 1. | Mississippi River, Thebes, Illinois | 1.82 | | 60 | 296 |
| 2. | Mississippi River, St. Louis, Missouri | 1.78 | | 58 | 283 |
| 3. | Illinois River, Hardin, Illinois | 2.60 | | 89 | 423 |
| 4. | Missouri River, Herman, Missouri | 2.87 | | 49 | 231 |
| 5. | Yellow River, Zhengzhou, Henan, China | 0.38 | | 63 | 300 |

*Source*: Data from Kile et al. (1995).

of CT and DCB on representative soils and bed sediments are presented in Figures 7.8 and 7.9. The virtual linearity of the isotherms is typical of the solute partition in SOM, as expected. With the reported uncertainty of $K_{oc}$ values being about ±8% and that of the organic-carbon mass fraction in soil/sediment ($f_{oc}$) being ±5%, the observed differences between soil $K_{oc}$ values (or between sediment $K_{oc}$ values) for both CT and DCB are relatively small. The mean $K_{oc}$ value for CT on 32 normal soils is 60 (SD = ±7) and the mean $K_{oc}$ value for DCB is 290 (SD = ±42). The $K_{oc}$ values for both CT and DCB on 36 normal bed sediments are generally higher and show about the same variation as the $K_{oc}$ values on soils. The mean $K_{oc}$ value on bed sediments for CT is 102 (SD = ±11) and for DCB is 502 (SD = ±66); they are greater by a factor of 1.7 than the mean $K_{oc}$ values for CT and DCB on soils. This difference is more than the standard deviation (SD) of the means and is illustrated graphically in Figure 7.10 for CT and in Figure 7.11 for DCB. The finding that the $K_{oc}$ values for DCB are a factor of 4 to 6 greater than respective $K_{oc}$ values for CT on all soils and sediments is essentially what Eq. (3.15) would predict, based on the different water solubilities of CT (800 mg/L) and DCB (154 mg/L) and the comparable solubilities of low-polarity liquids in SOM (Rutherford et al., 1992), as shown in Table 7.2.

The high degree of invariance of the CT and DCB $K_{oc}$ values between most soils or between most bed sediments is phenomenal. The normalized sorption coefficients ($K_{oc}$ values) for both solutes show little dependence on soil or sediment OC contents (e.g., 0.16 to 6.09% for soils) and on (dry) soil or sediment surface areas (e.g., 1.07 to 54.0 $m^2$/g for soils), as shown in Table 7.3. This finding reveals the similarity in SOM polarity/composition between soils and between

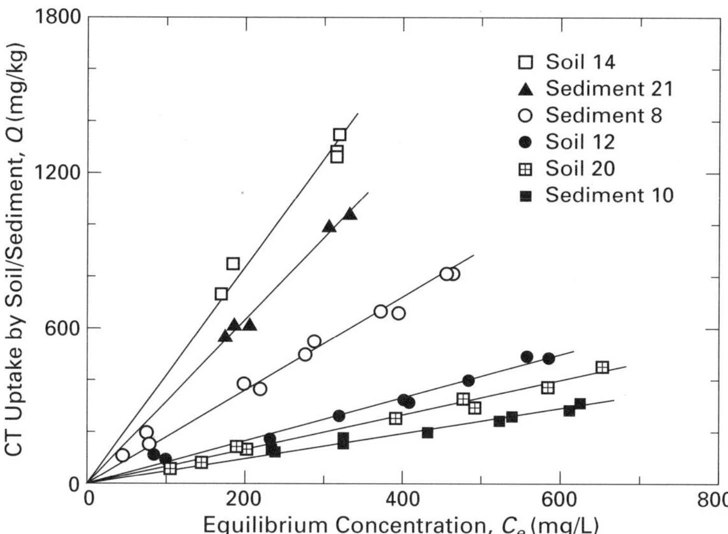

**Figure 7.8** Sorption of carbon tetrachloride (CT) on representative soils and bed sediments listed in Table 7.3. [Data from Kile et al. (1995). Reproduced with permission.]

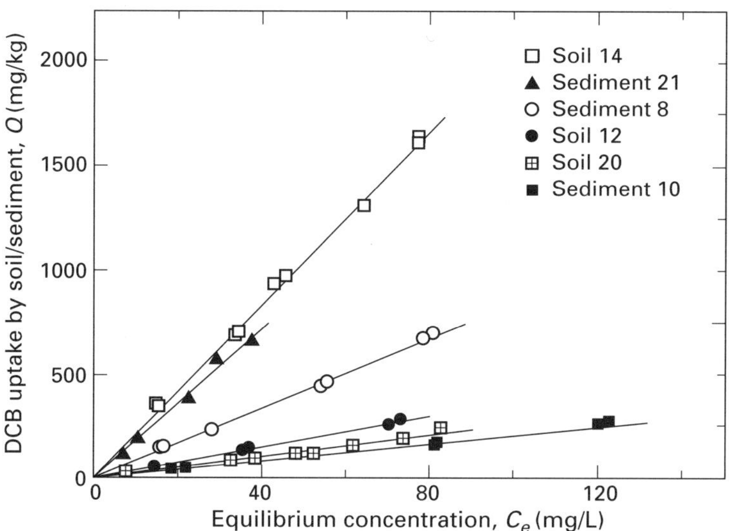

**Figure 7.9** Sorption of 1,2-dichlorobenzene (DCB) on representative soils and bed sediments listed in Table 7.3. [Data from Kile et al. (1995). Reproduced with permission.]

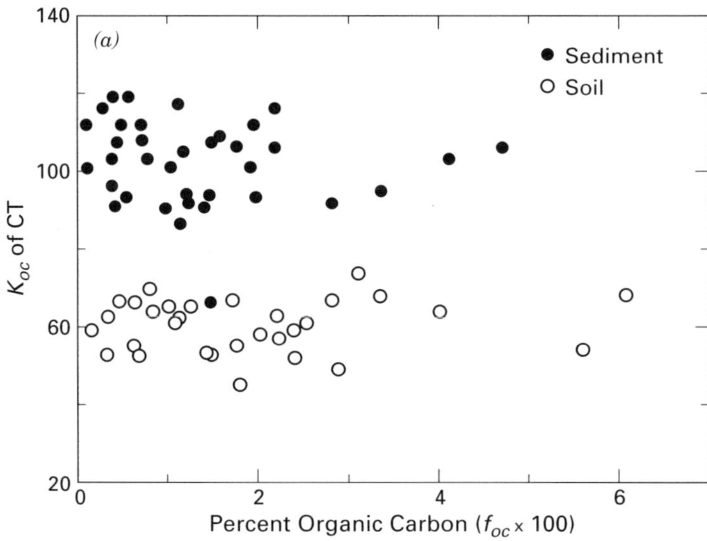


**Figure 7.10**    Plot of CT $K_{oc}$ versus soil/sediment $f_{oc}$ for 32 soils and 36 bed sediments listed in Table 7.3. [Data from Kile et al. (1995). Reproduced with permission.]

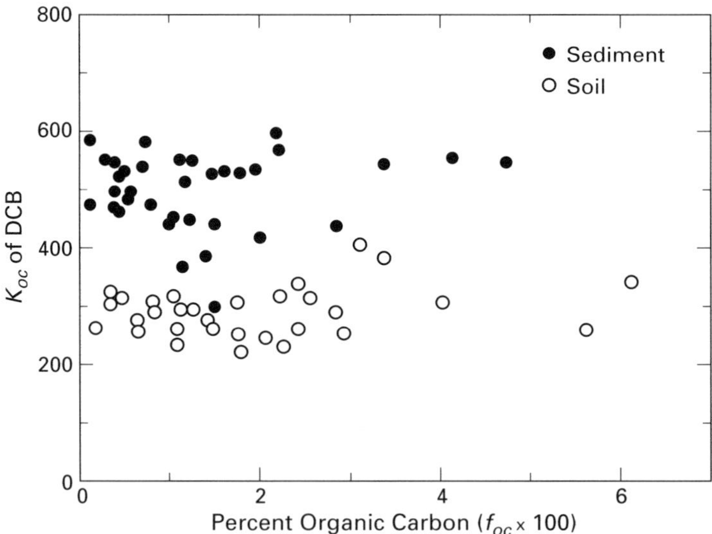


**Figure 7.11**    Plot of DCB $K_{oc}$ versus soil/sediment $f_{oc}$ for 32 soils and 36 bed sediments listed in Table 7.3. [Data from Kile et al. (1995). Reproduced with permission.]

bed sediments and the predominance of solute partitioning in SOM over mineral adsorption; if mineral adsorption were important, the sorption coefficient would be affected by the sorbent's SA, which is not observed. The range of variation for the soils is much smaller than that reported in other studies, with a smaller set of combined data analyzed by different analytical methods (Kenaga and Goring, 1980; Mingelgrin and Gerstl, 1983). The extreme $K_{oc}$ values for soils (or sediments) differ by less than a factor of 2. The relative invariance in $K_{oc}$ suggests that the properties of the humified SOM that mediate nonpolar solute solubility are quite similar for a wide variety of uncontaminated soils, and also likely for relatively pristine bed sediments. There does not appear to be a large variability in SOM polarity and composition between well-weathered soils from diverse geographic locations. In view of the relative invariance in $K_{oc}$ between soils or between sediments as illustrated, the use of average soil (or sediment) $K_{oc}$ values for assessing the sorption of low-polarity contaminants to different soils (or sediments) would seem sufficient in most environmental applications. However, an account of the difference in sorption to soil and sediment of low-polarity contaminants would seem warranted.

The fact that most soil $K_{oc}$ values are distinct from bed-sediment $K_{oc}$ values implies that the process that turns eroded soils into bed sediments brings about a noticeable change in the property of the organic constituent. A possible cause for this change is that the sedimentation process fractionates soil organic constituents such that the more polar and water-soluble organic components in SOM (e.g., fulvic and humic acid fractions), or that the soil particles with more polar organic components, are separated out to form dissolved organic matter and colloids in water, with the less-polar soil organic constituents preserved in the bed sediment. The time scale to bring about a complete soil-to-sediment conversion should be a function of hydrodynamics. Another possible cause for this change would be the biological influence. However, this effect would seem small since the difference between bed-sediment and soil $K_{oc}$ values, although statistically significant, is not large.

Part of the variation in $K_{oc}$ within bed sediments reflects the extent of conversion of the eroded soils to bed sediments. Consider, for example, the relatively low $K_{oc}$ values of CT and DCB with sediments 1, 6, and 7 in Table 7.3. Sediment 1 is a U.S. EPA sample taken from the mouth of Isaacs Creek at the junction with the Ohio River near Ripley, Ohio. The fact that the $K_{oc}$ values on sediment 1 are significantly lower than the rest but are very similar to soil $K_{oc}$ values suggests that this sample could be a recently eroded soil which retains most of its soil-organic-matter composition. The somewhat lower $K_{oc}$ values with sediments 6 and 7 relative to the average sediment $K_{oc}$ may again be a result of incomplete conversion of eroded soils to bed sediments. The $K_{oc}$ data suggest that bed sediments from most large rivers and lakes are relatively comparable in their SOM polarities and compositions, probably because they are more aged and contain less recently eroded soils.

The difference between soil and bed-sediment $K_{oc}$ values as detected by relatively nonpolar solutes lends a basis for identifying the source of suspended solids in rivers. In the study of Kile et al. (1995), the suspended solids from the Mississippi River, Missouri River, and Yellow River were collected during high river flows, and the sample from the Illinois River was collected during a low-to-normal river flow. Here the $K_{oc}$ values of CT and DCB are typical of those of soils for the former but are more representative of bed sediments for the latter (Table 7.3). One may infer from these data that the suspended solids during high water flows in these three rivers consist mainly of newly eroded soils and the suspended solids from the Illinois River under low-to-normal water flow consist largely of resuspended bed sediment. The Yellow River suspended solid, which shows its origin as an eroded soil, is in keeping with the river's high carrying load of eroded soils during the high-flow season. In contrast, bed sediment collected from the Yellow River (sediment 22) gives $K_{oc}$ values typical of those for other bed sediments. Thus the sorption data serves as a simple indicator of the source and time history of the suspended solids.

The relatively low $K_{oc}$ values of CT and DCB, about one order of magnitude lower than their respective $\log K_{ow}$ values, suggest that the SOM of soils (or sediments) must be fairly polar in nature to limit the partition (solubility) of these nonpolar organic solutes. To investigate the effect of SOM composition and polarity on solute partition, Rutherford et al. (1992) measured the partition coefficients of two relatively nonpolar solutes, benzene and CT, in relation to the elemental compositions of relatively ash-free natural organic matters: cellulose, muck, peat, and treated peat (peat washed by $0.1\,N$ NaOH to lower the oxygen content). The weight ratio of [oxygen + nitrogen] to carbon of the natural organic matter [i.e., the (O + N)/C value] was used as an approximate polarity index of the sample, which gives the relative polarity order: cellulose > muck > peat > treated peat. An inverse relation is evident between the $K_{oc}$ (or $K_{om}$) of both CT and DCB and the (O + N)/C of the organic matter sample, as shown in Figure 7.12.

The results for CT in Figure 7.12 illustrate several points of interest. First, the partition of a nonpolar solute to a natural organic matter is sensitive to the organic matter polarity (or composition). The small $K_{oc}$ value of CT with cellulose compared to that with humified materials (such as muck or peat) results from a poor match in polarity between a nonpolar solute and a highly polar organic phase, which makes cellulosic materials a poor partition phase for nonpolar contaminants. Second, the (O + N)/C values for normal soils should fall into a relatively narrow range, in view of relatively constant $K_{oc}$ values of CT on soils from diverse sources (mean $K_{oc}$ = 60; SD = ±7), which would place them somewhere between the (O + N)/C values of Houghton muck (0.777) and Florida peat (0.657). Third, the low (O + N)/C value of the treated peat (0.488) and the observed $K_{oc}$ value of 115 for CT with this sample, which exceeds the $K_{oc}$ values with normal soils but resembles the $K_{oc}$ values with bed sediments, substantiate the contention that the sediment OM has a

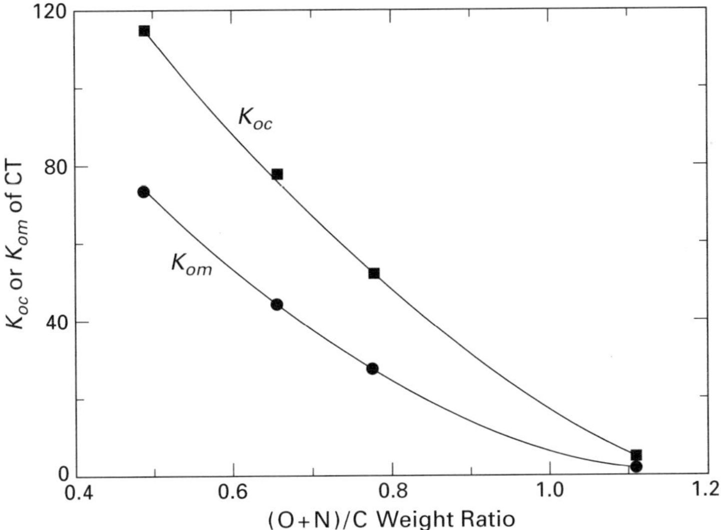

**Figure 7.12**  Plot of the CT $K_{om}$ or $K_{oc}$ value versus the (O + N)/C ratio of natural organic matters. [Data from Rutherford et al. (1992).]

generally lower polar group content than soil OM, as is substantiated by the $^{13}$C-NMR data (Kile et al., 1999).

### 7.3.3  Effect of Contaminant Water Solubility

We now look into the effect of solute water solubility ($S_w$) on the solute sorption (partition) coefficient ($K_{om}$) with SOM in relation to the model equation, (3.15), and the relation of $K_{om}$ to $K_{ow}$ for the solutes. Rewriting Eq. (3.15) with $K_{pw} = K_{om}$, one gets

$$\log K_{om} = -\log S_w \overline{V} - \log \rho - (1+\chi)/2.303 - \log\left(\gamma_w/\gamma_w^*\right) \tag{7.11}$$

where the meanings of all terms in Eq. (7.11) are defined earlier [see Eqs. (3.12–3.15)]. Presented in Table 7.4 is a list of the $\log K_{om}$ values of 12 substituted aromatic compounds with Woodburn soil ($f_{om} = 0.019$), the respective $\log S_w$ values, molar volumes ($\overline{V}$), and $\log K_{ow}$ values. A plot of $\log K_{om}$ versus $\log S_w \overline{V}$ is shown in Figure 7.13 along with the *ideal line*, which is obtained by assuming that $\rho = 1.2$ and $\chi_S = 0.25$ for SOM and $\log\left(\gamma_w/\gamma_w^*\right) = 0$. With the $\log S_w \overline{V}$ of a compound specified, the difference between $\log K_{om}^{\circ}$ from the ideal line and experimental $\log K_{om}$ equals the sum of $\chi_H/2.203$ and $\log\left(\gamma_w/\gamma_w^*\right)$. The magnitude of $\log\left(\gamma_w/\gamma_w^*\right)$ is generally small for compounds with $\log K_{om} \leq 3$, assuming that the amount of organic matter released from soil into water after soil–water equilibration is <100 mg/L (the actual value varies with the soil–water system) (Chiou et al., 1984; Gschwend and Wu,

**TABLE 7.4. Water Solubilities ($S_w$), Molar Volumes ($\overline{V}$), Octanol–Water Partition Coefficients ($K_{ow}$), and $K_{om}$ Values of Some Substituted Benzenes and Polychlorinated Biphenyls (PCBs) on Woodburn Soil**

| Compound | $\log S_w^a$ (mol/L) | $\overline{V}$ (L/mol) | $\log S_w\overline{V}$ | $\log K_{om}$ | $\log K_{ow}$ |
|---|---|---|---|---|---|
| Benzene | −1.64 | 0.0894 | −2.69 | 1.26 | 2.13 |
| Anisole | −1.85 | 0.109 | −2.82 | 1.30 | 2.11 |
| Chlorobenzene | −2.36 | 0.102 | −3.35 | 1.68 | 2.84 |
| Ethylbenzene | −2.84 | 0.123 | −3.75 | 1.98 | 3.15 |
| 1,2-Dichlorobenzene | −2.98 | 0.113 | −3.98 | 2.27 | 3.38 |
| 1,3-Dichlorobenzene | −3.04 | 0.114 | −3.98 | 2.23 | 3.38 |
| 1,4-Dichlorobenzene | (−3.03) | 0.118 | −3.96 | 2.20 | 3.39 |
| 1,2,4-Trichlorobenzene | −3.57 | 0.125 | −4.47 | 2.70 | 4.02 |
| 2-PCB | (−4.57) | 0.174 | −5.33 | 3.23 | 4.51 |
| 2,2′-PCB | (−5.08) | 0.189 | −5.57 | 3.68 | 4.80 |
| 2,4′-PCB | (−5.28) | 0.189 | −5.97 | 3.89 | 5.10 |
| 2,4,4′-PCB | (−5.98) | 0.204 | −6.67 | 4.38 | 5.62 |

*Source*: Data from Chiou et al. (1983).

[a] Values at 20 to 25°C. The numbers in parentheses are the supercooled liquid solubilities estimated according to Eqs. (3.9) and (3.25) with $\Delta \overline{H}_{fus}/T_m = 56.5\,\text{J/mol·K}$.

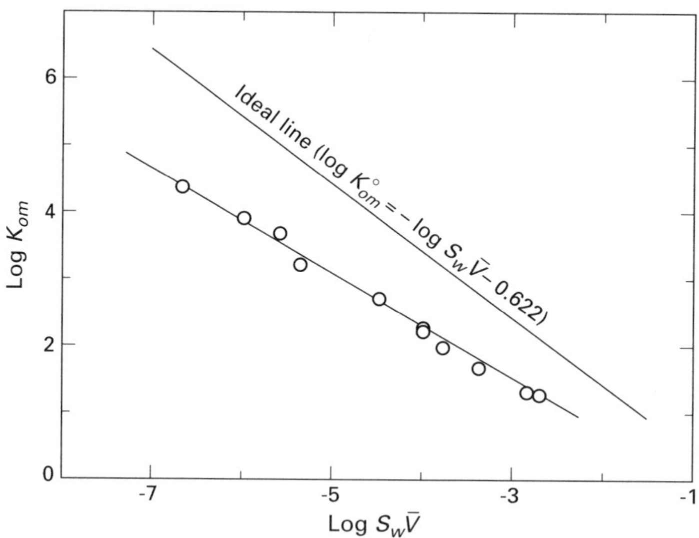


**Figure 7.13**  Plot of $\log K_{om}$ versus $\log S_w\overline{V}$ for substituted aromatic compounds in Table 7.4. [Data from Chiou et al. (1983). Reproduced with permission.]

1985). As an approximation, we assume that $\log (K^{\circ}_{om}/K_{om}) \simeq \chi_H/2.203$ in the present analysis.

Comparison of the magnitude of $-\log S_w \overline{V}$ with that of $\log (K^{\circ}_{om}/K_{om})$ indicates that $-\log S_w \overline{V}$ is the major determinant of $\log K_{om}$ for the organic compounds, which results in a highly linear correlation of $\log K_{om}$ with $\log S_w \overline{V}$ as shown in Figure 7.13. For the 12 compounds (mainly chlorinated benzenes and PCBs) on Woodburn soil, the regression equations gives

$$\log K_{om} = -0.813 \log S_w \overline{V} - 0.993 \tag{7.12}$$

with $r^2 = 0.995$, where the $S_w$ values (mol/L) for solid solutes are their estimated supercooled liquid solubilities [see Eq. (3.9)]. Here one sees a good correspondence between the correlation of $\log K_{om}$ versus $\log S_w \overline{V}$ in Eq. (7.12) and the correlation of $\log K_{ow}$ versus $\log S_w$ in Eq. (5.3) for substituted aromatic solutes. The increased deviation between $\log K^{\circ}_{om}$ and $\log K_{om}$ relative to that between $\log K^{\circ}_{ow}$ and $\log K_{ow}$ is ascribed to the increased incompatibility of these low-polarity solutes with relatively polar SOM over that with the less-polar octanol. For more-polar solutes, the differences between $\log K^{\circ}_{om}$ and $\log K_{om}$ should be less because of their enhanced partition in SOM, and the relation of $\log K_{om}$ to $\log S_w \overline{V}$ (or a related property) may change accordingly.

Since the variation of $\overline{V}$ among solutes is quite small compared to that of $S_w$, a linear relation should also exist between $\log K_{om}$ and $\log S_w$, as has been widely recognized (Chiou et al., 1979, 1983; Karickhoff et al., 1979; Kenaga and Goring, 1980; Means et al., 1980; Briggs, 1981; Hassett et al., 1981; Karickhoff, 1984). Using the data in Table 7.4, one finds a correlation equation of

$$\log K_{om} = -0.729 \log S_w + 0.001 \tag{7.13}$$

with $n = 12$ and $r^2 = 0.996$. Here the predominant effect of $S_w$ on $K_{om}$ is much anticipated for the partition equilibria of organic solutes in a partially miscible mixture of an organic phase and water, despite the fact that SOM is not nearly as good a solvent for nonpolar solutes as normal organic solvents such as octanol. For this reason, the $\log K_{om}$ is largely linearly related to $\log K_{ow}$, as illustrated for a wide variety of organic contaminants on various soils and sediments, as stated earlier. From the data in Table 7.4 one finds that

$$\log K_{om} = 0.904 \log K_{ow} - 0.779 \tag{7.14}$$

with $n = 12$ and $r^2 = 0.989$. For compounds with $\log K_{ow}$ in the range 2 to 5, the $K_{om}$ values are roughly one order of magnitude smaller than the corresponding $K_{ow}$ values. If the sorption coefficient is expressed in terms of soil organic carbon ($K_{oc}$), equivalent correlation equations may be established through $K_{oc} \simeq 1.72 K_{om}$. The conversion factor of 1.72 is derived with the assumption that SOM is about 58% in carbon (Hamaker and Thompson, 1972). (Based on the mean $K_{oc}$ of CT on soils and the relation of $K_{oc}$ to the organic matter composition in Figure 7.12, an assumed carbon content of about 54% in SOM would seem more appropriate. This would change the conversion between $K_{om}$ and $K_{oc}$ to $K_{oc} \simeq 1.85 K_{om}$.)

Since the magnitude of $K_{oc}$ (or $K_{om}$) reflects the difference in solute solubilities in SOM and water, which varies with the solute, the slopes and intercepts in Eqs. (7.13) and (7.14) would vary with the compounds studied. For a single correlation to suffice for different classes of solutes, the solutes with the same $S_w$ or $K_{ow}$ value must exhibit nearly the same (supercooled-liquid) solubility in SOM [i.e., about the same $S_{om}$ value in Eq. (7.10)]. Although the data for many low-polarity solutes, or for a class of solutes with similar molecular groups, may largely satisfy this requirement, it is not observed by all solutes. In such correlations, the slope reflects the (differential) change in $\log S_{om}$ with the change in the solute's $\log S_w$ or $\log K_{ow}$ value, while the intercept is the theoretical $\log K_{oc}$ (or $\log K_{om}$) value of a hypothetical reference solute having $\log S_w = 0$ or $\log K_{ow} = 0$. As the $S_{om}$ value changes with solute polarity in a different manner for different series of solutes, the slope and intercept of the correlation change accordingly. For example, Briggs (1981) measured the $\log K_{om}$ values of a wide variety of polar compounds and pesticides (anilines, anilides, phenols, nitrobenzenes, substituted ureas, carbamates, organophosphates, and others) on soil and found the correlation between $\log K_{om}$ and $\log K_{ow}$ as

$$\log K_{om} = 0.52 \log K_{ow} + 0.64 \tag{7.15}$$

with $n = 105$ and $r^2 = 0.90$. Most compounds in Briggs's study have $\log K_{ow} < 3$, however. As seen, Eq. (7.15) gives a much smaller slope but a much larger intercept than does Eq. (7.14). At $\log K_{ow} < 3$, as for most polar solutes, the $\log K_{om}$ values estimated by Eq. (7.15) are comparable in magnitude with the respective $\log K_{ow}$ values and thus considerably higher than the $\log K_{om}$ values estimated by Eq. (7.14). An example for the difference in $\log K_{om}$–$\log K_{ow}$ correlation between low- and high-polarity compounds is given in Figure 7.14, which compares the $K_{om}$ data for low-polarity compounds in Table 7.4 with the data of Briggs (1981) for high-polarity compounds shown in Table 7.5.

The smaller slope in Eq. (7.15) than in Eq. (7.14) indicates that whereas highly polar solutes (those with low $\log K_{ow}$ values) exhibit relatively large $S_{om}$ values, the $S_{om}$ decreases more rapidly with increasing $K_{ow}$ as the solute polarity (or $S_w$) decreases within the series selected when compared with the variation in $S_{om}$ for a series of relatively nonpolar solutes. It has been found that polar organic liquids exhibit a considerably higher partition (solubility) than nonpolar liquids in dry SOM, due presumably to more powerful molecular interactions by polar and H-bonding forces over that by London (dispersion) forces (Chiou and Kile, 1994). Such polar or specific forces for polar solutes with relatively polar SOM may result in a nonlinear sorption, and therefore the $K_{om}$ values commonly measured as single solutes could also be concentration dependent. Therefore, unlike the $\log K_{ow}$ versus $\log S_w$

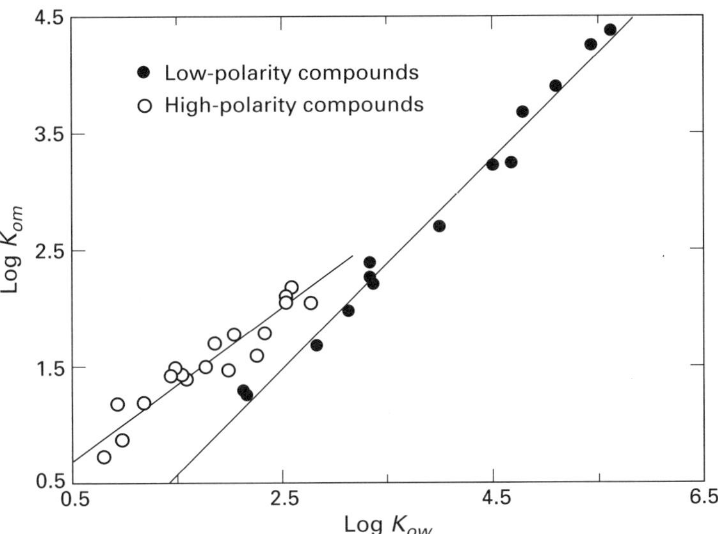


**Figure 7.14**   Plots of $\log K_{om}$ versus $\log K_{ow}$ for nonpolar solutes in Table 7.4 and polar solutes in Table 7.5.

**TABLE 7.5.  Log $K_{om}$ and log $K_{ow}$ Values of Some High-Polarity Compounds**

| Compound | $\log K_{om}$ | $\log K_{ow}$ |
|---|---|---|
| Aniline | 1.17 | 0.90 |
| *m*-Toluidine | 1.41 | 1.42 |
| 3,4-Dichloroaniline | 2.05 | 2.78 |
| Anilide | 1.19 | 1.16 |
| 3-Bromoanilide | 1.77 | 2.04 |
| 3,4-Dibromoanilide | 2.10 | 2.54 |
| Phenol | 1.48 | 1.46 |
| 4-Bromophenol | 2.17 | 2.59 |
| Nitrobenzene | 1.70 | 1.85 |
| 4-Bromonitrobenzene | 2.18 | 2.60 |
| Fenuron | 0.88 | 0.96 |
| Monuron | 1.46 | 1.98 |
| Methyl-*N*-phenylcarbamate | 1.49 | 1.76 |
| Ethyl-*N*-phenylcarbamate | 1.58 | 2.26 |
| Dimethoate | 0.72 | 0.79 |
| Simazine | 1.44 | 1.51 |
| Aldicarb | 1.39 | 1.57 |
| Carbaryl | 1.78 | 2.32 |
| Captan | 2.06 | 2.54 |

*Source*: Data from Briggs (1981).

correlation, which is relatively insensitive to solute polarity for the reasons stated in Chapter 5, the $\log K_{om}$ versus $\log K_{ow}$ (or $\log K_{om}$ versus $\log S_w$) correlation is much more sensitive to solute polarity. In this sense, weakly polar octanol (or $K_{ow}$) is not a good replicate model for relatively polar SOM (or $K_{om}$).

### 7.3.4  Behavior of PAHs versus Other Nonpolar Contaminants

Although the $K_{oc}$ values of many nonpolar contaminants can be estimated with sufficient accuracy from their physicochemical properties, such as $K_{ow}$ or $S_w$, the selection of a proper correlation for estimation is, however, complicated by the inconsistency of published correlations for selected nonpolar compounds on soils and sediments. A particular case in point is the inconsistency of the correlations for polycyclic aromatic hydrocarbons (PAHs) relative to other nonpolar organic compounds. For instance, with $K_{oc} = 1.85\,K_{om}$, the log $K_{oc}$ versus $\log K_{ow}$ correlation established by Chiou et al. (1983) [i.e., Eq. (7.14)] for substituted aromatic compounds (primarily, chlorinated benzenes and PCBs) on a soil over the range $\log K_{ow} = 2.11$ to $5.62$ gives

$$\log K_{oc} = 0.904 \log K_{ow} - 0.512 \tag{7.16}$$

which is drastically different from a similar correlation established by Karickhoff et al. (1979) for mainly PAHs and their derivatives on river sediments, where the compounds cover a comparable $\log K_{ow}$ range (2.11 to 6.34). The sediment $K_{oc}$ values for the latter compounds and their $K_{ow}$ values are given in Table 7.6, which give a correlation equation:

$$\log K_{oc} = 1.00 \log K_{ow} - 0.21 \tag{7.17}$$

**TABLE 7.6.  Sediment $\log K_{oc}$ Values and Corresponding $\log K_{ow}$ Values of Mainly Polycyclic Aromatic Hydrocarbons (PAHs) and Their Derivatives**

| Compound | $\log K_{oc}$ | $\log K_{ow}$ |
| --- | --- | --- |
| Benzene | 1.9 | 2.1 |
| Naphthalene | 3.1 | 3.4 |
| 2-Methylnaphthalene | 3.9 | 4.1 |
| Phenanthrene | 4.4 | 4.6 |
| Anthracene | 4.4 | 4.5 |
| 9-Methylanthracene | 4.8 | 5.1 |
| Pyrene | 4.9 | 5.2 |
| Methoxychlor | 4.9 | 5.1 |
| Tetracene | 5.8 | 5.9 |
| Hexachlorobiphenyl | 6.1 | 6.3 |

*Source*: Data from Karickhoff et al. (1979).

As noted, Eq. (7.17) predicts higher $K_{oc}$ than does Eq. (7.16) at the same $K_{ow}$ value; the difference between the $K_{oc}$ values estimated by the two equations increases significantly with increasing $K_{ow}$ value. In the range of $\log K_{ow} = 3$ to 6, the ratio of $K_{oc}$ from Eq. (7.17) for PAHs on sediment to that from Eq. (7.16) for chlorinated benzenes and PCBs on soil increases from 4 to 8.

In view of these two distinct $\log K_{oc}$–$\log K_{ow}$ correlations, it appears either that PAHs on sediments exhibit appreciably higher $K_{oc}$ values than do substituted aromatic solutes on soils or else that PAHs exhibit considerably smaller $K_{ow}$ values than do substituted aromatic solutes. The latter effect can be ruled out on the basis that plots of $\log K_{ow}$ versus $\log S_w$ (supercooled liquid solubility) for different nonpolar solutes, including substituted aromatic solutes and PAHs, fall virtually onto a single line (Kile et al., 1995). Similarly, since there are no strong polar groups in PAHs, the much higher $K_{oc}$ values of PAHs cannot be attributed to specific interactions with SOM or with soil minerals in water solution (Haderlein and Schwarzenbach, 1993; Chiou, 1995). Therefore, the higher $K_{oc}$ values of PAHs according to Eq. (7.16) result supposedly from two effects: (1) higher partition with sediments than with soils for a wide variety of nonpolar solutes, including PAHs; and (2) PAHs exhibiting higher $K_{oc}$ values with SOM than other nonpolar solutes. The first effect was noted earlier for the sorption data of CT and DCB on a wide range of soils and sediments and presumably is also pertinent to PAHs. The contributions of these two potential effects were investigated by Chiou et al. (1998) by analyzing the sorption of three PAHs: naphthalene (NAP), phenanthrene (PHN), and pyrene (PYR), on a series of "clean" soils as well as freshwater and coastal sediments to elucidate their partition effects with SOM. Similar experiments were performed on some "contaminated" coastal marine sediments. The solid-state $^{13}$C-NMR data of some soils and sediments were also obtained to substantiate the effect of SOM composition on the partition behavior of the PAHs.

Molecular properties of NAP, PHN, and PYR are given in Table 7.7. The sorption isotherms of these compounds on selected soils, river sediments, and coastal sediments in Massachusetts Bay and in Boston Harbor (at Spectacle Island, Peddocks Island, and Fort Point Channel) in Massachusetts are essentially linear over a significant range of relative concentrations ($C_e/S_w$). The experimental $\log K_{oc}$ values of the PAHs are compiled in Table 7.8.

To evaluate the impact of the original SOM polarity between soil and sediment on $K_{oc}$, the $\log K_{oc}$ values for the three PAHs on five soils and the corresponding values on four sediments (the three freshwater sediments and the coastal Massachusetts Bay marine sediment) are compared. The data for other coastal sediments of Boston Harbor are excluded from this evaluation because they are either known or suspected of being contaminated by organic wastes (McGroddy and Farrington, 1995). As shown in Table 7.8, the $K_{oc}$ values between soils or between sediments are relatively invariant, as previously noted for CT and DCB on a large set of soils and freshwater sediments. The

**TABLE 7.7. Molecular Properties of Naphthalene (NAP), Phenanthrene (PHN), and Pyrene (PYR) at Room Temperature**

| Property | NAP | PHN | PYR |
|---|---|---|---|
| Molecular weight | 128 | 178 | 202 |
| Melting point (°C) | 80 | 101 | 156 |
| Heat of fusion (kJ/mol)[a] | 19.3 | 18.6 | 17.6 |
| Log $K_{ow}$ (octanol–water)[a] | 3.36 | 4.57 | 5.18 |
| Log $S_w$ of solid solute (mol/L) | −3.61 | −5.14 | −6.18 |
| Solubility parameter, $\delta$[b] | 9.9 | 9.8 | 10.6 |

*Source*: Data from Chiou et al. (1998) and references therein.

[a] Values as cited in Chiou et al. (1982b).

[b] In units of $(\text{cal/cm}^3)^{0.5}$. Values for NAP and PHN from Hildebrand et al. (1970) and for PYR from Acree (1998).


**TABLE 7.8. Organic-Carbon-Normalized Partition Coefficients ($K_{oc}$) of NAP, PHN, and PYR on Selected Soils and Sediments**

| | | $\log K_{oc}$ | | |
|---|---|---|---|---|
| Soil/Sediment Sample | Sample $f_{oc}$ | NAP | PHN | PYR |
| Woodburn soil | 0.0126 | 2.61 | 4.27 | 4.99 |
| Elliot soil | 0.0290 | 2.63 | 4.27 | 4.98 |
| Marlette soil | 0.0180 | 2.68 | 4.12 | 4.96 |
| Piketon soil | 0.0149 | 2.77 | 4.27 | 4.97 |
| Anoka soil | 0.0107 | 2.76 | 4.10 | 4.97 |
| Lake Michigan sediment | 0.0402 | 2.91 | 4.38 | 5.14 |
| Mississippi River sediment (St. Fransville, LA) | 0.0040 | 2.86 | 4.45 | 5.22 |
| Mississippi River sediment (Helena, AK) | 0.0160 | 2.88 | 4.53 | 5.23 |
| Massachusetts Bay sediment | 0.0163 | 2.87 | 4.33 | 5.12 |
| Spectacle Island sediment | 0.0334 | 2.89 | 4.42 | 5.04 |
| Peddocks Island sediment | 0.0312 | 2.95 | 4.62 | 5.24 |
| Fort Point Channel sediment | 0.0519 | 3.07 | 4.64 | 5.45 |

*Source*: Data from Chiou et al. (1998) and references therein.


$K_{oc}$ values on Massachusetts Bay sediment are similar to those on freshwater sediments, in reflection of the supposedly terrestrial origin of this coastal sediment and its lack of serious organic contamination.

The mean $\log K_{oc}$ values of NAP, PHN, and PYR on the five soils are 2.69 (±0.073 SD), 4.21 (±0.088), and 4.98 (±0.009) and the values on the four sediments are 2.88 (±0.022), 4.42 (±0.087), and 5.18 (±0.056). The sediment $\log K_{oc}$ values of the three PAHs agree fairly well with the predictions of Eq. (7.17). From these mean $\log K_{oc}$ values, the partition of PAHs with sediment OM is about 1.6 times as effective as that with soil OM, which is greater than the standard deviations of the mean $K_{oc}$ values for the soils and for the sediments. The difference in PAH partition observed between the soil and sediment (i.e., by about a factor of 1.6) is comparable in magnitude to that (by about a factor of 1.7) found for CT and DCB on a large set of soils and freshwater sediments described earlier. The result suggests that the unequal solute partition efficiencies with soil versus sediment OM accounts roughly for a factor of 2 for the discrepancy in $K_{oc}$ as exhibited by Eq. (7.16) for soils and by Eq. (7.17) for sediments. The remaining gap stems presumably from the unequal solubilities of PAHs and substituted aromatic compounds in the SOM of soils or sediments.

The disparate partition effects between PAHs and substituted aromatic solutes on SOM may be illustrated by comparing the mean soil $K_{oc}$ values of PHN ($\log K_{oc} = 4.21$; $\log K_{ow} = 4.57$) and PYR ($\log K_{oc} = 4.98$; $\log K_{ow} = 5.18$) in Table 7.8 with the reported soil $\log K_{oc}$ values of 2-PCB ($\log K_{oc} = 3.50$; $\log K_{ow} = 4.51$) and 2,4′-PCB ($\log K_{oc} = 4.16$; $\log K_{ow} = 5.10$) in Table 7.4, where the PAH and PCB solutes have comparable $\log K_{ow}$ values. As seen, there is a sharp decline in $K_{oc}$ for substituted aromatic solutes (PCBs) relative to that for the PAHs with increasing molecular weight (or $K_{ow}$); the $K_{oc}$ values for PHN and PYR are 5.1 and 6.6 times, respectively, those of 2-PCB and 2,4′-PCB. These differences are comparable in magnitude to those by Eqs. (7.16) and (7.17), when a correction is made for the soil versus sediment effect. In addition, the enhanced reduction in $K_{oc}$ with increasing $K_{ow}$ for chlorinated solutes (PCBs) relative to PAHs indicates that the partition effect of the former with SOM decreases more rapidly with increasing molecular weight.

It is of interest to explore reasons for the enhanced partition of PAHs in SOM over that of other nonpolar solutes. Since the SOM contains various polar and nonpolar moieties (Baldock et al., 1992), including aromatic and aliphatic groups, it is pertinent to compare the solubility behaviors of PAHs in apolar aromatic (e.g., benzene) and aliphatic (e.g., $n$-hexane) solvents and in a moderately polar solvent (e.g., octanol). The solubilities of NAP, PHN, and PYR in benzene at room temperature are 3.26 mol/L (Acree and Rytting, 1983), 2.27 mol/L (Hildebrand et al., 1917), and 0.697 mol/L (Judy et al., 1987), respectively; they are about 2.5, 4.9, and 7.4 times higher than their respective solubilities in $n$-hexane. The mole fraction solubilities of these PAHs in benzene are in fact very close to their ideal solubilities by Raoult's law [Eq. (5.12)], with PYR showing the greatest deviation from its ideal solubility by only about a factor of 2 (i.e., the PAH molecular size has no strong bearing on the solubility in benzene relative to the ideal solubility). The solubilities of relatively nonpolar PAHs in $n$-hexane should be comparable with those in octanol (see related partition data in Table 5.2). The increased deviation of PAH solubilities in benzene and $n$-hexane (or octanol) with increasing molecular size suggests that there is an increasing incompatibility of PAHs with an aliphatic phase (or a moderately polar phase) as the PAH molecular size increases. This result is analogous to the increased disparity in $K_{oc}$ between PAHs and other low-polarity solutes (e.g., chlorinated benzenes and PCBs) with increasing molecular weight or $K_{ow}$. From these results, PAHs should exhibit much greater partition with the SOM's aromatic components (especially, those without polar substituents) than with its other components or molecular segments.

To relate the $K_{oc}$ data to solute partition with the SOM's components, it is necessary to know the aromatic and other group contents in SOM. Based on $^{13}$C-NMR data for whole soils, the organic carbon in SOM is about $20 \pm 5\%$ aromatic, $25 \pm 6\%$ alkyl, $40 \pm 10\%$ $O$-alkyl (e.g., carbohydrate), and $15 \pm 5\%$ carboxyl + amide + ester (Baldock et al., 1992). The aromatic content includes hydrogen- and carbon-substituted aromatics, oxygenated aromatics (e.g., those with —OH and —COOH), and unsaturated carbons (Baldock et al., 1992); the separation of these components is technically difficult, however. Using the above-estimated carbon fractions in SOM, if one assumes that (1) the PAHs partition to SOM's aromatic components as effectively as they do to a pure aromatic solvent (as in benzene); (2) the partition to SOM's aliphatic components is about the same as to $n$-hexane; and (3) the partition to the remaining SOM components is not significant, the $K_{oc}$ values calculated for NAP, PHN, and PYR by the assumed solubilities in SOM and in water should be about the same as the values measured. However, the $K_{oc}$ values calculated, even discounting the PAH partition to SOM's aliphatic fraction, are significantly higher than the values measured, to be explained shortly; the partition to aliphatic fraction alone is not sufficient to account for the measured $K_{oc}$ values. If an equivalent "nonpolar aromatic carbon" fraction of about 0.10 in SOM is assumed, the calculated $\log K_{oc}$ values for NAP (3.12), PHN (4.50), and PYR (5.02) are then comparable with their values measured with sediments in Table 7.8 (2.88, 4.42, and 5.18, respectively) and with other sediment $\log K_{oc}$ values reported by Karickhoff et al. (1979) in Table 7.6 (3.11, 4.36, and 4.92, respectively). The similarly calculated $\log K_{oc}$ for anthracene (4.55) using its solubilities in benzene ($8.90 \times 10^{-2}$ mol/L) (Acree and Rytting, 1983) and in water ($2.51 \times 10^{-7}$ mol/L) (Chiou et al., 1982b) also agrees well with the $\log K_{oc}$ value of 4.41 measured with sediments (Karickhoff et al., 1979).

Whereas a quantitative account of the $K_{oc}$ values requires well-defined SOM molecular structures that are presently unavailable, the results above support the assumed influence of the SOM's aromatic components on PAH partition. The observation that the calculated $K_{oc}$ values based on the total aromatic content in SOM are much higher than the actual values suggests that most aromatic structures in SOM are substituted with some polar and nonpolar groups, which reduce their overall compatibilities with PAHs. Since sediments exhibit slightly higher $K_{oc}$ values than soils, the sediment organic matter should have either a lower polar-group content or a higher aromatic content. The approximate aromatic fractions in SOM for some of the present soils and sediments will be estimated later using the solid-state $^{13}$C-NMR spectra of the whole samples.

The enhanced affinity of PAHs over other nonpolar solutes for natural organic matter has been attributed to the PAH's planar structures that facilitate their gaining a closer approach to the aromatic components in natural organic matter (Chin et al., 1997) or their intermolecular attractions through $\pi$–$\pi$ electron interactions (Gauthier et al., 1987). From the partition standpoint, the favored interaction between aromatic components seems more likely to be related to similarity in their cohesive energy densities (CEDs) (or their solubility parameters, $\delta$). As shown in Table 2.2, for example, the $\delta$ values, in $(cal/cm^3)^{1/2}$, for benzene (9.2), NAP (9.9), PHN (9.8), and anthracene (9.9) are consistently high compared with those of other nonpolar solutes. The $\delta$ value estimated for PYR from its solubilities in hydrocarbons is about 10.6 (Acree, 1998). Thus, on going from benzene to PYR, the $\delta$ values for PAHs stay relatively constant with increasing molecular size. The $\delta$ values for most short-chain halogenated compounds are in the range 8.5 to 10 (e.g., $\delta = 8.6$ for CT) and the values for aliphatic hydrocarbons (e.g., $n$-hexane) are much lower, mainly in the range 7 to 8. The $\delta$ values for low-molecular-weight alkyl-substituted aromatic compounds (e.g., toluene and xylenes) are about 8.6 to 9.0 (Hildebrand et al., 1970). The $\delta$ data for halogenated aromatic compounds (e.g., chlorinated benzenes and PCBs) are essentially unavailable. The $\delta$ for SOM is reported to range from 10.3 with relatively nonpolar solutes to 17.2 with highly polar and H-bonding solutes (Chiou and Kile, 1994). As noted, since the aromatic compounds have similar structures and presumably comparable CEDs, the two mechanisms considered to favor the partition of PAHs with SOM are interrelated.

On the basis that the SOM is a natural *heterogeneous polymer*, where the various components (or molecular segments) may be present in sizable domains, it is possible that a given class of solutes would interact more favorably with certain SOM components of comparable CEDs, such as for PAHs with SOM's aromatic regions. This consideration is consistent with the variable $\delta$ value of SOM, as stated earlier. The solubility of a compound in a polymer medium may thus depend more on the polymer's functional-group content than on its overall polarity, the latter being more important for low-molecular-weight organic substances. Since the $\delta$ values of PAHs are fairly constant, their solubilities (as supercooled liquids) in SOM should be approximately the same and largely independent of the molecular weight; this is contrary to the significant solubility decrease of other nonpolar solutes in SOM with increasing molecular weight. This dissimilarity provides an a priori account of the increased deviation between the $K_{oc}$ values of PAHs and other nonpolar solutes with an increase in $K_{ow}$.

The relative amounts of different structural carbons in SOM, as revealed by the solid-state $^{13}C$-NMR spectra of whole soils and sediments, provide added insights into the effect of SOM composition on solute $K_{oc}$. However, such assignment of organic carbon types is only semiquantitative because of the overlapping chemical shift of certain functional-group carbons and of the imprecise quantification of some functional groups (Baldock et al., 1992). The $^{13}C$-NMR spectra of Marlette soil, Helena Mississippi River sediment, Spectacle Island sediment, and Fort Point Channel sediment are selected for

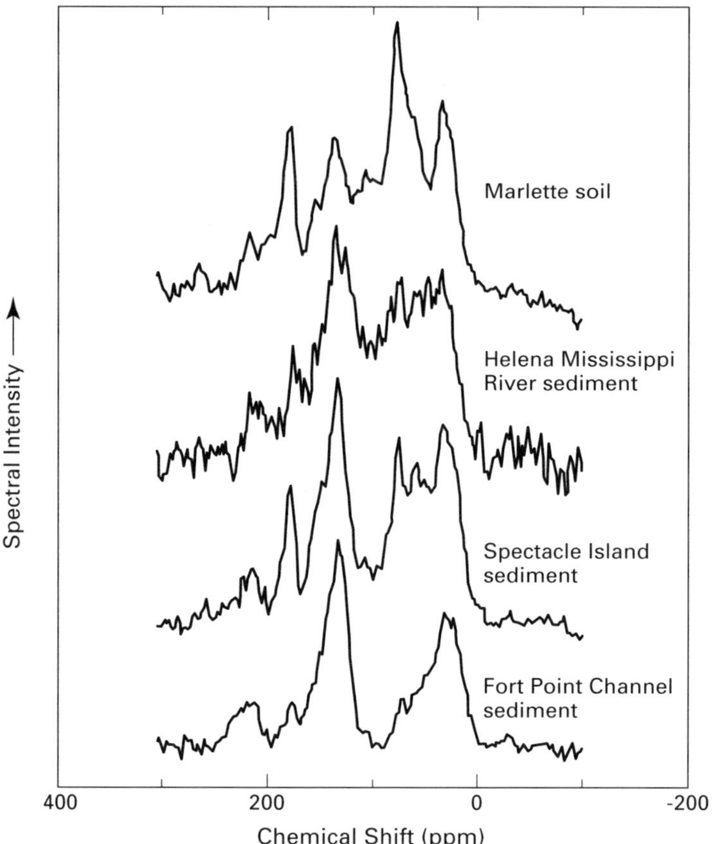



**Figure 7.15**    Solid-state $^{13}$C-NMR spectra of selected soil and sediment samples. [Data from Kile et al. (1999).]

comparisons (Figure 7.15). Analyses of the spectra of a number of "clean" soils and freshwater sediments (including those not reported here) indicate that the soils generally have higher combined fractions of polar $O$-alkyl and carboxyl-amide-ester components than the sediments. The aromatic carbon fractions observed (110 to 160 ppm) fall largely into the range of $26 \pm 8\%$ for both soils and sediments; there is no clear pattern that the sediments exhibit higher aromatic fractions. For Marlette soil and Helena Mississippi River sediment, the aromatic carbon fractions are about 20% and 33%, respectively, which include all aromatic carbons with polar and nonpolar groups and other unsaturated carbons. The respective carbon fractions for polar $O$-alkyl (45 to 110 ppm) and carboxyl-amide-ester (160 to 200 ppm) groups other than phenols (140 to 160 ppm) are 56% and 35%, respectively. The difference in alkyl carbon content between the samples is relatively minor (25% versus 33%). The higher $K_{oc}$ values of PAHs on Helena Mississippi River sediment than on Marlette soil correlates more closely with the sediment's lower polar-group content. Thus, while the aromatic content in SOM appears to enhance the $K_{oc}$ values for PAHs, the difference between soil and sediment $K_{oc}$ values is related, at least in part, to their different polar-group contents.

We now examine how the $\log K_{oc}$ data of PAHs with contaminated coastal marine sediments compare to those with "clean" river sediments. The Fort Point Channel sediment from Boston Harbor is known to be severely contaminated by highly aromatic organic wastes that resemble a coal tar (McGroddy and Farrington, 1995), whereas Spectacle Island sediment and Peddocks Island sediment are suspected of being contaminated to lesser extents by similar organic wastes. For these samples, one notes that the $K_{oc}$ values of the PAHs on Spectacle Island sediment are virtually the same as those of "clean" freshwater sediments and Massachusetts Bay marine sediment. This suggests that the level of contamination in this sediment has not significantly affected the overall organic composition of the sample. The carbon fractions for alkyl (29%), aromatic (32%) and $O$-alkyl and carboxyl-amide-ester components (39%) of Spectacle Island sediment are about the same as those of Helena Mississippi River sediment (Figure 7.15). The moderately high $K_{oc}$ data with Peddocks Island sediment indicate that the sediment is contaminated by a significant level of organic wastes (McGroddy and Farrington, 1995), which either reduces the polar-group content or increases the aromatic content of the sediment (NMR spectra not available). The severely contaminated Fort Point Channel sediment exhibits a high aromatic carbon content (46%); the combined carbon content of $O$-alkyl and carboxyl-amide-ester components is low (21%), excluding the oxygenated aromatic carbons (e.g., the phenolic carbons at 140 to 160 ppm). The alkyl carbon content (33%) is the same as that of Helena Mississippi River sediment. The spectral data for this sediment correlate semiquantitatively with the much higher than average $K_{oc}$ values of the PAHs. A comparison of the NMR and $K_{oc}$ data between Helena Mississippi River sediment and Fort Point Channel sediment further illustrates the point that the polar-group content in SOM is a significant factor affecting the $K_{oc}$ values of PAHs and other solutes.

### 7.3.5  Estimation of Sorption Coefficients for Nonpolar Contaminants

The foregoing analysis of the SOM compositional effect on the $K_{oc}$ values of nonpolar solutes with soils and sediments and of the enhanced partition of PAHs versus other nonpolar solutes with SOM offers an improved algorithm for estimating the $K_{oc}$ values of various nonpolar contaminants with soils and sediments by applying appropriate $\log K_{oc}$–$\log K_{ow}$ correlations. Starting with Eq. (7.16) for estimating the $K_{oc}$ values of nonpolar non-PAH compounds on soils, the corresponding $K_{oc}$ values for the compounds on sediments can be obtained by assuming that the sediment $K_{oc}$ is twice the soil $K_{oc}$, that is,

$$\log K_{oc} = 0.904 \log K_{ow} - 0.211 \tag{7.18}$$

Similarly, the correlation for PAHs on soils may be achieved by assuming the soil $K_{oc}$ value to be one-half the sediment $K_{oc}$ value, which can in turn be obtained by adjusting Eq. (7.17) to give

$$\log K_{oc} = 1.00 \log K_{ow} - 0.51 \tag{7.19}$$

The merits of Eqs. (7.16) through (7.19) have been largely substantiated by Xia (1998) and Allen-King et al. (2002) in a statistical analysis of a large set of nonpolar compounds on a wide range of soils and sediments if the SOM content is more than about 0.1%. A more extensive testing of the generality of these equations is warranted.

For soils and sediments having a SOM content less than about 0.1%, the measured $K_{oc}$ values for given nonpolar solutes tend to be greater than predicted (i.e., higher than the values with high-SOM soils and sediments). A logical explanation is that the occurrence of a weak and linear solute uptake by soil minerals, due to solute concentration near (rather than condensation on) mineral surfaces, as discussed earlier, may become quite significant for low-SOM soils and sediments relative to the concurrent linear solute partition to SOM. In this case, since the overall solute sorption would be virtually linear, normalizing the sorption coefficient with SOM leads to a higher $K_{oc}$ than predicted. With a given low-SOM soil or sediment, the magnitude of this deviation should be greater for more water-soluble solutes because of their smaller partition coefficients with SOM, making the mineral effect more important. Finally, one would not expect the soil versus sediment and PAH versus nonPAH effects, as dealt with here for nonpolar solutes, to be equally evident and important for relatively polar compounds because their partition to SOM would be mediated strongly by more powerful polar interactions.

### 7.3.6  Sorption to Previously Contaminated Soils

We have so far considered primarily the sorptive behavior of contaminants from water to relatively clean soils or sediments. In many organic-contaminated sites, the sorption of a contaminant may be greatly influenced by the preexisting contaminants in soil or sediment. If a local environment contains an exotic organic plume (i.e., an excess organic phase), there is little doubt that it will exert a large effect on contaminant uptake, as it will act effectively as an excess partition phase to sequester the contaminant. However, if the contaminated soil or sediment contains no clearly discernible separate phase, the result would depend on the level of soil contamination, or more exactly on whether the quantity in the soil exceeds the saturation limit in SOM, because an excess phase will not appear before this limit is exceeded. Often, the excess phase is not readily visible because it may be microscopic in size. On the other hand, if the amount of contaminants in SOM is below saturation, the effect on contaminant sorption should be insignificant (Sun and Boyd, 1991). This is because the substance partitioned into SOM cannot function as a partition phase. For a substance to maintain a stable separate phase in soil with time, it must be relatively refractory and insoluble in water, such as petroleum hydrocarbons, PCB oils, coal tars, and some chlorinated solvents.

When a separate organic phase exists in a soil, the sorption of a contaminant from water into this soil consists of a concurrent partition into SOM and into the exotic organic phase. In their treatment of contaminant sorption to soils contaminated by a PCB oil, Sun and Boyd (1991) expressed the apparent contaminant sorption coefficient ($K_d^*$) as

$$K_d^* = f_{om}K_{om} + f_{oil}K_{oil} = K_d + f_{oil}K_{oil} \qquad (7.20)$$

where $f_{om}$, $K_{om}$, and $K_d$ are as defined previously, $f_{oil}$ the mass fraction of the oil in soil, and $K_{oil}$ the contaminant partition coefficient between the oil and water. Note that in Eq. (7.20), $f_{oil}$ accounts in principle only for the oil content that exists as a separate phase, not including the mass partitioned into SOM, as described below. It may be safely assumed that the small amount of oil partitioned into SOM does not significantly alter the medium properties of the SOM.

Since $K_{oil}$ is expected to be orders of magnitude greater than $K_{om}$ for nonpolar organic compounds, the $K_d^*$ value of a nonpolar contaminant with a soil containing an excess oil phase will be greatly higher than $K_d$ if $f_{oil}$ is a sizable quantity relative to $f_{om}$. The measured $K_d^*$ and $K_d$ values for 2-chlorobiphenyl (2-PCB) on four excess-PCB-oil-loaded soils are given in Table 7.9. From the data for $K_d^*$, $K_d$, and $f_{oil}$, a $K_{PCB\text{-}oil}$ value of 114,000 was derived for 2-PCB as compared to $K_{om} = 1700$, the difference being 67 times (Sun and Boyd, 1991). Here the $K_{PCB\text{-}oil}$ value is about 3.5 times greater than $K_{ow}$ (32,400). This difference is well expected because 2-PCB should form a nearly ideal solution in the PCB oil because of their similar compositions, while the solution of 2-PCB in octanol would not be as close to being ideal because of some compositional differences.

In the plot of $K_d^* - K_d$ against total $f_{oil}$ for 2-PCB on a soil with $f_{om} = 0.015$ contaminated by different contents of a PCB oil, Sun and Boyd (1991) found that at $K_d^* - K_d = 0$, $f_{oil}$ has a value of 0.001 instead of 0. This means that when the PCB oil content in this soil is less than 0.1%, no enhanced sorption of 2-PCB occurs. This threshold value corresponds to 6 to 7% of $f_{om}$. It suggests that the saturation capacity of the PCB oil in SOM is about 6 to 7% by weight, and thus a separate PCB-oil phase will emerge in the soil only after the PCB-

**TABLE 7.9. Partition Coefficients of 2-Chlorobiphenyl on Soils Contaminated by Commercial PCB Oils and on Respective Soils after Removal of PCB Oils**

| Soil No. | SOM (%) | PCB Oil (%) | Total OM (%) | $K_d$ | $K_d^*$ | $K_{om}^*/K_{om}$ |
|---|---|---|---|---|---|---|
| 1 | 2.0 | 0.22 | 2.22 | 34 | 162 | 4.3 |
| 2 | 1.6 | 0.74 | 2.34 | 27 | 747 | 19 |
| 3 | 1.6 | 0.77 | 2.37 | 27 | 788 | 20 |
| 4 | 0.21 | 0.38 | 0.59 | 3.6 | 319 | 32 |

*Source*: Data from Sun and Boyd (1991).

oil level exceeds this threshold capacity. This saturation effect is generally applicable to any soils contaminated by oils or other organic liquids. Soils with high SOM contents will thus take up a relatively large quantity of oils before a separate oil phase emerges, and vice versa for soils with low SOM contents. The observed PCB-oil saturation capacity in SOM, when expressed in liquid volume, is comparable in magnitude to the values for other nonpolar liquids, as given in Tables 7.2 and 7.21. Information on whether a separate organic phase exists in a contaminated soil is essential to the behavior and fate of both preexisting and incoming contaminants and to the strategy to be taken for soil remediation.

A more straightforward means to ascertain whether a separate organic phase exists in a soil (or sediment) is to compare the measured $K^*_{om}$ (or $K^*_{oc}$) value of a model nonpolar solute against its intrinsic $K_{om}$ (or $K_{oc}$) value, since we found earlier that the $K_{om}$ (or $K_{oc}$) values for nonpolar solutes (e.g., CT and DCB) are relatively invariant between normal soils. To do so, one normalizes the $K^*_d$ and $K_d$ in Eq. (7.20) to the total organic matter content ($f_{tom}$) or to the total organic carbon content ($f_{toc}$) of the contaminated soil. The following general equations can then be derived after manipulation of the related terms:

$$K^*_{om} = K^*_d/f_{tom} = K_{om} + (K_{hom} - K_{om})(f_{hom}/f_{tom}) \qquad (7.21)$$

or

$$K^*_{oc} = K^*_d/f_{toc} = K_{oc} + (K_{hoc} - K_{oc})(f_{hoc}/f_{toc}) \qquad (7.22)$$

in which $K_{hom}$ is the partition coefficient of a model nonpolar solute between the hydrocarbon phase in soil and water, $f_{hom}$ is the hydrocarbon organic-matter fraction in the soil, and $f_{tom}$ is the total organic-matter fraction in the soil (i.e., $f_{tom} = f_{om} + f_{hom}$). The $K_{hoc}$, $f_{hoc}$, and $f_{toc}$ are the corresponding terms based on the organic-carbon content of the contaminated soil. For a nonpolar solute, the $K_{hom}$ (or $K_{hoc}$) value is usually orders of magnitude higher than $K_{om}$ (or $K_{oc}$), as observed with 2-PCB. In Eqs. (7.21) and (7.22), it should be recognized that while $K_{om}$ and $K_{oc}$ are interrelated by a nearly constant factor because the carbon content in SOM is relatively constant, as discussed before (see pages 135–136), $K^*_{om}$ and $K^*_{oc}$ are not related to each other by a constant factor, as the carbon content in total organic matter of a contaminated soil is not fixed.

By use of Eqs. (7.21) and (7.22) and the measured $f_{tom}$, $K^*_{om}$, and $K_{om}$ (or $f_{toc}$, $K^*_{oc}$, and $K_{oc}$), the resulting $K^*_{om}/K_{om}$ or $K^*_{oc}/K_{oc}$ value offers a sensitive test for an excess hydrocarbon phase in contaminated soils. Here a finding of $K^*_{om}/K_{om} \gg 1$ or $K^*_{oc}/K_{oc} \gg 1$ indicates the presence of a separate hydrocarbon phase (e.g., oils), the magnitude being proportional to $f_{hom}/f_{tom}$ (or $f_{hoc}/f_{toc}$). If a model nonpolar solute with known $K_{om}$ (or $K_{oc}$) is employed for the test, only $f_{tom}$ and $K^*_{om}$ (or $f_{toc}$ and $K^*_{oc}$) are required for completing the analysis. Moreover, if the

$K_{hom}$ (or $K_{hoc}$) of the model solute is known or can be estimated with fair accuracy, the value of $f_{hom}$ (or $f_{hoc}$) can then readily be estimated. The $K_{om}^*/K_{om}$ values for 2-PCB with four PCB-oil contaminated soils (Sun and Boyd, 1991) are shown in the last column of Table 7.9. Similarly, using Eq. (7.22) with CT and DCB as model nonpolar solutes, very large $K_{oc}^*/K_{oc}$ values are observed for soils and sediments in which a petroleum hydrocarbon phase exists (Kile et al., 1995). In Table 7.8 we also find that the measured $K_{oc}$ values of PAHs for some highly contaminated sediments are appreciably higher than those of relatively clean sediments, due to the presence of a separate hydrocarbon phase.

### 7.3.7  Deviations from Linear Sorption Isotherms

Some recent studies on the sorption of single contaminants (solutes) from water on some soils and sediments indicate that the measured sorption at low relative concentrations ($C_e/S_w$) may often be nonlinear with enhanced sorption coefficients compared to the upper linear sorption range. Young and Weber (1995) found that the sorption of a nonpolar solute (phenanthrene) on some soils and shales exhibits a significant nonlinearity with a concave-downward shape at low concentrations. Spurlock and Biggar (1994) observed nonlinear sorption of relatively polar substituted ureas (herbicides) on soils at low concentrations, with the nonlinear sorption coefficient increasing with decreasing solute concentration. Xing et al. (1996) also found deviations from linear sorption at low concentrations for some polar pesticides (triazines) and, to a lesser extent, for relatively nonpolar trichloroethylene (TCE) on selected soil and organic-matter samples.

It is of practical interest to deliberate on the cause of such nonlinear sorption for organic solutes at low $C_e/S_w$, since a wide variety of relatively soluble organic contaminants may fall into this range in natural systems. Although the unsuppressed adsorption of polar solutes on certain clay fractions of low-organic-content soils (Laird et al., 1992; Haderlein and Schwarzenbach, 1993; Weissmahr et al., 1997) could result in nonlinear sorption at low $C_e/S_w$, the effect as noted for polar and nonpolar solutes on soils with relatively high SOM contents points instead to the occurrence of a strong nonpartition effect (e.g., adsorption or specific interaction) of solutes with either a small amount of active SOM groups or with a small amount of nonmineral soil fraction.

A number of conceptual models have been postulated to account for the nonlinear solute sorption on soils of significant SOM contents: (1) the different equilibrium rates of the solute with the assumed two structural entities of the SOM, one in a rubbery state and the other in a glassy state (Young and Weber, 1995; Weber and Huang, 1996) where the solute sorption to rubbery SOM is linear in reflection of partition and that to glassy SOM is nonlinear in reflection of a surface adsorption; (2) the presence of a small amount of high-surface-area carbonaceous material (HSACM) (such as charcoal or soot) that exhibits a greater nonlinear adsorption at low relative concentrations than the linear partition to SOM (Chiou, 1995; Gustafsson et al., 1997; Chiou et al., 1998); (3) the availability of compound-specific "internal holes" (or internal pores) in SOM for adsorption of specific solutes in addition to solute partition into the water-saturated SOM (Pignatello and Xing, 1996; Xing et al., 1996; Xing and Pignatello, 1997); and (4) the occurrence of specific interactions for polar solutes with limited active sites in SOM in addition to solute partition to SOM, the former effect approaching saturation at lower solute concentrations (Spurlock and Biggars, 1994). Some pertinent experimental data are presented below to give readers a brief overview of the problem involved.

The sorption data on some reference samples (a soil, peat, and soil humic acid) from Xing et al. (1996) indicate that the nonlinear sorption tends to be more pronounced for polar solutes (e.g., atrazine and prometon) than for low-polarity solutes (e.g., TCE), the data being based on their Freundlich (or log-log) plots. On the Cheshire fine sandy loam used by Xing et al. (1996), TCE exhibits essentially no nonlinearity, with concentrations ranging from <0.1 mg/L (<$10^{-4}$ in $C_e/S_w$) to >100 mg/L (>$10^{-1}$ in $C_e/S_w$). By comparison, the sorption of phenanthrene on some soils and shales reported by Young and Weber (1995) exhibits significant nonlinearity at low concentrations. It thus appears that the extent of nonlinear sorption for nonpolar solutes tends to depend on the soil source and to be smaller in magnitude relative to that for polar solutes. In addition, Xing et al. (1996) noted that in the binary-solute systems a coexisting polar solute (prometon) strongly suppresses the nonlinear sorption of the nominal polar solute (atrazine), whereas a coexisting nonpolar solute (TCE) exerts only a small suppressing effect. To account for these observations, Xing et al. (1996) proposed that different sets of compound-specific internal holes exist in SOM for adsorption of different compounds in addition to their conventional partition into bulk SOM.

To contemplate on the sources of sorption nonlinearity, Chiou and Kile (1998) presented extensive sorption data utilizing several polar and nonpolar compounds on a peat (organic) soil and a mineral soil. The compounds studied and their physicochemical properties are listed in Table 7.10. To minimize complications from the interactions of polar solutes with minerals, most of the

**TABLE 7.10. Physicochemical Properties of Selected Organic Compounds Used for Detection of Nonlinear Sorption to Florida Peat and Woodburn Soil**

| Compound | Abbreviation | $S_w$ (mg/L) | $\log K_{ow}$ | $pK_a$ |
|---|---|---|---|---|
| Phenol | PHL | 87,000 | 1.46 | 9.89 |
| 3,5-Dichlorophenol | DCP | 8,050 | 3.23 | 7.85 |
| Monuron | MON | 275 | 1.98 | <−1 |
| Diuron | DUN | 38 | 2.68 | <−1 |
| Ethylene dibromide | EDB | 3,520 | 1.99 | — |
| Trichloroethylene | TCE | 1,100 | 2.53 | — |
| Lindane | LND | 7.8 | 3.75 | — |

*Source*: Data from Chiou et al. (1998) and references therein.

sorption data were obtained on the peat soil ($f_{oc} = 0.493$), with supplemental data on a mineral (Woodburn) soil ($f_{oc} = 0.0126$). Polar solutes from two chemical classes (phenols and substituted ureas) and three low-polarity solutes (ethylene dibromide, TCE, and lindane) were employed in sorption experiments. In addition to single-solute isotherms, the isotherms of nominal solutes in many binary-solute mixtures were also determined, with the competing solutes (co-solutes) taken from either the same class or from a different class. Binary-solute sorption studies examine the competitive sorption of polar solutes both between and within chemical classes, considering that the earlier binary-solute studies were confined mainly to solutes from the same or similar class (Xing et al., 1996). Results on solute competition are critical to assessing the effect of various co-solutes on the behavior of a given solute (contaminant) in multisolute natural systems. A comparison of single-solute and binary-solute isotherms enables one to separate the relative effects of linear partition to SOM and nonlinear sorption to soil.

Typical single-solute sorption isotherms of TCE, ethylene dibromide (EDB), diuron (DUN), and 3,5-dichlorophenol (DCP) at room temperature ($24 \pm 1°C$) on the peat are shown in Figures 7.16 to 7.19 and on Woodburn soil in Figures 7.20 and 7.21. The isotherms are plotted on a linear scale of the solute uptake per unit mass of soil ($Q$) against the relative concentration of the solute in water ($C_e/S_w$) to enable a better distinction of individual solute behaviors. Unique characteristics exist between the solutes. In all cases, the isotherms display nonlinearity with concave-downward curvatures at low

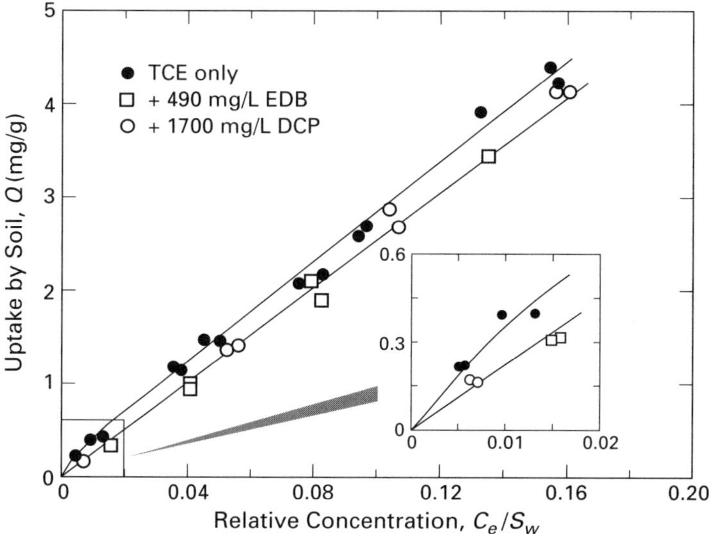


**Figure 7.16** Sorption of TCE as a single solute and as a binary solute on peat soil with EDB and DCP as co-solutes at specified equilibrium concentrations. [Data from Chiou and Kile (1998).]

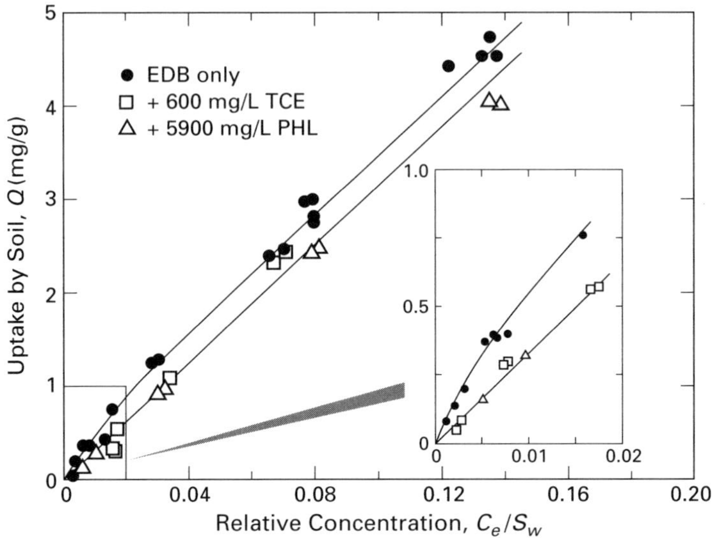

**Figure 7.17**   Sorption of EDB as a single solute and as a binary solute on peat soil with TCE and PHL as co-solutes at specified equilibrium concentrations. [Data from Chiou and Kile (1998).]

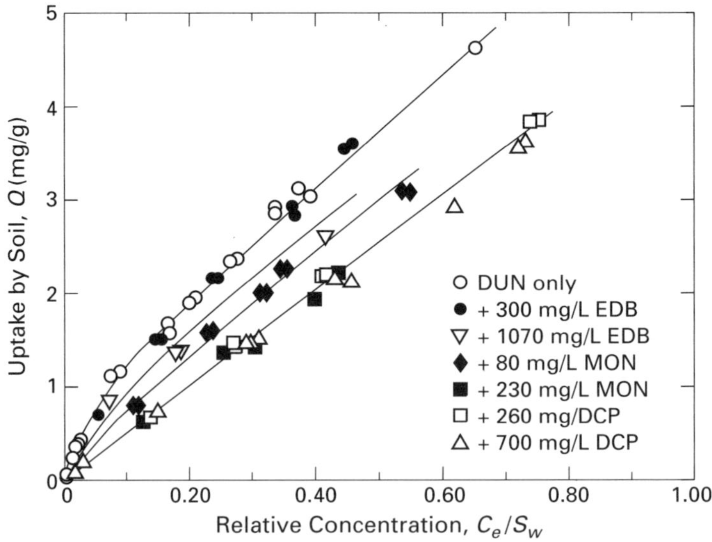

**Figure 7.18**   Sorption of DUN as a single solute and as a binary solute on peat soil with EDB, MON, and DCP as co-solutes at specified equilibrium concentrations. [Data from Chiou and Kile (1998).]

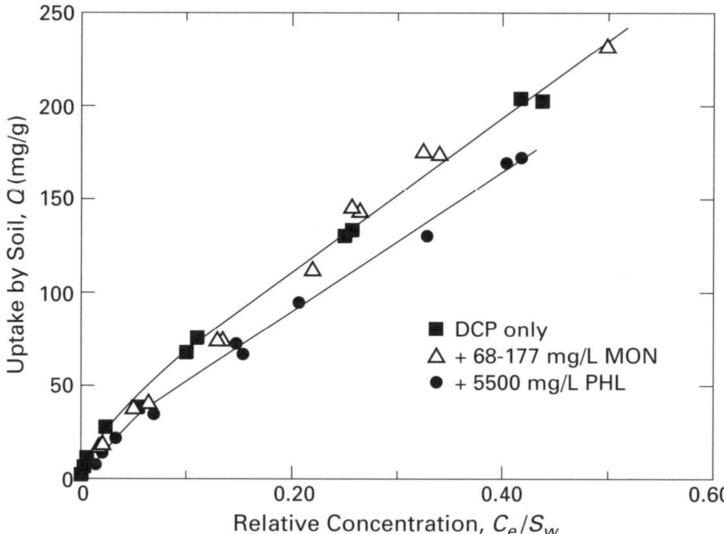


**Figure 7.19** Sorption of DCP as a single solute and as a binary solute on peat soil with MON and PHL as co-solutes at specified equilibrium concentrations. [Data from Chiou and Kile (1998).]

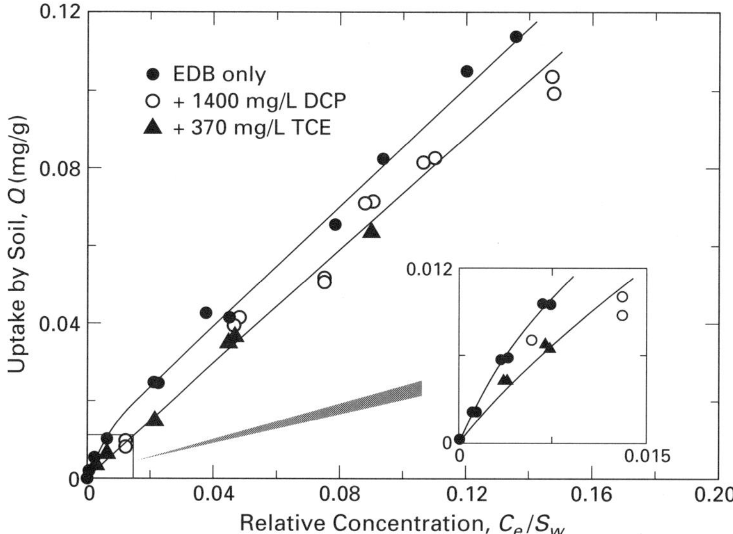


**Figure 7.20** Sorption of EDB as a single solute and as a binary solute on Woodburn soil with TCE and DCP as co-solutes at specified equilibrium concentrations. [Data from Chiou and Kile (1998).]

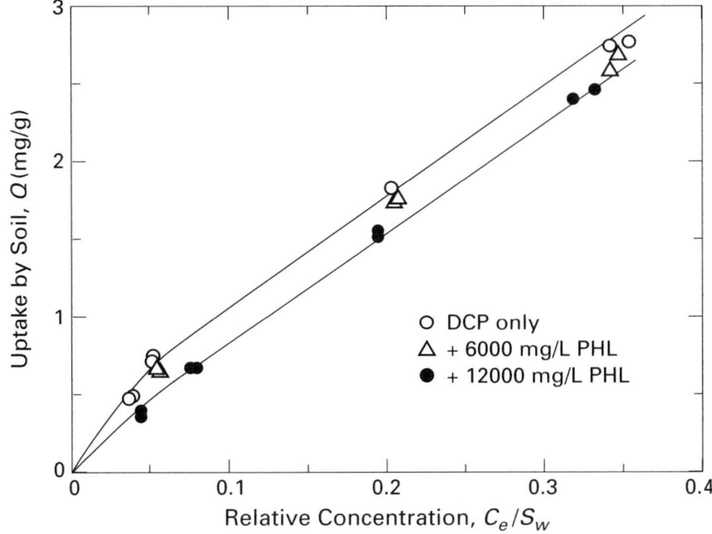


**Figure 7.21**   Sorption of DCP as a single solute and as a binary solute on Woodburn soil with PHL as co-solute at specified equilibrium concentrations. [Data from Chiou and Kile (1998).]

$C_e/S_w$ values but exhibit a practically linear shape at moderate to high $C_e/S_w$ values. The nonlinear effects are relatively more visible for the polar solutes (e.g., DCP and DUN) than for the nonpolar solutes (e.g., EDB and TCE). Such an overall isotherm shape bears no resemblance to the conventional adsorption shapes shown in Figure 4.1. Moreover, whereas the apparent nonlinear saturation capacities, as measured by the extrapolated intercepts of the upper linear lines to the $Q$ axis, are about the same for the nonpolar solutes, they vary appreciably among the polar solutes both on peat soil and on Woodburn soil. The apparent nonlinear saturation capacities ($Q_{ans}$), the approximate $C_e/S_w$ values at which the nonlinear capacities approach apparent saturation, and the organic-carbon-normalized partition coefficients for the linear sections of the isotherms are presented in Table 7.11.

As seen, DCP exhibits the highest $Q_{ans}$ value on either soil. The $Q_{ans}$ value decreases progressively with decreasing solute polarity for other solutes: monuron (MON), DUN, EDB/TCE, and lindane (LND). On the peat soil (Figures 7.16 through 7.19), the $Q_{ans}$ is about 0.15 mg/g for TCE, 0.18 mg/g for EDB, 0.60 mg/g for DUN, 0.82 mg/g for MON, and 25 mg/g for DCP. The corresponding $Q_{ans}$ value with Woodburn soil is <0.008 mg/g for EDB, <0.005 mg/g for LND, and about 0.38 mg/g for DCP (Figures 7.20 and 7.21). The nonlinear sorption approaches apparent saturation at about $C_e/S_w = 0.010$ to 0.015 for EDB, TCE, and LND and at about 0.10 to 0.13 for DCP, MON, and DUN on both soils. The present DUN isotherm shape is much like the one for DUN

**TABLE 7.11. Apparent Nonlinear Saturation Capacities ($Q_{ans}$), Apparent Nonlinear Saturation Points [$(C_e/S_w)_{ans}$], and Organic-Carbon-Normalized Linear Partition Coefficients ($K_{oc}$) of the Solutes on Florida Peat and Woodburn Soil**

| Sorption System | $Q_{ans}$ (mg/g) | $(C_e/S_w)_{ans}$ | $\log K_{oc}$ |
| --- | --- | --- | --- |
| TCE/peat | 0.15 | 0.012 | 1.69 |
| EDB/peat | 0.18 | 0.010 | 1.28 |
| DUN/peat | 0.60 | 0.10 | 2.43 |
| MON/peat | 0.82 | 0.13 | 1.45 |
| DCP/peat | 25 | 0.12 | 2.03 |
| EDB/Woodburn | <0.008 | 0.015 | 1.23 |
| LND/Woodburn | <0.005 | 0.010 | 2.92 |
| DCP/Woodburn | 0.38 | 0.080 | 1.87 |

*Source*: Data from Chiou and Kile (1998).

with other soils reported by Hance (1965) when Hance's data in log-log plots are converted into linear scales.

In binary-solute systems, the sorption isotherms of TCE, EDB, DUN, and DCP with various polar and nonpolar co-solutes (competitors) on peat soil at fixed co-solute concentrations are shown in Figures 7.16 through 7.19. The isotherms for EDB and DCP with other co-solutes on Woodburn soil are given in Figures 7.20 and 7.21. On either soil, the isotherms of nominal nonpolar solutes (e.g., TCE and EDB) in binary-solute systems exhibit significantly lower capacities than those of their respective single-solute isotherms only at very low $C_e/S_w$ values (mostly, <0.02) (Figures 7.16, 7.17, and 7.20). That is, with the uncertainties of the sorption data being about 10%, the slopes of the nominal solutes at high $C_e/S_w$ values in single-solute and binary-solute experiments are not statistically different. At the applied co-solute concentrations, the TCE and EDB isotherms become relatively linear at low $C_e/S_w$ values. For example, the small nonlinear EDB capacities on both peat soil and Woodburn soil (Figures 7.17 and 7.20) are greatly suppressed by nonpolar TCE as the co-solute at 370 mg/L ($C_e/S_w = 0.34$), by polar DCP at 1400 mg/L ($C_e/S_w = 0.17$), and by polar PHL at 5900 mg/L ($C_e/S_w = 0.068$). In contrast, the nominal solute isotherms of the polar solutes vary considerably among the co-solute/soil systems.

As noted, the sorption of DCP on peat is largely unaffected by MON as the co-solute even with MON present at 177 mg/L ($C_e/S_w = 0.64$) (Figure 7.19). However, when PHL is used as the co-solute at 5500 mg/L ($C_e/S_w = 0.063$), it suppresses a fair amount of the DCP nonlinear capacity, although the suppression is not complete. The relative competitive powers of various co-solutes are exhibited most distinctly by the binary-solute DUN sorption isotherms (Figure 7.18). In this case, EDB as the co-solute imposes little effect on DUN sorption at the EDB concentration of 300 mg/L ($C_e/S_w = 0.085$), while it exhibits a small depression of DUN sorption at the EDB concentration of 1070 mg/L ($C_e/S_w = 0.30$). MON at 80 mg/L ($C_e/S_w = 0.29$) exhibits a significant but incomplete suppression of the DUN nonlinear capacity while MON at 230 mg/L ($C_e/S_w = 0.84$) and DCP at 260 mg/L and 700 mg/L ($C_e/S_w = 0.032$ and 0.087, respectively) suppress most of the nonlinear capacity. In the latter case, the resultant DUN isotherms are relatively linear with slopes about equal to that of the upper linear DUN single-solute isotherm. The response of DCP to co-solute PHL on Woodburn soil (Figure 7.21) is similar to that with peat soil (Figure 7.19), in which PHL as the co-solute at 12,000 mg/L ($C_e/S_w = 0.14$) erases most of the DCP nonlinear capacity.

The results above illustrate that the nonlinear behavior of a nominal solute in binary-solute systems is influenced by both the co-solute type and its concentration and that a polar co-solute (e.g., DCP) of one chemical class may effectively suppress the nonlinear sorption of a nominal polar solute of a different class (e.g., DUN). In this respect, phenolic compounds are more powerful sorbates and competitors than are substituted ureas, which is consistent with their relative nonlinear sorption capacities. If the nominal solute is a polar compound, suppression by various co-solutes occurs in a highly selective manner. Here the large nonlinear capacities of polar solutes (e.g., DCP and DUN) are not strongly affected by nonpolar co-solutes at relatively low $C_e/S_w$ values; a large suppression occurs if the co-solute is of high polarity even at relatively low $C_e/S_w$ values, as illustrated by DCP on DUN (Figure 7.18). For the polar solutes studied, the relative suppressive power follows the order PHL $\geq$ DCP > MON > DUN, which is essentially the order of their $S_w$ values. By contrast, the small nonlinear capacities of nonpolar solutes (e.g., EDB and TCE) are more effectively depressed by either polar or nonpolar co-solutes if the $C_e/S_w$ of the co-solute is appreciably higher than the upper $C_e/S_w$ limit (0.010 to 0.015) of the sorption nonlinearity for nonpolar solutes.

Some important isotherm features for solutes at low $C_e/S_w$ values, as revealed explicitly by Chiou and Kile (1998) and less explicitly by Xing et al. (1996) and Xing and Pignatello (1997), include (1) the smaller nonlinearity effects for nonpolar than for polar solutes; (2) the relatively small suppression of the sorption of a polar solute (e.g., atrazine) by a nonpolar co-solute (e.g., TCE) over a range of co-solute concentrations; and (3) the significant suppression of the sorption of a polar solute (atrazine) by other polar co-solutes (e.g., prometon and other triazines). Here the linear plots, as employed by Chiou and Kile (1998), are better adapted than the log-log plots used in related studies for displaying the different extents of nonlinear sorption in the different systems.

The unique isotherm shape for single-solute systems (i.e., nonlinear at low $C_e/S_w$ but virtually linear at other $C_e/S_w$ values) suggests that more than one mechanism is operative over the entire concentration range. Moreover, since the (apparent) nonlinear capacity and the point of nonlinear saturation are not the same for polar and nonpolar solutes, the data suggest that the primary causes for their nonlinear behaviors at low concentrations are not the same.

We consider first the probable source of the sorption nonlinearity for nonpolar solutes, where the nonlinear capacity is relatively small and approaches apparent saturation at very low $C_e/S_w$ (0.010 to 0.015). On peat soil, the nonlinear capacities of about 0.18 mg/g for EDB and 0.15 mg/g for TCE are well within the allowed monolayer adsorption capacity of the soil based on its BET-$N_2$ surface area of 1.4 m$^2$/g. The same is true for EDB and LND on Woodburn soil, where the nonlinear capacities of EDB and LND are <0.008 mg/g and the surface area of the soil is 11.2 m$^2$/g. Since there is little tendency for nonpolar solutes to engage in specific interaction with SOM and since the peat soil has a very low mineral content, these features might be ascribed to strong solute adsorption at low $C_e/S_w$ values on a small amount of HSACM (e.g., charcoal-like materials) in soil (Chiou, 1995), on which water exhibits a weak competitive adsorption (see Chapter 6, section 6.5, on the adsorption of water on activated carbon). At moderate to high $C_e/S_w$ values, this adsorption is largely saturated and the partition in SOM predominates to make the isotherm essentially linear.

The HSACM hypothesis is consistent with the characteristics of solute adsorption on activated carbon. For adsorbates with a density of about 1 g/mL on a typical activated carbon, 1 m$^2$/g of carbon surface area corresponds to about 0.25 mg/g for the adsorbate monolayer capacity, the saturation capacity is about twice as high. Adsorption on activated carbon rises sharply at low $C_e/S_w$ values (see the related discussion in Chapter 6). At $C_e/S_w$ = 0.01 to 0.02, the adsorbed capacity is about 40 to 50% of the saturation capacity (Manes and Hofer, 1969; Chiou and Manes, 1974). From this point up to $C_e/S_w$ = 1, adsorption approaches full saturation more gradually. With a small quantity of assumed HSACM and a significant amount of SOM in soil, the isotherms for nonpolar solutes at $C_e/S_w$ > 0.01 to 0.02 would therefore become relatively linear as the (linear) partition into SOM outweighs the adsorption on HSACM. On peat soil, the nonlinear EDB and TCE capacities (0.15 to 0.18 mg/g) are consistent with the HSACM hypothesis, if most of the soil surface area (1.4 m$^2$/g) comes from a small amount of HSACM. For Woodburn soil, the surface area is much higher and the nonlinear capacity for nonpolar EDB is much smaller. The large surface area results presumably from mineral surfaces, on which the strong interaction with water minimizes the solute adsorption (Chiou and Shoup, 1985; Chiou et al., 1985). The lower nonlinear capacity of EDB on Woodburn soil may be attributed to a trace amount of HSACM in the soil. Since adsorption on activated carbon occurs primarily by London forces as discussed in Chapter 4, and the same is expected for HSACM, the solute polarity would not be relevant for competitive adsorption. This is corroborated by the nonspecific suppression of the EDB sorption by both polar and nonpolar co-solutes on peat.

The nonlinear sorption characteristics of nonpolar solutes on peat are internally consistent with the $N_2$ adsorption data on this sample. The $N_2$ adsorption exhibits a similar sharp rise and a downward concavity at low relative pressures ($P/P°$) ($\leq$0.02) with a monolayer capacity of 0.36 mg/g (Chiou et al., 1993;

see Figure 6.6). This behavior suggests that a small amount of high-affinity adsorption sites is present, as pictured by the HSACM postulate. The HSACM concept is thus in accord with the widespread natural occurrence of charcoal-like materials that are commonly produced by biomass burning. The ubiquity of low levels of charcoal-like materials in sediments has been well documented (Smith et al., 1973; Griffin and Goldberg, 1983; Masiello and Druffel, 1998). Many soils may thus be naturally blended with small amounts of charcoal-like substances. Although the HSACM postulate does not rule out the possibility of a small quantity of high-affinity adsorption sites in a specific fraction of the SOM, it seems unlikely that the SOM contains such unique sites. In this respect, the HSACM is better viewed as an extraneous substance in soil rather than as a portion of SOM, even though it is often counted as part of the soil organic carbon in SOM analysis by high-temperature combustion methods. More work is needed to establish clearly whether a direct relation exists between nonlinear capacity and SOM content.

We now consider the sorption data of polar solutes on peat soil and Woodburn soil. For polar solutes, the greater nonlinear capacity requires an additional nonlinear model. For DCP on peat, for instance, the nonlinear capacity observed (25 mg/g) greatly exceeds the adsorption capacity accountable by the small surface area of the soil. The nonlinear capacities for MON and DUN on peat are smaller but still higher than can be reconciled with the soil surface area. These findings imply that the relatively large nonlinear sorption of polar solutes at low $C_e/S_w$ values is strongly related to solute polarity and occurs within the interior network of SOM. The data are compatible with the specific-interaction (SI) model of Spurlock and Biggar (1994), which captures the nonlinear features of polar pesticides at low (relative) concentrations. The model postulates that the specific interaction of polar solutes with highly active SOM sites approaches saturation at a much lower concentration than does the concurrent partition to SOM, and therefore the isotherm is nonlinear at low (relative) concentrations.

Since the assumed specific-interaction (SI) model involves the polar groups of solute and SOM, it makes sense that the magnitude of nonlinear sorption and the solute competitive power depend on the solute polarity, as manifested by the experimental data. In this respect, a nonpolar co-solute is unable to suppress the large nonlinear uptake of a polar solute. The small reduction of DUN uptake on peat by co-solute EDB (with $C_e/S_w = 0.30$), as shown in Figure 7.18, and that of atrazine uptake at low concentrations on soil by co-solute TCE (Xing et al., 1996) may be attributed to adsorptive competition on a small amount of HSACM in soil. The finding that the nonlinear sorption capacities of a polar solute (e.g., DCP) on peat and Woodburn soil (see Table 7.11) correlate largely with respective SOM contents is in keeping with assumed specific interactions of polar solutes with the interior active sites of SOM. Chiou and Kile (1998) also found that the nonlinear sorption of DCP on peat does not disappear by lowering the solution pH to 2.0, suggesting that the active sites in SOM are not confined to ionizable groups. Overall, the combination of the specific-interaction model of Spurlock and Biggar (1994) for polar solutes and the HSACM model for nonpolar solutes reconcile the outstanding features of the nonlinear and competitive sorption of the solutes with certain soil samples.

Other substantial supporting evidence for the HSACM model emerged soon after the work of Chiou and Kile (1998). Kleineidam et al. (1999) used a petrographic method to identify charcoal-like particles in natural solids (limestone) and showed that the low-concentration sorption isotherm of a contaminant (phenanthrene) exhibits the greatest nonlinear sorption on solids that contain the highest amount of charcoal particles and reworked vitrinite. The mineral matter of these geosorbents contributed little to phenanthrene uptake, as expected, because of the adsorptive suppression by water. Karapanagioti et al. (2000) used the same method to identify similar charcoal-like particles in alluvial sediments. Karapanagioti et al. (2001) also achieved an independent confirmation of the charcoal-like substances in the (Florida) peat and Woodburn soil samples employed earlier by Chiou and Kile (1998). A photograph showing the charcoal-like particles in the peat sample is presented in Figure 7.22; the sample contains approximately 8% charcoal-like substance, as determined by manual counting of the opaque particles. This sug-

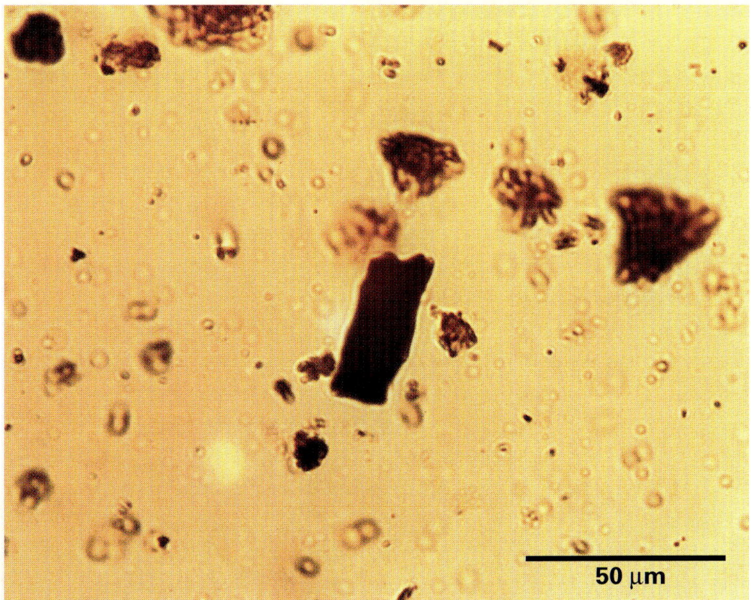


**Figure 7.22**  Photomicrograph exhibiting the presence of sharp-edged opaque particles in the untreated peat soil characteristic of charcoal-like material. The photomicrograph was produced with a transmitted white light (field of view 0.32 mm wide) to a thin sample layer on a strew slide. The amount of opaque particles in the sample by hand counting is about 8% by weight. (Courtesy of Dr. David Sabatini and Dr. Jeff Childs, University of Oklahoma, Norman.)

gests that the surface area of the charcoal-like substance in the peat sample is relatively small (estimated to be about $15\,m^2/g$). Only a trace amount of charcoal-like content was found in the Woodburn soil, as expected.

Additional support for the HSACM model was given by Xia and Ball (1999), who applied the Polanyi adsorption potential theory to evaluate the nonlinear sorption components of nine nonpolar single solutes (benzene, chlorinated benzenes, and PAHs) on an aquitard solid. A plot of the adsorbed solute volume versus the solute adsorption potential per unit molar volume (see Chapter 4) for all liquid solutes yielded essentially a single curve and thus about the same limiting adsorption volume (see the data with activated carbon in Chapter 6). A similar plot for solid solutes displayed reduced adsorption capacities and limiting volumes, due supposedly to a less efficient packing of solid compounds within the porous adsorbent structure. These unique characteristics are typical of the organic-solute adsorption onto activated carbon (Manes, 1998). In a subsequent binary-solute study, Xia and Ball (2000) showed that suppression of the nonlinear sorption of a solute by other co-solutes is similar to that described earlier.

The complete isolation of small amounts of charcoal-like particles (HSACM) from ordinary soils and sediments is extremely difficult because of the small HSACM mass and the lack of effective isolation methods. However, it is possible to remove these particles by a density-fractionated technique, since the charcoal particle, which resembles activated carbon, should have a higher density than SOM, the latter estimated to be about $1.3\,g/mL$ (Chiou et al., 1983). On this basis, Chiou et al. (2000) prepared relatively pure (i.e., HSACM-free) humic acids (HAs) and a presumably HSACM-enriched humin (HM) from the (Florida) peat sample used in their sorption studies by a density-fractionated method. A similar HA was extracted from a Michigan muck. The raw HA was first extracted by base, followed by centrifugation to precipitate and remove fine charcoal-like particles. The solution was neutralized with acid to precipitate HA and the suspensions were removed. The HA fraction was then redissolved with base and centrifuged to remove the solids precipitated. By repeating this procedure several times, a relatively pure HA was obtained. The base-unextracted portion of the peat, the HM fraction, should thus be enriched with charcoal-like particles. The idea here is that if the HSACM is the source of nonlinear sorption for nonpolar solutes on the whole peat, the sorption of the same solutes on the purified HA is expected to be essentially linear. Conversely, the sorption isotherm with HM should exhibit a greater nonlinear effect than that for the whole peat sample. The organic carbon contents and the surface areas of the prepared HAs and HM are shown in Table 7.12.

The sorption isotherms of EDB from water on the two HA sorbents are depicted in Figure 7.23, and the isotherms for EDB on whole peat and peat-derived HM are shown in Figure 7.24. The EDB isotherms on both HAs exhibit virtually no visible nonlinearity from low to high $C_e/S_w$; the high linearity indicates that solute partition into SOM is the governing sorption

**TABLE 7.12. BET-$N_2$ Surface Areas (SA) and Percents of Organic Carbon (OC) of Peat and Fractionated Organic Matters**

| Sample | OC (%) | SA (m²/g) |
|---|---|---|
| Peat | 49.3 | 1.4 |
| Peat HA | 47.3 | 0.16 |
| Muck HA | 46.3 | 0.17 |
| HM | 51.4 | 4.5 |

*Source*: Data from Chiou et al. (2000).

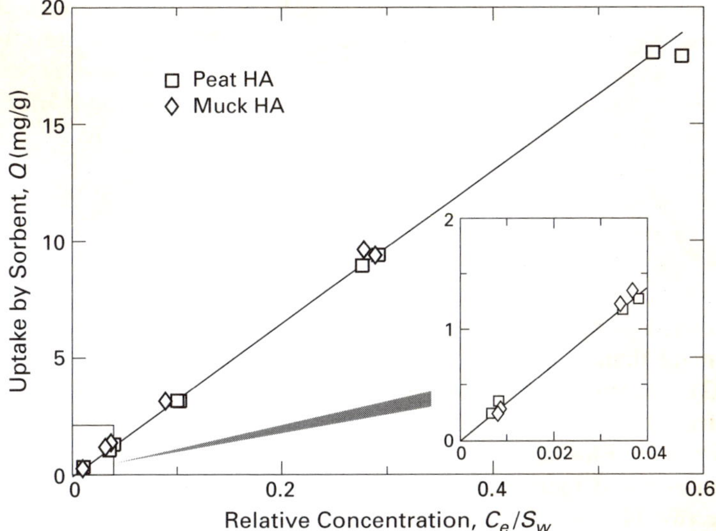


**Figure 7.23**  Sorption of EDB from water on peat HA and muck HA. [Data from Chiou et al. (2000). Reproduced with permission.]

mechanism. In contrast, the EDB isotherms on whole peat and (peat-derived) HM display a noticeable nonlinearity at low $C_e/S_w$ values. The apparent nonlinear capacity ($Q_{ans}$) of EDB with HM is much greater than with whole peat; the extent of EDB sorption nonlinearity at low $C_e/S_w$, as characterized by the Freundlich $n$ exponent in Table 7.13, is also somewhat higher with HM than with whole peat. The increased linearity at moderate to high $C_e/S_w$ reflects the dominance of solute uptake by partition into SOM. The partition capacities of EDB on the two HAs are practically equal. The uptake capacity of EDB on whole peat at all but low $C_e/S_w$ values is also about the same as that on peat HA.

The sorption isotherms of moderately polar DUN on the same four sorbents are illustrated in Figure 7.25, in which the sorption to HM exhibits

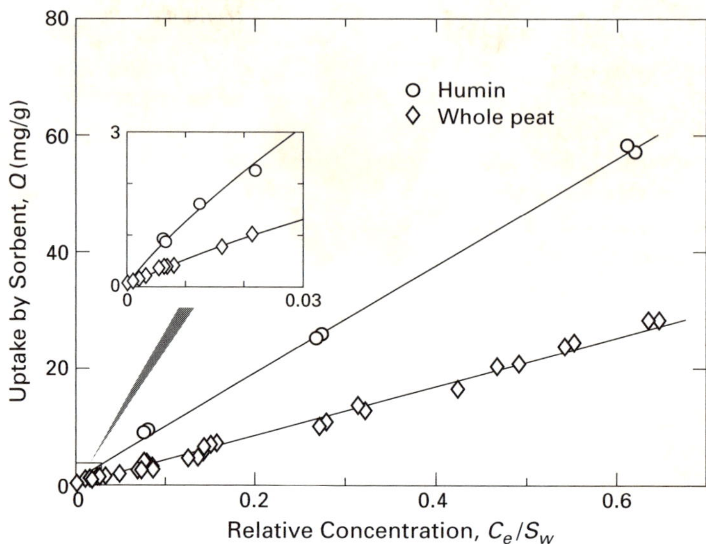


**Figure 7.24** Sorption of EDB from water on whole peat and peat HM. [Data from Chiou et al. (2000). Reproduced with permission.]

**TABLE 7.13. Apparent Nonlinear Saturation Capacities ($Q_{ans}$), Nonlinear Saturation Points [$(C_e/S_w)_{ans}$], and Freundlich Exponents ($n$) for the Sorption of Selected Solutes on Sorbents**

| Solute/Sorbent | $Q_{ans}$ (mg/g) | $(C_e/S_w)_{ans}$ | $n_1{}^a$ | $n_2{}^a$ |
|---|---|---|---|---|
| EDB/peat | 0.18 | 0.010 | 0.91 | 0.99 |
| EDB/peat HA | ~0 | ~0 | 1.00 | 1.00 |
| EDB/muck HA | ~0 | ~0 | 1.00 | 1.00 |
| EDB/HM | 0.50 | 0.018 | 0.88 | 1.00 |
| DUN/peat | 0.60 | 0.10 | 0.63 | 1.00 |
| DUN/peat HA | 0.40 | 0.12 | 0.72 | 0.99 |
| DUN/muck HA | 0.50 | 0.10 | 0.72 | 0.99 |
| DUN/HM | 2.0 | 0.17 | 0.51 | 0.94 |
| DCP/peat | 25 | 0.12 | 0.56 | 0.98 |
| DCP/peat HA | 16 | 0.10 | 0.61 | 1.00 |
| DCP/muck HA | 16 | 0.12 | 0.61 | 1.00 |
| DCP/HM | 45 | 0.13 | 0.41 | 1.00 |

*Source*: Data from Chiou et al. (2000).

[a] $n_1$ is the best-fit $n$ exponent at $P \geq 0.05$ in a Freundlich plot, $Q = K_f(C_e/S_w)^n$, for the sorption data within the stated nonlinear range [i.e., at $C_e/S_w \leq (C_e/S_w)_{ans}$]; $n_2$ is the corresponding best-fit $n$ value at $P \geq 0.05$ for the sorption data at $C_e/S_w \geq (C_e/S_w)_{ans}$.

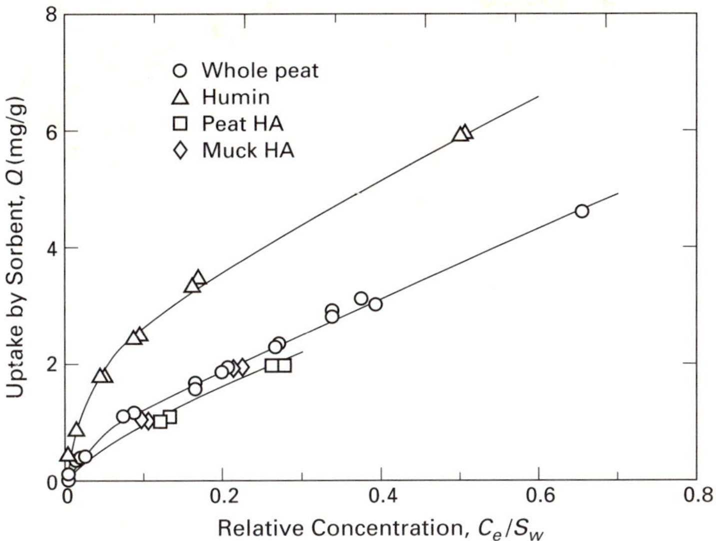


**Figure 7.25**   Sorption of DUN from water on whole peat, peat HA, muck HA, and peat HM. [Data from Chiou et al. (2000). Reproduced with permission.]

the highest uptake and nonlinear capacity relative to the sorption to other sorbents. The DUN uptake capacities on the two HAs are comparable in magnitude, being somewhat lower than that on whole peat. The sorption isotherms of more polar DCP on the four sorbents are shown in Figure 7.26. Again, the sorption to HM exhibits the highest capacity and nonlinear effect, and the isotherms with whole peat, peat HA, and muck HA are relatively comparable. The apparent nonlinear capacities and the approximate apparent nonlinear saturation points, $(C_e/S_w)_{ans}$ for the studied solutes on all sorbent samples are given in Table 7.13. As noted graphically in Figures 7.23 through 7.26, and substantiated by the Freundlich $n$ exponents in Table 7.13, the sorption isotherms at $C_e/S_w > (C_e/S_w)_{ans}$ are essentially linear.

Let us consider first the sorption data of EDB. As shown in Table 7.12, the peat HA exhibits a much lower BET-$N_2$ surface area than the original peat ($0.16\,m^2/g$ versus $1.4\,m^2/g$), due presumably to the removal of HSACM from the peat when the HA was prepared by density fractionation. In the study with peat previously, the observed sorption nonlinearity for nonpolar solutes (e.g., EDB) at low $C_e/S_w$ values was ascribed to adsorption on a small amount of HSACM, concurrent with solute partition into SOM. The absence of nonlinear sorption at low $C_e/S_w$ values with density-fractionated HA confirms this hypothesis. Further, the petrographic analysis revealed no charcoal-like particles in the HA sample. The base-insoluble peat HM fraction contains presumably a greater proportion of HSACM per unit mass than the original peat, as manifested by the approximately threefold-higher BET-$N_2$ surface area of the HM ($4.5\,m^2/g$ versus $1.4\,m^2/g$). The comparable organic-carbon contents

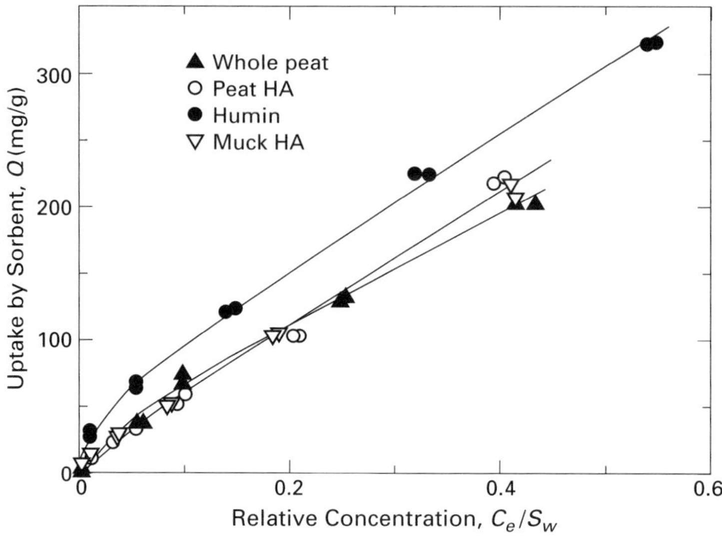


**Figure 7.26**  Sorption of DCP from water on whole peat, peat HA, muck HA, and peat HM. [Data from Chiou et al. (2000). Reproduced with permission.]

with HM and whole peat (Table 7.12) suggest that the increased surface area of HM over that of the whole peat is not related to the mineral or ash content of the sample. As noted in Table 7.13, the nonlinear capacity of EDB on HM (about 0.50 mg/g) is approximately three times the capacity on the peat (0.18 mg/g), in support of the HSACM hypothesis. In all cases, the BET-$N_2$ surface areas of the sorbents are sufficient to account for the EDB nonlinear capacities. The much higher EDB linear uptake on HM than on other sorbents at moderate to high $C_e/S_w$ suggests that the organic medium of HM is less polar than that of the peat or peat-derived HA. The partition of nonpolar solutes to organic media is sensitive to the polarity of the medium, as discussed in Chapter 5 and earlier in this chapter.

The nonlinear capacity of DUN on the peat, as noted before, is about 0.60 mg/g, which results presumably from both its specific interaction with active SOM groups and its adsorption on a small quantity of HSACM. On this basis, the nonlinear capacity of DUN on peat HA is expected to be only slightly lower than that on the peat. The value observed (about 0.40 mg/g) is consistent with the estimation. Conversely, the nonlinear capacity of DUN on HM is expected to be greater than that on the peat because HM has a much higher surface area (and presumably more HSACM), if the active sites in peat and HM are not vastly different. The nonlinear capacity observed for DUN on HM (about 2.0 mg/g) agrees semiquantitatively with this expectation; the surface area of HM (or the HSACM contribution) is not sufficient to account for the nonlinear capacity of DUN on HM. The slope of the upper DUN–HM isotherm, which reflects the DUN partition effect is similar to that of DUN on peat (Figure 7.25). This indicates that the reduced HM polarity has less impact on the partition uptake of a moderately polar solute (DUN) than that of a nonpolar solute (EDB).

The sorption of DCP on peat, HA, and HM exhibits trends similar to those observed for the DUN sorption. For relatively polar DCP, the nonlinear effect probably results predominantly from its specific interaction with active SOM groups, as mentioned before. Here the nonlinear capacity of DCP is about 25 mg/g with the peat, 16 mg/g with HA, and 45 mg/g with HM (Table 7.13). The disparate nonlinear capacities of DCP on these sorbents appear to reflect some changes in either the affinity or the abundance of the active sites in HA and HM materials when prepared from the peat soil. Nonetheless, the nonlinear capacities for DCP on all sorbents are far too high to be reconciled with the measured BET-$N_2$ surface areas alone. Thus the solute–SOM specific interactions appear to predominate over the much weaker solute adsorption on a small amount of HSACM for the nonlinear sorption of relatively polar solutes. The presumed different sources for the nonlinear effects of polar and nonpolar solutes is further illustrated by their different nonlinear-sorption ranges [i.e., the observed $(C_e/S_w)_{ans}$ values are considerably greater for polar DCP and DUN than for nonpolar EDB] (Table 7.13). As with the DUN sorption, the slope of the upper DCP–HM isotherm is comparable with that of DCP on the peat (Figure 7.26), suggesting that the contents of relatively polar sorbents do not significantly affect the partition uptake of a relatively polar solute.

The nonlinear characteristics of EDB, DUN, and DCP on HA, HM, and peat samples are inherently consistent with the expectations of the HSACM-SI model, in which the nonlinear capacities observed for nonpolar EDB are well related to the BET-$N_2$ surface areas (or to the presumed amounts of HSACM) of the sorbents, whereas those for polar DUN and DCP call for additional specific interactions with SOM. By contrast, the diversity of the data cannot readily be reconciled with the glassy–rubbery SOM model nor with the internal-hole model without much additional ad hoc hypothesis. The glassy–rubbery SOM model does not consider specifically the disparate nonlinear effects for polar and nonpolar solutes. The results with density-fractionated HA and base-insoluble HM would force this model to further hypothesize that the impact of the glassy component in SOM on sorption nonlinearity also depends on the solute polarity; that is, the glassy component (or its effect) exists only in HM but not in HA for nonpolar solutes, whereas it occurs in both HA and HM for polar solutes. Similarly, the internal-hole model would have to further assume that the compound-specific internal holes accessible to nonpolar solutes are located only in HM but those accessible to polar solutes exist in both HA and HM. It is difficult, however, to rationalize the inconsistency on the origin and effect of the glassy component or internal holes in SOM, since both polar and nonpolar solutes should have equal access to the presumed glassy SOM or internal holes.

In light of the mutual consistency of the sorption, surface area, and petrographic data, the existence of small amounts of HSACM in soils or natural solids is at least one of the primary causes of the nonlinear effect of nonpolar compounds at low $C_e/S_w$ values. Although the available information does not exclude the possibility of other nonlinear sources for nonpolar compounds (e.g., other materials in natural solids with carbon-like adsorptive properties), the HSACM model for nonpolar solutes is currently supported by more experimental evidence. For the nonlinear effect of polar solutes, such as that shown for DCP on HA, where the nonlinear capacity greatly surpasses the amount attributable to surface adsorption, the solute–SOM specific interaction, as suggested by Spurlock and Biggar (1994), seems to be a reasonable explanation of the result, although the mode of specific interaction remains to be substantiated. At this point, greater effort is needed to explore all possible nonlinear sources and causes for nonionic compounds with soils and natural solids.

We now shift our attention to another potential source of sorption nonlinearity for polar solutes, which has to do with their adsorption on certain minerals. We recall from our deliberation in Chapter 6 that a montmorillonite (SAz-1) exchanged with $K^+$ exhibits a lower water uptake than benzene uptake at low $P/P^\circ$ values, as a result of the weak cation hydration and the low affinity of water for siloxane surfaces. This suggests that very polar solutes may compete effectively against water for adsorption on such clay minerals in soil. Although it would be very unusual for ordinary soils to have a high content of montmorillonite with predominantly weak hydrating cations (e.g., $K^+$ and $Cs^+$), one cannot rule out this potential adsorptive effect for very polar contaminants at low $C_e/S_w$ values with such soils or sediments that exhibit unusually high montmorillonite contents.

Laird et al. (1992) studied atrazine uptake from water solution by 13 individual Ca-saturated smectitic clays, where the cation exchange capacity (CEC) varies from 79 to 134 cmol/kg. They found that the net atrazine uptake ranged widely from 0 to nearly 100%. The extent of atrazine uptake decreased generally with increasing clay surface charge density or largely with increasing CEC of the clay, since the siloxane plane areas of all smectites are relatively comparable. Haderlein and Schwarzenbach (1993) and Haderlein et al. (1996) further investigated the clay uptake of polar contaminants from water. The latter study compared the sorption data of a series of nitroaromatic compounds (NACs) (e.g., nitrobenzenes, nitrotoluences, nitroanilines, and nitrophenols) on relatively pure kaolinite (Cornwall, UK), illite (Tokay, Hungary), and montmorillonite (SAz-1) in different cationic forms. On a given clay type, the uptake of NACs was relatively high when the clay was exchanged with $K^+$, $Cs^+$, or $NH_4^+$ ion, and the isotherms are notably nonlinear, characteristic of adsorption; the uptake became negligible when the clay was exchanged with strongly hydrating $Na^+$, $Ca^{2+}$, $Mg^{2+}$, and $Al^{3+}$ ions. The discrepant results were attributed to the powerful hydration of the latter set of cations on clay's siloxane surfaces, which reduces their water-unoccupied surfaces accessible to polar solutes (Weissmahr et al., 1997). On this basis, the high uptake of atrazine by low-CEC Ca-saturated smectites, as observed by Laird et al. (1992), may be attributed to the abundance of hydration-unaffected siloxane surfaces available for atrazine adsorption.

Among the K-saturated clays, the adsorption powers observed follow the order montmorillonite > illite > kaolinite, which reflects largely the amounts of siloxane surfaces present in these clays. Here illite is less effective than montmorillonite, presumably because the former has a much higher surface charge (or charge density), making the surface more hydrophilic. An example of the nonlinear (adsorption) isotherm of 1,3,5-trinitrobenzene (TNB) on K-saturated montmorillonite, illite, and kaolinite is presented in Figure 7.27. Among the NACs studied, the planar solutes with several electron-withdrawing substituents exhibit the highest adsorption. The adsorptive uptake is considered to take place by certain electron donor–acceptor inter-actions between clay's siloxane oxygens and NAC's oxygens (Haderlein et al., 1996); the spectroscopic data show no evidence for significant H-bonding between siloxane oxygens and NAC's hydrogens (Weissmahr et al., 1997). Although the polar-solute adsorption on certain clay minerals is evidently sig-nificant, the significance of this adsorption with ordinary soils and the related sorption nonlinearity have not hitherto been well documented. This is due at least in part to the fact that most soils and sediments rarely have sufficient amounts of these clay minerals to make the effect clearly dominant.

The observed nonlinear sorption for single solutes at low $C_e/S_w$ and its sup-pression by other coexisting solutes have important environmental implica-

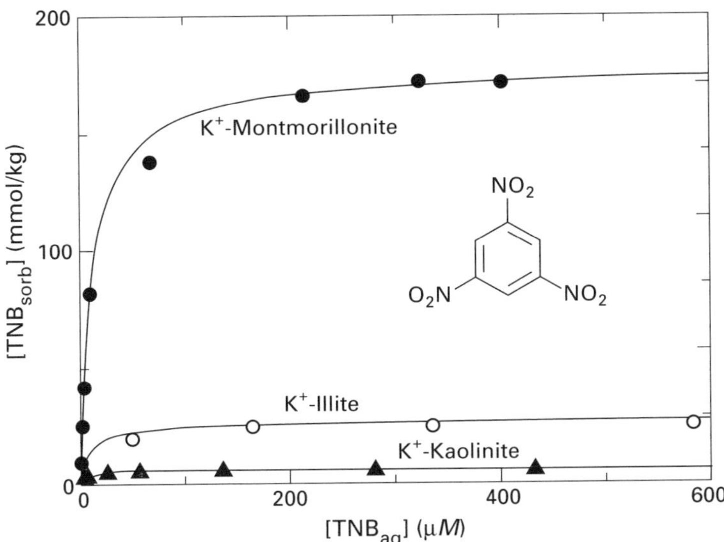


**Figure 7.27**   Adsorption of 1,3,5-trinitrobenzene by homoionic K⁺-exchanged clays in aqueous suspensions. [Data from Haderlein et al. (1996). Reproduced with permission.]

tions. Whereas the single-solute sorption data apply to most laboratory studies, the behavior of a contaminant in natural systems at low $C_e/S_w$ will depend strongly on system conditions. The principal deciding factors for nonpolar contaminants are the amount of HSACM relative to SOM, the contaminant solubility, and the number and amount of other coexisting contaminants. If the HSACM content is small, the sorption coefficients of most nonpolar solutes should be relatively constant over the typical environmental concentration range (i.e., at subparts per million or higher), as the linear partition to SOM would outweigh the nonlinear adsorption on HSACM. This is expected to be especially the case for all nonpolar contaminants with very low $S_w$ values (e.g., chlorinated hydrocarbons, PCBs, and PAHs) because at environmentally significant concentrations they will be placed at relatively high $C_e/S_w$ values (i.e., they will easily fall into the partition-dominated range). In the evaluation of the sorption coefficient ($K_{om}$ or $K_{oc}$) of a contaminant at low $C_e/S_w$ in natural systems, it is also imperative to keep in mind the effect of mutual solute competition on the sorption behavior of individual solutes. The possible outcome of this effect is outlined as follows.

If the system contains multiple solutes and the HSACM content is low, the nonlinear sorption would be most significant when there are only few co-existing solute species and/or if the solutes are at extremely low $C_e/S_w$ values. However, if the system contains a dominant component (of high $C_e/S_w$ value), the sorption of all nonpolar solutes, irrespective of the number of solute species, at low $C_e/S_w$ values should become relatively linear because of adsorptive suppression by the dominant species on HSACM. In this case, the sorption coefficients of all contaminants, including the dominant species, should conform largely to their respective linear partition coefficients with SOM. The same should apply for nonpolar solutes if the system contains a large number of solute species, each at about the same $C_e/S_w$ value. For polar solutes in multiple-solute systems, where the major cause for sorption nonlinearity is the SOM content, or the specific clay content on rare occasions, the sorption coefficients could be subject to greater variation with solute concentration, since suppression of the nonlinear sorption in this case requires higher co-solute $C_e/S_w$ values and more specific co-solute polarity. Nonetheless, if a powerful polar contaminant (e.g., phenol) dominates, the nonlinear sorption effects of the less-polar contaminants should again be greatly diminished. The observed suppression of the nonlinear sorption of a given polar solute by polar co-solutes from other classes reduces the complexity of the multiple-solute system.

### 7.3.8  Influence of Dissolved and Suspended Natural Organic Matter

For many highly water-insoluble compounds, the term $\log (\gamma_w/\gamma_w^*)$ often has a more significant impact on the $\log K_{om}$ value in soil–water systems [Eq. (7.11)] than, say, on $\log K_{ow}$ values in octanol–water systems [Eq. (5.1)]. In soil–water mixtures, the term $\gamma_w/\gamma_w^*$ expresses the enhancement of solute concentration or solubility by a given level of dissolved and/or suspended (colloidal) organic matter in water derived from soil/sediment or other sources (i.e., $\gamma_w/\gamma_w^* = C_e^*/C_e = S_w^*/S_w$), as described later. Even when present in trace quantities, dissolved or suspended high-molecular-weight humic material is known to be able to significantly enhance the water solubility of otherwise extremely insoluble organic compounds. Wershaw et al. (1969) observed that the apparent solubility of DDT in 0.5% soil sodium–humate solution is more than 200 times greater than in pure water (5.5 µg/L). Carter and Suffet (1982) found that the added sediment humic acid in water solution significantly enhances the concentration of DDT over that in pure water. This solubility enhancement effect is attributed to a partition interaction of solutes with colloidal organic matter (Gschwend and Wu, 1985) or to a partitionlike interaction with the microscopic organic environment of the dissolved organic matter (Chiou et al., 1986). A useful relationship between apparent solute concentration (or solubility) in water with a given level of dissolved and/or suspended organic matter ($C_e^*$) and solute concentration (or solubility) in pure water ($C_e$) has been established by Chiou et al. (1986), which gives

$$C_e^* = C_e + XK_{dom}C_e = C_e(1 + XK_{dom}) \tag{7.23}$$

or

$$S_w^* = S_w(1 + XK_{dom}) \tag{7.24}$$

where $X$ is the total mass of dissolved and suspended organic matter per unit weight (or volume) of water (usually in a dimensionless unit), which, for simplicity, is operationally termed the concentration of dissolved organic matter (DOM); $K_{dom}$ is the enhancement (or partition) coefficient of the solute between DOM and water (dimensionless), which is a function of the type of solute and the composition of DOM; $S_w^*$ is the apparent solute solubility in water with $X$ amount of DOM; and $S_w$ is the solute water solubility in pure water at the same temperature. If the term $X$ in Eqs. (7.23) and (7.24) is expressed alternatively in terms of the dissolved organic carbon (DOC) mass per unit weight of water, the term $K_{dom}$ in these equations is replaced by $K_{doc}$. Relating $C_e^*$ with solute concentration in soil ($Q$) at equilibrium, one obtains an apparent soil–water distribution coefficient for the solute ($K_d^*$) as

$$K_d^* = \frac{Q}{C_e(1 + XK_{dom})} = \frac{K_d}{1 + XK_{dom}} \tag{7.25}$$

where $K_d = f_{om}K_{om}$, as defined before. Normalization of $K_d^*$ to $f_{om}$ leads to

$$K_{om}^* = \frac{K_{om}}{1 + XK_{dom}} \tag{7.26}$$

in which $K_{oc}^*$ and $K_{oc}$ may be substituted for $K_{om}^*$ and $K_{om}$, respectively, if $K_d^*$ and $K_d$ are normalized to soil (or sediment) organic carbon content ($f_{oc}$).

The magnitude of $K_{dom}$ (or $K_{doc}$), from the standpoint of DOM, should depend on the polarity and molecular size of DOM. For truly dissolved organic matter (not suspended organic matter), the DOM molecules must in principle be sufficiently large and must possess a sizable intramolecular nonpolar moiety in order to promote partitionlike interaction with the solute. A dissolved low-molecular-weight organic matter (e.g., an organic solvent) is not expected to be equally effective for promoting this type of interaction because of its size limitation. For the solutes, the important properties for a large solubility enhancement are very low water solubility and significant compatibility with the organic phase.

A critical analysis of the functional relationship of $K_{dom}$ (or $K_{doc}$) with solute $S_w$ and with the source and composition of DOM in truly dissolved form was given by Chiou et al. (1986, 1987) using a series of solutes with vastly different $S_w$ values and of DOMs with varied compositions and structures. The solutes used and their $S_w$ values at room temperature are: $p,p'$-DDT, $S_w = 5.5\,\mu g/L$; $2,4,5,2',5'$-PCB, $S_w = 10\,\mu g/L$; $2,4,4'$-PCB, $S_w = 115\,\mu g/L$; lindane, $S_w = 7.8\,mg/L$; and 1,2,3-trichlorobenzene (TCB), $S_w = 16.3\,mg/L$. The various DOMs used are: Sanhedrin soil humic acid (SSHA), Sanhedrin soil fulvic acid (SSFA), Suwannee River humic acid (SRHA), and Suwannee River fulvic acid (SRFA), extracted from soil and stream, and human-made phenylethanoic acid and polyacrylic acid. The high purities of extracted natural DOM samples provide sufficiently accurate elemental data (shown in Table 7.14) as indices of their polarities. Plots of the $S_w^*$ values against DOM concentrations for selected solutes with SSHA as DOM according to Eq. (7.24) are shown in Figure 7.28. The corresponding plots with SRHA, phenylethanoic acid, and polyacrylic acid as DOMs are presented in Figures 7.29, 7.30, and 7.31, respectively.

The results in Figures 7.28 through 7.31 indicate that SSHA is most effective in enhancing the solute water solubility, while SSFA, SRHA, SRFA, phenylethanoic acid, and polyacrylic acid exhibit less or no enhancing effects. For the solutes, the effect decreases progressively from $p,p'$-DDT to $2,4,5,2',5'$-PCB and to $2,4,4'$-PCB with increasing $S_w$, and becomes negligible for relatively water-soluble lindane and TCB over the DOM concentrations studied (0 to 100 mg/L for SSHA, SSFA, SRHA, SRFA, and polyacrylic acid; 0 to 700 mg/L for phenylethanoic acid). The linear relation between $S_w^*$ and DOM concentration, the decrease in solubility enhancement with increasing $S_w$, and

**Figure 7.28** (*a*) Apparent water solubility of $p,p'$-DDT (●), $2,4,5,2'5'$-PCB (■,□), and $2,4,4'$-PCB (▲,△) as a function of SSHA concentration at 24 to 25°C. Solid symbols are for single solutes; open symbols for the two PCBs are for their binary mixtures. (*b*) Apparent water solubility of lindane and 1,2,3-trichlorobenzene as a function of SSHA concentration. [Data from Chiou et al. (1986).]

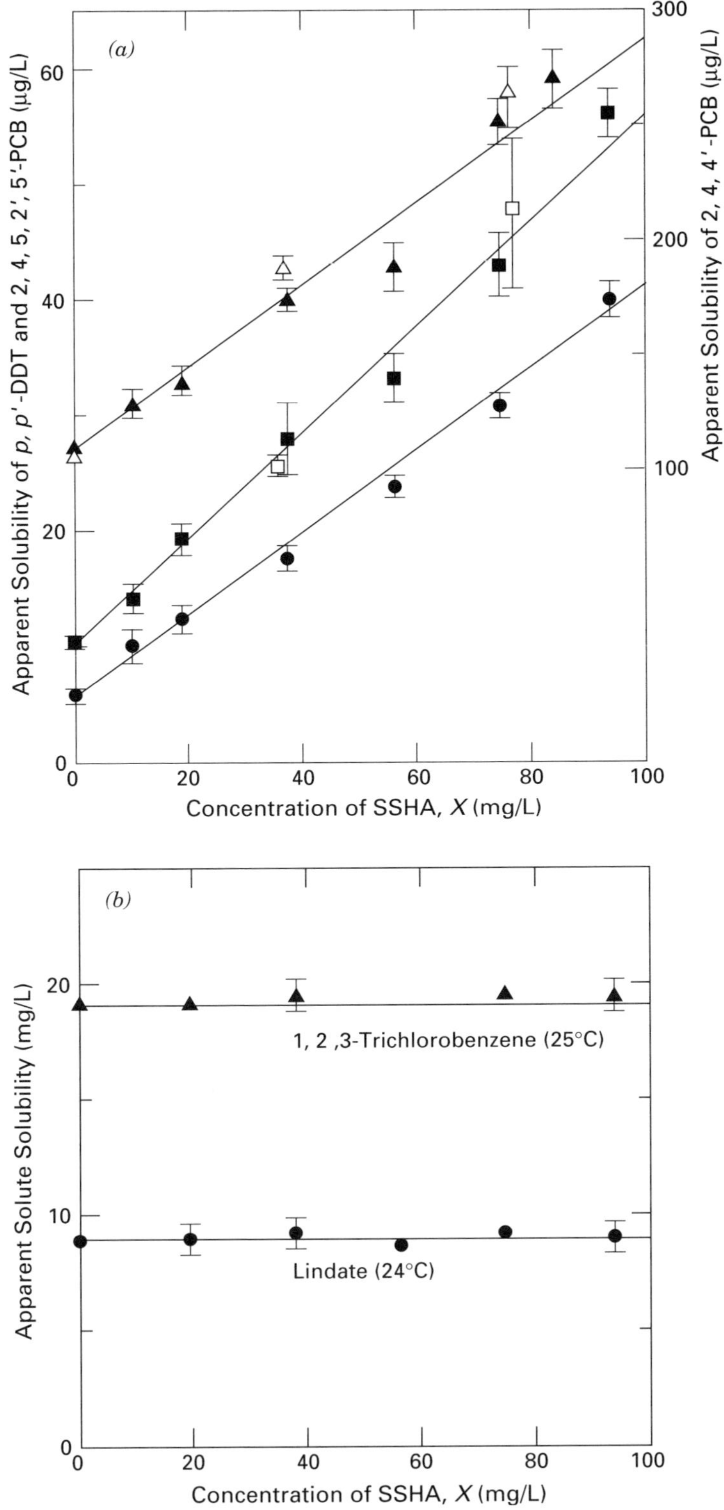

(a)
Apparent Solubility of p, p'-DDT and 2, 4, 5, 2', 5'-PCB (µg/L)
Apparent Solubility of 2, 4, 4'-PCB (µg/L)
Concentration of SSHA, X (mg/L)
(b)
Apparent Solute Solubility (mg/L)
1, 2 ,3-Trichlorobenzene (25°C)
Lindate (24°C)
Concentration of SSHA, X (mg/L)

**TABLE 7.14. Percent Elemental Contents of Soil and Stream Organic Matter Extracts on a Moisture-free, Ash-free Basis**

| Organic Matter Extract | C | H | O | N | S | P | Ash |
|---|---|---|---|---|---|---|---|
| Sanhedrin soil humic acid (SSHA) | 58.0 | 3.64 | 33.6 | 3.26 | 0.47 | 0.10 | 1.19 |
| Sanhedrin soil fulvic acid (SSFA) | 48.7 | 4.36 | 43.4 | 2.77 | 0.81 | 0.59 | 2.25 |
| Suwannee River humic acid (SRHA) | 54.2 | 4.14 | 39.0 | 1.21 | 0.82 | 0.01 | 3.18 |
| Suwannee River fulvic acid (SRFA) | 53.8 | 4.24 | 40.3 | 0.65 | 0.60 | 0.01 | 0.68 |

*Source*: Data from Chiou et al. (1986).

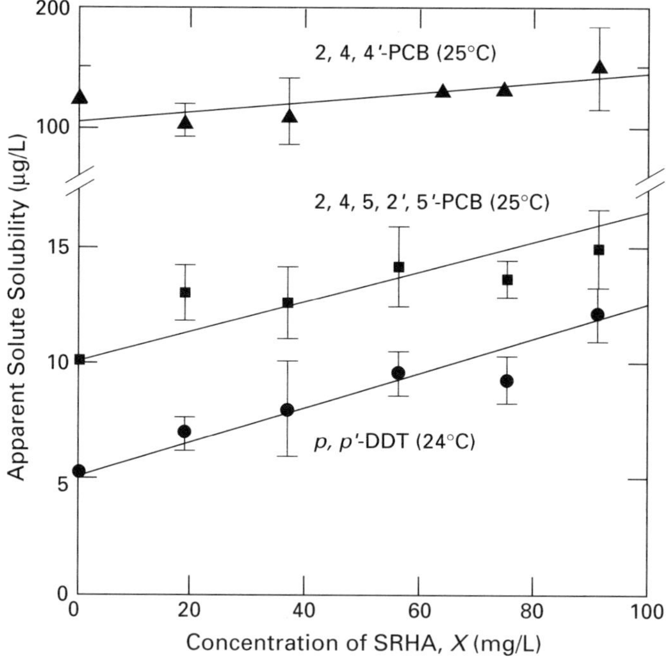


**Figure 7.29**    Apparent water solubility of *p,p'*-DDT, 2,4,5,2'5'-PCB, and 2,4,4'-PCB as a function of SRHA concentration. [Data from Chiou et al. (1986).]

the lack of interference between binary solutes (for 2,4,5,2',5'-PCB and 2,4,4'-PCB with SSHA) are in keeping with the postulated partitionlike interaction between the solute and DOM for the solubility enhancement observed. The different effects for a group of solutes with a given DOM are attributed to their different $K_{dom}$ (or $K_{doc}$) values. The calculated $\log K_{dom}$ and $\log K_{doc}$ values of the solutes with SSHA, SSFA, SRHA, and SRFA are tabulated in Table 7.15. Since, for a given DOM, the magnitude of $K_{dom}$ (or $K_{doc}$) should be

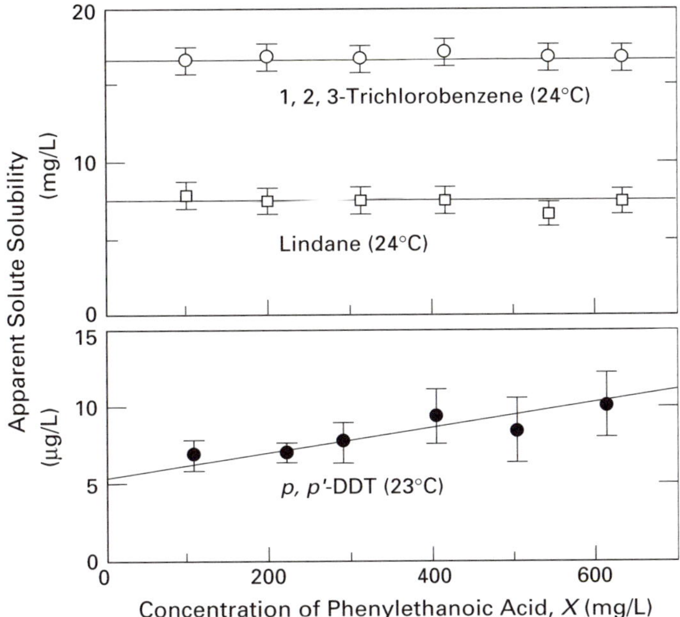

**Figure 7.30**  Apparent water solubility of $p,p'$-DDT, lindane, and 1,2,3-trichlorobenzene as a function of phenylethanoic acid concentration. [Data from Chiou et al. (1986).]


**TABLE 7.15.  Comparison of $\log K_{dom}$ and $\log K_{doc}$ Values of Selected Organic Solutes on SSHA, SSFA, SRHA, and SRFA with Respective Octanol–Water Partition Coefficients ($\log K_{ow}$) of the Solutes**

| Compound | $\log K_{ow}$ | $\log K_{dom}/\log K_{doc}$ | | | |
|---|---|---|---|---|---|
| | | SSHA | SSFA | SRHA | SRFA |
| $p,p'$-DDT | 6.36 | 4.82/5.06 | 4.27/4.58 | 4.12/4.39 | 4.13/4.40 |
| 2,4,5,2′,5′-PCB | 6.11 | 4.63/4.87 | 3.81/4.12 | 3.80/4.07 | 3.83/4.10 |
| 2,4,4′-PCB | 5.62 | 4.16/4.40 | 3.58/3.89 | 3.27/3.54 | 3.30/3.57 |
| 1,2,3-Trichlorobenzene[a] | 4.14 | ~2.8/3.0 | ~2.0/2.3 | ~1.7/2.0 | ~1.7/2.0 |
| Lindane[a] | 3.70 | ~2.5/2.7 | ~1.5/1.8 | ~1.2/1.5 | ~1.2/1.5 |

*Source*: Data from Chiou et al. (1986).

[a] Approximate $\log K_{dom}$ and $\log K_{doc}$ values are obtained from linear extrapolations of $\log K_{dom}$ and $\log K_{doc}$ against $\log K_{ow}$ of $p,p'$-DDT, 2,4,5,2′,5′-PCB, and 2,4,4′-PCB.


inversely related to $S_w$ of the solute, the $\log K_{dom}$ (or $\log K_{doc}$) values are essentially linear with respective $\log K_{ow}$ values.

The finding that small DOMs (e.g., phenylethanoic acid) are unable to produce as strong a solubility enhancement effect as humic and fulvic acids agrees with the required DOM size and polarity in such a partitionlike inter

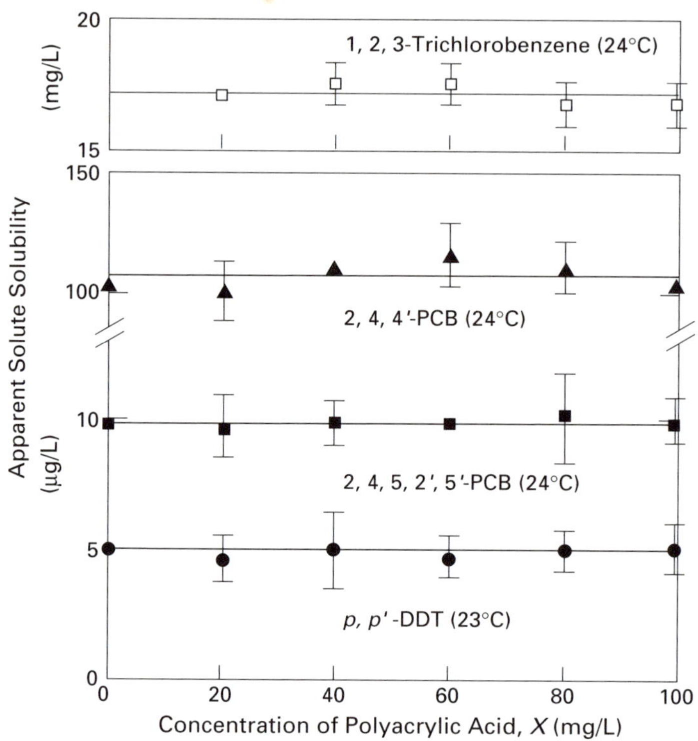


**Figure 7.31**  Apparent water solubility of $p,p'$-DDT, 2,4,5,2'5'-PCB, 2,4,4'-PCB, and 1,2,3-trichlorobenzene as a function of polyacrylic acid (MW = 2000) concentration. [Data from Chiou et al. (1986).]

action. Note that if the size of DOM were not important, phenylethanoic acid should exhibit a greater enhancement effect than other DOMs because of its less-polar molecular composition (70.6% C, 5.9% H, and 23.5% O). In situations where the solubility enhancement is caused by a specific interaction, one would expect such an effect to be more related to the acidity or polarity of the DOM, and therefore phenylethanoic acid should be a strong solubility enhancer.

The relatively large solubility enhancement for DDT and PCBs with SSHA suggests that SSHA must have a substantially higher molecular weight than those of the solutes, which is supported by the proposed molecular weights of 2000 to 20,000 for soil humic acids (Schnitzer and Kahn, 1972). The partition-like interaction (i.e., by van der Waals forces), as postulated for a relatively insoluble solute with high-molecular-weight DOMs, is mechanistically similar to solute solubilization in surfactant micelles, where a microscopic organic phase is formed through the aggregation of surfactant monomers. The much smaller enhancement effects with soil fulvic acid (SSFA), aquatic humic acid (SRHA), and aquatic fulvic acid (SRFA) may be ascribed to their smaller molecular sizes and, perhaps to a greater extent, to their greater polarities (which enhance their interactions with water) as measured by (O + N)/C. The suggested molecular weights are 1000 to 5000 for soil fulvic acids (Buffle et al., 1978; Underdown et al., 1981), 1000 to 10,000 for aquatic humic acids (Thurman et al., 1982), and 500 to 1000 for aquatic fulvic acids (Aiken and Malcolm, 1987).

The comparable effects as found with SRHA, SRFA, and SSFA, despite some differences in their molecular sizes, suggest that the DOM polarity has a greater impact on solubility enhancement in this situation. Further, poly-acrylic acid (with MW = 2000 and 90,000) as a linear polyelectrolyte shows no enhancement effect at all, notwithstanding that its C (50%), H (5.6%), and O (44.4%) contents are comparable with those of SSFA. Here the inability of polyacrylic acid to enhance solute solubility is attributed to the frequent and orderly attachment of hydrophilic carboxyl groups to the carbon chain and to an extended chain structure, which prohibit the formation of a sizable intramolecular nonpolar environment. Thus, although the molecular size of DOM is essential, this property is not the sole deciding factor in solubility enhancement. To be a strong solubility enhancer, the DOM must possess both a favorable size of nonpolar moiety and a sufficiently large molecular weight.

By the results in Table 7.15, the $K_{dom}$ values with dissolved soil humic acid (SSHA) are about half as large as the $K_{om}$ values with bulk soil organic matter (SOM) in concentrating organic compounds on a unit weight basis. Compar-atively, the particulate or colloidal organic matter derived from suspended soil or sediment particles should give $K_{dom} \simeq K_{om}$. In soil–water mixtures, the total amount of dissolved and suspended organic matter in water $(X)$ usually increases with increasing soil-to-water ratio. With high $X$ value, the resulting $K_d^*$ value may be significantly smaller than the $K_d$ value for certain solutes because of their enhanced water solubility. Hence, if the aqueous phase con-tains, say, $X = 30\,mg/L$ of soil humic acid, the apparent $K_m^*$ value of DDT with $\log K_{dom} = 4.8$ would be about three times lower than the intrinsic $K_{om}$ value. Gschwend and Wu (1985) measured the $K_d^*$ values of highly water-insoluble 2,4,5,2′,5′-PCB and 2,3,4,5,6,2′,5′-PCB with sediments in relation to total organic carbons of suspended sediment microparticles. By use of Eq. (7.25), they showed that when precautions were taken either to eliminate or to account for the suspended microparticles in water the calculated $\log K_{oc}$ (or $\log K_{om}$) values remained essentially constant for both solutes over a wide range of sediment-to-water ratios, whereas the $\log K_{oc}^*$ value decreased with the amount of suspended microparticles.

With the preceding account, the observed large variation in $\log K_{om}$ for highly water-insoluble solutes with the solid/water ratio in the sorption experiment (Means et al., 1982; Karickhoff, 1984) could be attributed in part to the different extents of solute solubility enhancement by dissolved and/or suspended organic matter. It would also appear that the significant increase in the apparent $K_{om}^*$ value of DDT with soils following the soil extraction with ether and alcohol (Shin et al., 1970) might be more reasonably explained by the reduction of soluble or suspended soil organic matter rather than by the postulate that the applied solvents remove lipids from organic matter and hence enhance the accessibility of DDT to soil organic matter.

By Eqs. (7.23) to (7.26), the effect of DOM on solute solubility, and hence on the solute sorption coefficient, would be less significant for solutes with higher water solubility because their $K_{dom}$ or $K_{doc}$ values are smaller. Consequently, more water-soluble compounds, such as lindane and TCB, show no detectable solubility increases even at relatively high DOM concentrations (Caron et al., 1985; Chiou et al., 1986). This consequence is predicated by Eq. (7.24). For example, for a solute–DOM pair to achieve $S_w^* = 2S_w$, it is required that $X = 1/K_{dom}$. Since $K_{dom}$ increases with decreasing (supercooled-liquid) $S_w$, a higher $X$ is needed for a more water-soluble solute than for a less-soluble solute to produce the same result. Thus, with soil humic acid as DOM, DDT with $K_{dom} = 6.6 \times 10^4$ would give $S_w^* = 2S_w$ at $X = 1.5 \times 10^{-5}$ (or 15 mg/L), while the more soluble TCB with $K_{dom} = 6.3 \times 10^2$ can achieve this result only at $X = 1.6 \times 10^{-3}$ (or 1600 mg/L). Other relatively soluble solutes (most polar solutes) would likewise show no discernible solubility enhancement at $X <$ 100 mg/L.

At $S_w^* = 2S_w$, if one writes $X = S_w/(S_w K_{dom})$, with $S_w K_{dom}$ in units of g solute/g DOM (i.e., dimensionless), a more explicit relation between $S_w$ and required $X$ can be obtained for water solubility enhancement. For nonpolar solutes, the term $S_w K_{dom}$ is small because the corresponding $S_w K_{om}$ term, which expresses the solute $S_{om}$ [Eq. (7.10)] is small (see Table 7.2), and because $K_{om} \geq K_{dom}$ for soil-derived DOM. The dimensionless $S_{om}$ values vary to some extent among nonpolar liquids and supercooled liquids (i.e., about 0.03 for benzene and <0.01 for supercooled DDT). Hence, for a solute to achieve $S_w^* = 2S_w$ with soil humic acid as DOM ($K_{dom} \simeq 0.5 K_{om}$), it is required that $X \geq 60 S_w$, where $S_w$ is for liquids or supercooled liquids [see Eqs. (3.7) and (3.9)]. This factor should increase with increasing polarity of DOM (as with SRHA and SRFA) because of their reduced $K_{dom}$. For most nonpolar solutes, except possibly for PAHs, as discussed below, an empirical rule is that the level of the truly dissolved natural organic matter must be at least about two orders of magnitude greater than the solute's supercooled-liquid $S_w$ in order to produce an enhancement factor of 2. Since the natural DOM level seldom exceeds 100 mg/L, only those contaminants with $S_w < 1$ mg/L as supercooled liquids could possibly exhibit a significant increase in their concentrations by DOM in natural water. However, as discussed later, some human-made DOM, such as surfactant micelles, is able to produce a much greater effect than natural DOM at low concentrations.

We now attend specifically to the potential impact of DOM on the behavior of PAHs. As we recall, PAHs are a unique class of nonpolar solutes in respect to their enhanced partition to SOM, which leads to higher $K_{om}$ or $K_{oc}$ values, due to their aromatic moieties, as discussed earlier. Since the DOM derived from a natural source also contains aromatic moieties, although its content varies with the source (soil, river, lake, etc.), the DOM's aromatic content could thus significantly influence the apparent solubility (i.e., the $K_{dom}$ or $K_{doc}$ value) of a PAH in addition to the DOM's polarity and molecular size. Alternatively stated, one expects a PAH solute to exhibit a higher $K_{doc}$ value than a non-PAH solute with a given DOM, if the two solutes have similar $K_{ow}$ or supercooled $S_w$ value. Chin et al. (1997) found that the $K_{doc}$ value of pyrene with dissolved Aldrich humic acid, a commercial humic acid surrogate with a very high aromatic content, is only slightly smaller than that of DDT (by about a factor of 2), whereas the $K_{ow}$ value of DDT is about 15 times as large as that of pyrene. The measured $K_{doc}$ values for pyrene with a series of dissolved river and lake humic materials, which exhibit minor differences in oxygen content, appear to correlate well with the DOM's aromatic content as well as its molecular weight (Chin et al., 1997). This suggests that DOM in natural water might have a far greater enhancing effect on the concentration and mobility of some PAHs.

In natural systems, the $K_{dom}$ (or $K_{doc}$) values for all low-solubility solutes would vary with the DOM source. Based on the limited data available, the $K_{dom}$ values for natural (not human-made) DOM values in rivers and streams appear to fall between values for soil humic acid (SSHA) and for highly acidic aquatic fulvic acid (SRFA) (Chiou et al., 1987), which vary by a factor of about 5. A more comprehensive account of the $K_{dom}$ variance requires further studies with purified DOM samples. The earlier difficulties in extracting relatively pure aquatic humic materials for solute–DOM interaction studies have led some researchers to the use of commercial humic acids as surrogates for aquatic humic materials. Since the elemental compositions of certain commercial humic acids are vastly different from those of aquatic and soil-derived humic substances (Malcolm and MacCarthy, 1986), the data derived from these samples would not serve as a realistic reference for aquatic humic materials. For example, the $K_{dom}$ (or $K_{doc}$) values of DDT and PCBs with Aldrich humic acid are about three times higher than with soil humic acid and some 4 to 20 times higher than with aquatic humic acid and fulvic acid extracts (Chiou et al., 1987). The exceptionally high $K_{dom}$ (or $K_{doc}$) values with Aldrich humic acid are much a result of its exceptionally high carbon (65.3%) and low oxygen (25.1%) contents on an ash-free basis of the sample (Malcolm and MacCarthy, 1986). Thus the $K_{dom}$ (or $K_{doc}$) data with commercial humic acids could grossly overestimate the actual impact of aquatic humic materials on solute behavior in natural water.

As noted, both suspended and dissolved organic matter can affect the apparent solute solubility or solute sorption coefficient. Whereas the effect of suspended organic matter in soil–water mixtures on $K_d$ may be viewed effectively as the *third-phase effect* (Gschwend and Wu, 1985), the weaker effect as exhibited by truly dissolved organic matter of both natural and human-made origins cannot be well reconciled with a phase concept according to our convention. Rather, it would be more appropriate to treat a water solution containing dissolved organic matter as a mixed solvent, in which the solvency of water for solutes is altered by the dissolved organic substance. In natural systems, both dissolved and suspended organic matters would influence the behavior of sparingly soluble solutes to varying degrees, depending on their sources, concentrations, and compositions.

The enhanced solute solubility by DOM offers a logical explanation to the $K_d^*$ dependence on the soil/sediment–water ratio of solutes observed in soil–water mixtures. Alternatively, Di Toro (1985) hypothesized that collisions between solute-loaded suspended soil/sediment particles induce desorption of the solute from the particles, the extent being related to the suspended particle concentration, particle organic carbon fraction ($f_{oc}$) and solute $K_{oc}$. However, the idea that the solute sorption coefficient is affected by particle collision is unprecedented, as it violates the principle of equilibrium that the net amount of solute sorbed ought to be independent of particle collisions. From kinetic theory, the root-mean-square velocity of a substance (suspended particle or solute) is inversely related to the square root of its molecular weight; therefore, the important collision would be between suspended particles and solute molecules rather than between suspended particles themselves. Solute–particle collision affects solute concentrations (in both solution and particle phases) only before the system reaches equilibrium; it poses no net effect on concentration once equilibrium is reached. Thus, although the particle collision model gives a mathematical fit to Eq. (7.25), it lacks a physical basis to explain the influence of dissolved and suspended organic matter on solute solubility and the subsequent effect on solute sorption coefficient.

### 7.3.9  Influence of Surfactants and Microemulsions

The discharge of household and industrial organic wastes into the environment may produce complex mixtures of dissolved and suspended organic matter in natural water. A class of human-made dissolved and/or colloidal organic matter of special interest is that of surface-active agents (surfactants) due to their huge discharge quantities. The unique molecular structures of surfactants enable them to form stable aggregates (called *micelles*) above certain concentrations in water. Thus, whereas some surfactants or their degradation products could become potential pollutants in surface water or groundwater, their presence may also affect the solubility and partition behavior of other pollutants. The ability of a surfactant to form micelles is characterized by its *critical micelle concentration* (CMC). At concentrations below CMC, the dissolved surfactant is all in monomeric form; above the CMC, the amount of surfactant in excess of the CMC forms micelles. Unlike truly dissolved ordinary chemicals, which exist in monomeric form, the surfactant micelle offers a relatively large microscopic nonpolar environment to allow for solute partition (i.e., solubilization) (Rosen, 1978). This effect could greatly promote the (apparent) water solubility of otherwise relatively insoluble solutes, as compared to the effect of a dissolved natural organic matter on solute water solubility. As such, surfactants also provide a potential means to remove contaminants from contaminated soils (Sabatini et al., 1995), provided that the surfactant sorbed to the soil does not strongly enhance the contaminant uptake.

A general equation similar to Eq. (7.24) has been formulted to account for the solubility enhancement of a solute by a surfactant solution at a given temperature (Kile and Chiou, 1989):

$$S_w^*/S_w = 1 + X_{mn}K_{mn} + X_{mc}K_{mc} \qquad (7.27)$$

where $S_w^*$ and $S_w$ are as defined before; $X_{mn}$ is the concentration of the surfactant as monomers in water (dimensionless); $X_{mc}$ is the concentration of the surfactant in micellar form in water (dimensionless); $K_{mn}$ is the partition (enhancement) coefficient of the solute between surfactant monomers and water; and $K_{mc}$ is the partition coefficient of the solute between micelles and water. At a given temperature, $S_w$, $K_{mn}$, and $K_{mc}$ are constants. The $X_{mn}$ and $X_{mc}$ values are assigned as follows. If the surfactant concentration $X \le$ CMC, then $X_{mn} = X$; if $X >$ CMC, $X_{mn} =$ CMC and $X_{mc} = X -$ CMC.

By Eq. (7.27), a plot of the apparent solute solubility ($S_w^*$) against the surfactant concentration ($X$), extended over the CMC, will be bilinear, yielding a straight line with a slope of $S_wK_{mn}$ from $X = 0$ to $X =$ CMC, followed by another straight line of a much higher slope, $S_wK_{mc}$, at $X \ge$ CMC, provided that the surfactant is molecularly homogeneous and has a single monomer–micelle transition point (i.e., a single CMC). If the surfactant is molecularly heterogeneous, a succession of micelle formation may take place for different molecular fractions of the surfactant and the resulting plot of $S_w^*$ versus $X$ will not exhibit a single sharp transition. [In this case, the measurement of surface tension versus $X$, as commonly used to detect CMC, gives a breadth of monomer–micelle transition rather than a single sharp break.]

Kile and Chiou (1989) applied Eq. (7.27) to analyze the apparent water solubilities of extremely water-insoluble DDT and relatively soluble 1,2,3-trichlorobenzene (TCB) as influenced by several nonionic surfactants (Triton series and Brij 35), an anionic surfactant [sodium dodecyl sulfate (SDS)], and a cationic surfactant [cetyltrimethylammonium bromide (CTAB)]. The structures and properties of these surfactants are given in Table 7.16. Triton series and Brij 35 surfactants are molecularly heterogeneous in that the lengths of polar head groups [i.e., the number of ethylene oxides (EO) units] are average rather than single fixed values, whereas SDS and CTAB are virtually molecularly homogeneous. In related studies, Edwards et al. (1991) measured the apparent water solubilities of nathphalene, phenanthrene, and pyrene in water solutions of TX100, Brij 30, and other nonionic surfactants, and Jafvert et al. (1994) measured the apparent solubilities of hexachlorbenzene with similar surfactant types.

The solute solubility enhancement by a surfactant was treated by Edwards et al. (1991) in terms of molar solubilization ratio (MSR) (i.e., the number of moles of solute solubilized per mole of surfactant in micellar form) and by Jafvert et al. (1994) as a dimensional *equilibrium constant* of solute and micelle

**TABLE 7.16. Structures and Properties of Selected Commercial Surfactants**

| Surfactant | Structure | Molecular Weight | CMC (mg/L) |
|---|---|---|---|
| Triton series: | | | |

$$CH_3-\underset{\underset{CH_3}{|}}{\overset{\overset{CH_3}{|}}{C}}-CH_2-\underset{\underset{CH_3}{|}}{\overset{\overset{CH_3}{|}}{C}}-\langle\bigcirc\rangle-(OCH_2CH_2)_n-OH$$

| Surfactant | Structure | Molecular Weight | CMC (mg/L) |
|---|---|---|---|
| Triton X-100 (TX100) | $n = 9.5$ (average) | 628 | 130 |
| Triton X-114 (TX114) | $n = 7.5$ (average) | 536 | 110 |
| Triton X-405 (TX405) | $n = 40$ (average) | 1966 | 620 |
| Brij 35 (BJ35) | $C_{12}H_{25}(OCH_2CH_2)_{23}OH$ | 1200 | 74 |
| Sodium dodecyl sulfate (SDS) | $C_{12}H_{25}-OSO_3^- \cdot Na^+$ | 288 | 2100 |
| Cetyltrimethylammonium bromide (CTAB) | $C_{16}H_{33}-N(CH_3)_3^+ \cdot Br^-$ | 364 | 361 |

*Source*: Data from Kile and Chiou (1989).

concentrations in water. Although these values can be converted to $K_{mc}$, the surfactant effect as characterized by the $K_{mc}$ (and $K_{mn}$) in Eq. (7.27) is more convenient for our purposes, as it enables one to compare directly the magnitude of $K_{mc}$ (or $K_{mn}$) with the solute's solvent–water partition coefficients (e.g., $K_{ow}$) or with its partitionlike coefficient with a DOM (e.g., $K_{dom}$). This is especially useful for characterizing the behavior of micelle, which has been viewed as a *pseudo phase* in the surfactant literature.

For DDT, marked solubility enhancements and sharply rising slopes are observed, as the concentrations of SDS and CTAB exceed their (nominal) CMCs. Solubility enhancements of DDT with TX100 and TX114 manifest less sharp transitions and those with BJ35 and TX405 show a gradual transition, due presumably to their different degrees of molecular heterogeneity. TCB exhibits a similar solubility transition but a considerably lower enhancement, due to its much higher water solubility. The results for DDT are shown in Figures 7.32 and 7.33. In all cases, the enhancements as observed for DDT at concentrations above the CMC are tremendously large, manifesting the power of pseudophase micelles to solubilize highly insoluble nonpolar solutes. More soluble TCB shows considerably less solubility enhancement over the same surfactant concentration range. The $K_{mc}$ values calculated for DDT and TCB given in Table 7.17 are comparable in magnitude to the corresponding $K_{ow}$ values (see Table 5.1). Edwards et al. (1991) and Jafvert et al. (1994) show similar results for other solutes with micelles when their data are expressed in terms of the $K_{mc}$ values. The $K_{mn}$ values calculated for DDT, using solubility data at surfactant concentrations near zero, are some 40 times lower than

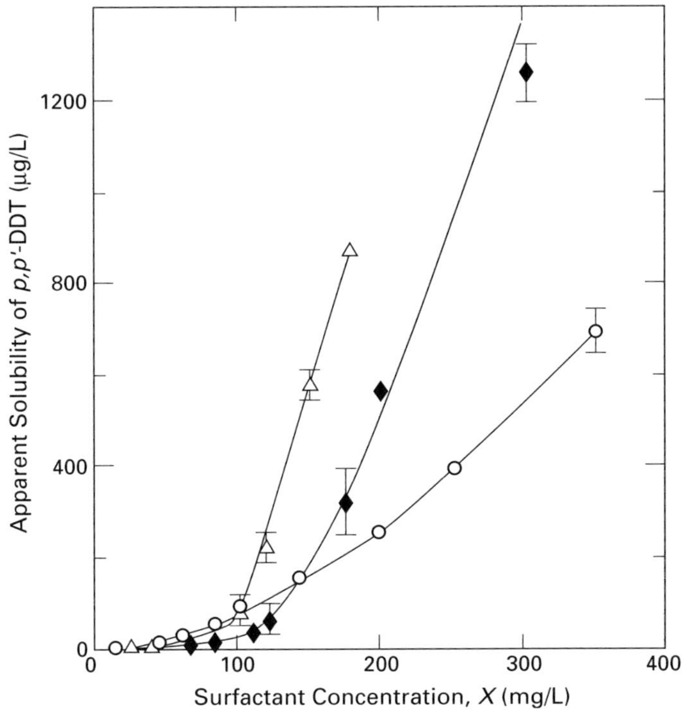



**Figure 7.32**   Apparent water solubility of $p,p'$-DDT as a function of TX100 ($\blacklozenge$), TX114 ($\triangle$), and BJ-35 ($\bigcirc$) concentration. [Data from Kile and Chiou (1989).]

**TABLE 7.17.  Log$K_{mn}$ and log$K_{mc}$ Values of $p,p'$-DDT and 1,2,3-Trichlorobenzene (TCB) Measured on Selected Surfactants**

| | log $K_{mn}$ | | log $K_{mc}$ | |
|---|---|---|---|---|
| Surfactant | $p,p'$-DDT | 1,2,3-TCB | $p,p'$-DDT | 1,2,3-TCB |
| Triton X-100 (TX100) | 4.26 | ND[a] | 6.15 | 3.82 |
| Triton X-114 (TX114) | 4.59 | ND | 6.18 | 3.95 |
| Triton X-405 (TX405) | 3.92 | ND | 5.56 | — |
| Brij 35 (BJ35) | 4.18 | ND | 5.75 | 3.31 |
| Sodium dodecyl sulfate (SDS) | 2.68 | ND | 5.38 | 3.54 |
| Cetyltrimethylammonium bromide (CTAB) | 3.54 | ND | 5.88 | 3.80 |

*Source*: Data from Kile and Chiou (1989).

[a] ND, not detectable.

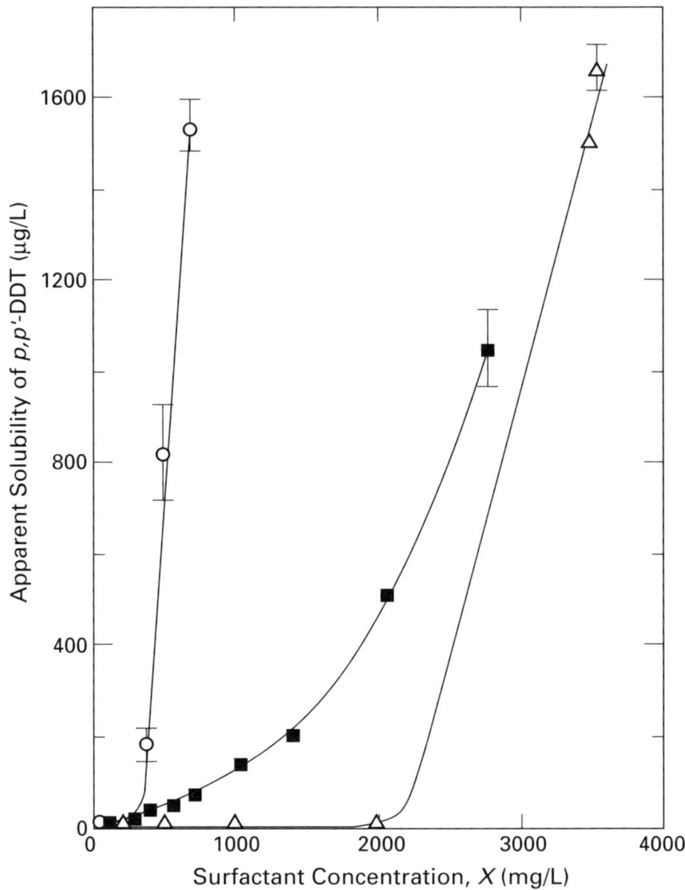

**Figure 7.33**    Apparent water solubility of $p,p'$-DDT as a function of TX405 (■), SDS (△), and CTAB (○) concentration. [Data from Kile and Chiou (1989).]

corresponding $K_{mc}$ values. The $K_{mn}$ values are comparable in magnitude with the $K_{doc}$ values of the solutes with dissolved aquatic humic substances. Whereas micelles of all surfactants would greatly enhance the solubility of highly insoluble contaminants, the effective surfactants from an economical standpoint are those that form micelles at low concentrations (i.e., the ones with low CMCs).

When surfactants are combined with oils in certain ratios to form stable microscopic mixtures (called *microemulsions*), they become especially effective in enhancing the water solubility of nonpolar organic contaminants, due to solute solubilization into the resulting suspended microscopic oil–surfactant phase. The microemulsions are stabilized through the interactions with water of the polar head groups of surfactants attached to the oil particles. Whereas such stable microemulsions are sometimes also labeled commercially as sur-

**TABLE 7.18. Composition and Properties of Commercial Petroleum Sulfonate Surfactants**

| Composition/Property | Petronate L | Petronate HL | Pyronate 40 |
|---|---|---|---|
| Percent petroleum sulfonate | 62 | 62 | 41 |
|    Fraction as hydrocarbon | 46.9 | 48.0 | 28.6 |
|    Fraction as -NaSO$_3$ | 15.1 | 14.0 | 12.4 |
| Percent free mineral oil | 33.0 | 32.5 | 12.0 |
| Percent inorganic salts | 0.5 | 0.5 | 8.5 |
| Percent water | 4.5 | 4.5 | 38.5 |
| Percent total hydrocarbon (water-free basis)[a] | 83.7 | 84.3 | 66.0 |
| MW of petroleum sulfonate | 415–430 | 440–470 | 330–350 |

*Source*: Data from Kile et al. (1990).

[a] Sum of hydrocarbons in petroleum sulfonate and free mineral oil on a moisture-free basis.

factants because of their abilities to reduce the surface tension of water, they exhibit virtually no monomer–micelle transitions (i.e., no CMCs), and hence possess a separate-phase property even at very low concentrations. An example of emulsion-forming commercial products is the class of petroleum sulfonate–oil (PSO) surfactants, formulated by mixing free mineral oil with surface-active petroleum sulfonate in certain proportions, as shown in Table 7.18. The solute solubility enhancement in water by microemulsions can be described in a form similar to Eq. (7.27) as

$$S_w^*/S_w = 1 + X_{em}K_{em} \qquad (7.28)$$

where $X_{em}$ is the concentration of microemulsion in water (dimensionless) and $K_{em}$ is the solute partition coefficient between the emulsified phase and water. Thus a plot of $S_w^*$ versus $X_{em}$ for a solute should produce a single straight line with a slope of $S_w K_{em}$, which is characteristically different from the plot with a normal surfactant. The measured $K_{em}$ values for DDT and TCB with PSO microemulsions have about the same magnitude as the $K_{mc}$ values, confirming the separate-phase property of microemulsions (Kile et al., 1990). In comparison with normal surfactants, however, the emulsified oil phase enhances the solute solubility at low concentrations more effectively than do normal surfactants at concentrations below their CMC values. In natural water, the possible formation of an emulsified oil phase from components of untreated organic wastes near waste-discharge sites could have a strong impact on the fate of certain organic pollutants.

The strong enhancement of the solute solubility by surfactants offers a potential means of remediating contaminated soils or natural solids by application of surfactant solutions. If the solution of a normal surfactant (or a microemulsion) is applied to remove contaminants in soil that exist as a separate nonaqueous-phase liquid (NAPL), the efficiency of the NAPL removal may be described by Eqs. (7.27) and (7.28), provided that the surfactant solution readily reaches the surfaces of soil particles and penetrates into their void spaces. For the NAPL-saturated soil, the sorption of surfactants to the soil should be small compared to the amount of NAPL retained by soil and therefore should not significantly alter the overall NAPL sorption by the soil. On the other hand, if the contaminants exist at subsaturated levels in the soil or a natural solid (i.e., no NAPL or excess contaminant phase is formed), the applied surfactant solution may produce disparate results, depending on the system, as described below.

Let us consider the situation for a contaminant at subsaturated level in a soil–water system to which a surfactant is applied. To begin with, one must recognize that there are fundamental differences in molecular properties between surfactants and common organic compounds that affect their interactions with soils or other solids. Most commercial and certain natural surfactants contain powerful ionic or polar head groups (as needed to keep them stabilized in water) that are attached to relatively long nonpolar tails. The high polarity and molecular weight of many surfactants (e.g., nonionic and cationic surfactants) enable them to compete more efficiently against water for adsorption onto certain soil minerals, besides their partition into SOM. In the absence of a separate contaminant phase, adsorption of the surfactant on minerals may induce the uptake of other contaminants onto or into the mineral-adsorbed surfactant phase and consequently affect the distribution coefficient of a contaminant. The simultaneous effects of soil-sorbed surfactants on contaminant sorption and unsorbed aqueous surfactants on contaminant dissolution have been addressed by Sun and Boyd (1993), Chiou (1998), and Lee et al. (2000) to delineate the consequent change in contaminant sorption coefficient. By this account, the apparent sorption coefficient ($K_d^*$) of a subsaturated contaminant in a soil–water mixture with a surfactant is given as

$$K_d^* = \frac{K_d + f_{sf} K_{sf}}{1 + X_{mn} K_{mn} + X_{mc} K_{mc}} \tag{7.29a}$$

or, in a more useful alternative form, as

$$\frac{K_d^*}{K_d} = \frac{1 + f_{sf} K_{sf}/K_d}{1 + X_{mn} K_{mn} + X_{mc} K_{mc}} \tag{7.29b}$$

where $K_d$ is as defined before; $f_{sf}$ the mass fraction of the sorbed surfactant in soil; and $K_{sf}$ is the solute (contaminant) sorption coefficient between sorbed surfactant and water. The term $f_{sf}$ is for the sum of adsorbed and partitioned surfactant. Here $K_{sf}$ is not a direct function of $f_{sf}$, but rather, a function of the aggregation state of the sorbed surfactant molecules. In principle, only the adsorbed surfactant can form a molecular aggregation, the extent being related to the amount adsorbed and mineral surface properties (Gu et al., 1992; Rutland and Senden, 1993; Nayyar et al., 1994). The surfactant partitioned into SOM is not subject to aggregation, in which case, $K_{sf} \approx 0$. Although surfactants partitioned into SOM may modify the SOM properties, the amount partitioned to SOM is usually not substantial, and thus the intrinsic SOM properties are probably not changed significantly. The magnitude of $(1 + X_{mn}K_{mn} + X_{mc}K_{mc})$ in Eq. (7.29) is determined by surfactant type and its concentration in water as well as by contaminant properties.

By the same reasoning as above, the relation between $K_d^*$ and $K_d$ for a (subsaturated) contaminant in a soil–water mixture with a microemulsion may be described as

$$\frac{K_d^*}{K_d} = \frac{1 + f_{em} K_{em}^{(s)} / K_d}{1 + X_{em} K_{em}^{(w)}} \tag{7.30}$$

where $f_{em}$ is the fraction of sorbed emulsified material in soil, $K_{em}^{(s)}$ the solute sorption coefficient between sorbed microemulsion and water, and $K_{em}^{(w)}$ the corresponding coefficient between aqueous-phase microemulsion and water. It is evident from Eqs. (7.29) and (7.30) that the applied surfactant (or microemulsion) imposes two opposing effects: The sorbed surfactant (or microemulsion) increases contaminant sorption, whereas the unsorbed surfactant promotes contaminant dissolution in water. The prerequisite condition for soil remediation ($K_d^* < K_d$) requires that $(1 + f_{sf}K_{sf}/K_d)$ be smaller than $(1 + X_{mn}K_{mn} + X_{mc}K_{mc})$ if a surfactant is used, or that $(1 + f_{em}K_{em}^{(s)}/K_d)$ be smaller than $(1 + X_{em}K_{em}^{(w)})$ if a microemulsion is involved.

To characterize $K_{sf}$ with respect to $f_{sf}$ (or $K_{em}^{(s)}$ in respect to $f_{em}$), detailed sorption information must be made available for various surfactants on various soils and solids in order to account for the extent and state of the sorbed surfactant. This requirement would seem too formidable a task to be accomplished realistically. Instead, we choose an empirical approach to develop the knowledge from a large set of relevant data. Because the molecular properties of surfactants vary considerably, the sorption of surfactants on different soils or solids is expectedly more complicated than that of nonionic organic compounds.

The sorption of cationic surfactants on clays and soils that contain significant cation exchange sites is known to take place primarily by a cation exchange process (Theng et al., 1967; Barrer, 1978; Lee et al., 1989; Xu and Boyd, 1994). Incorporation of a significant amount of long-chain alkylammonium cations onto charge exchange sites of either clays or soils creates an additional microscopic nonpolar phase for solutes to partition (i.e., to give large $f_{sf}K_{sf}$), which makes $K_d^* \gg K_d$ for nonpolar solutes (Boyd et al., 1988; Smith et al., 1990b). Therefore, the addition of cationic surfactants to soil–water systems would normally increase (rather than decrease) the uptake of organic pollutants on soil and hence reduce pollutant migration through a soil profile.

This would have a negative effect on soil remediation and a positive effect on soil contaminant uptake or immobilization. Anionic surfactants with large charges usually show relatively weak adsorption on mineral soils (Rouse et al., 1993). This would seem to make anionic surfactants more desirable for soil remediation. Currently, however, available data are insufficient to compare the effects of anionic surfactants and other surfactants on the sorption coefficients of organic contaminants in soil–water systems.

The influence of emulsified material on contaminant sorption coefficient has been investigated using a PSO surfactant (Petronate L) as reference material (Sun and Boyd, 1993). With naphthalene ($S_w$ = 31 mg/L), phenanthrene ($S_w$ = 1.3 mg/L), and 2,2′,4,4′,5,5′-hexachlorobiphenyl (2,2′,4,4′,5,5′-PCB) ($S_w$ = 1.0 μg/L) as model solutes and Oshtemo silt loam as the reference soil ($f_{om}$ = 0.0017), Sun and Boyd (1993) found that with applied $X_{em}$ = 50 mg/L the $K_d^*/K_d$ drops to about 0.02 for 2,2′,4,4′,5,5′-PCB and to about 0.75 for phenanthrene, but remains at about 1 for more water-soluble naphthalene (Figure 7.34). Thus, even at low $X_{em}$, the data indicate that $X_{em}K_{em}^{(w)}$ is much larger than $f_{em}K_{em}^{(s)}/K_d$ for highly insoluble solutes (high $K_d$), but the difference decreases rapidly with increasing solute $S_w$ or with decreasing $K_d$. For an emulsified material in soil–water systems, it is reasonable to assume that $K_{em}^{(s)} \simeq K_{em}^{(w)}$, since the sorbed material on solid surfaces would probably be in a separate phase similar to that in water. Therefore, if the emulsified material exhibits only a low-to-moderate sorption on a solid (i.e., if $f_{em}/X_{em}$ is not very large), $f_{em}K_{em}^{(s)}/K_d$ may then be less than $X_{em}K_{em}^{(w)}$, or $f_{em}/K_dX_{em} < 1$, for solutes with large $K_d$. For solutes of low $K_d$ value, the $f_{em}/K_dX_{em}$ increases and appears to approach 1 for naphthalene with a mineral soil. Overall, the addition of emulsified materials to water for remediation of subsaturated contaminants in soil would be most effective for highly insoluble contaminants and presumably for soils with high SOM content, both leading to high $K_d$.

The influence of normal surfactants on $K_d^*$ is expectedly more complicated. A few studies with Triton X-100 (TX100), a nonionic surfactant, have been reported. Sun et al. (1995) found that the $K_d^*/K_d$ of 1,2,4-trichlorobenzene (TCB) ($S_w$ = 18 mg/L) on Oshtemo silt loam ($f_{om}$ = 0.0017) increases initially at $X \ll$ CMC, reaches a maximum of about 5 at $X \approx 2$ CMCs and then declines slowly thereafter with $X$; the $K_d^*/K_d$ values for highly insoluble 2,2′,4,4′,5,5′-PCB ($S_w$ = 1.0 μg/L) and DDT ($S_w$ = 5.5 μg/L) on the same soil exhibit a maximum of 2.4 and 3.3, respectively, at $X \approx$ CMC and then decline sharply at higher $X$, with $K_d^*/K_d \approx 0.5$ at $X \approx 2$ CMCs (Figure 7.35) (see Table 7.16 for the CMC value). In a similar study, Deitsch and Smith (1995) found that with Triton X-100 the $K_d^*$ of relatively soluble trichloroethylene (TCE) ($S_w$ = 1100 mg/L) on an organic-rich soil ($f_{oc}$ = 0.24) exhibits no significant change over $K_d$ up to moderately high $X$ (300 mg/L) and that the $K_d^*$ shows a significant reduction relative to $K_d$ only at $X \gg$ CMC. An obvious consequence of the $K_d^* - K_d$ relation is that $K_d^* < K_d$ for all contaminants when $X$ exceeds greatly the CMC. This is because the sorption of surfactant to a soil reaches the saturation level at some point while the surfactant concentration in water ($X$) increases continuously past the CMC, making $(1 + X_{mn}K_{mn} + X_{mc}K_{mc}) \gg$

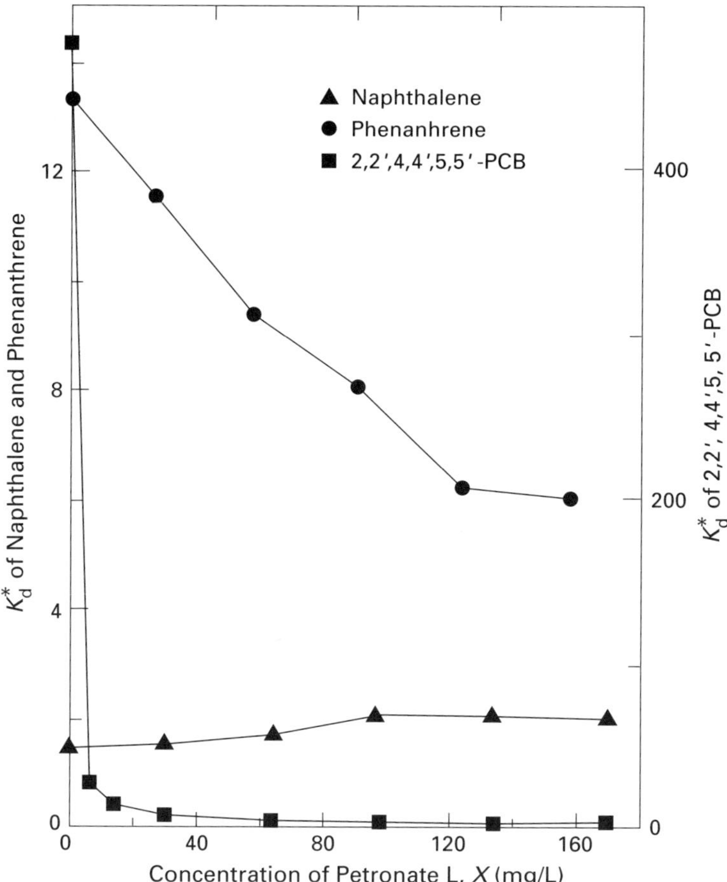



**Figure 7.34**    Apparent $K_d^*$ values of 2,2',4,4',5,5'-PCB, phenanthrene, and naphthalene with a soil ($f_{om}$ = 0.0017) as a function of petroleum sulfonate (Petronate L) concentration in water at 22°C. [Data from Sun and Boyd (1993). Reproduced with permission.]

$(1 + f_{sf}K_{sf}/K_d)$. Our primary interest here is for systems with low and moderate $X$, where the relative magnitudes of $(1 + X_{mn}K_{mn} + X_{mc}K_{mc})$ and $(1 + f_{sf}K_{sf}/K_d)$ may change sensitively from one system to another.

The later more extensive study with TX100 by Lee et al. (2000) elucidated conspicuously the roles of the surfactant when adsorbed on minerals and when partitioned into SOM, in addition to its better known effect on contaminant solubilization in water. The study consisted of four relatively water-soluble BTEX liquids (benzene, toluene, ethylbenzene, and $p$-xylene) and three chlorinated solid compounds [lindane, $\alpha$-BHC, and heptachlor epoxide (HPOX)], which cover a wide range of water solubility; the solid sorbents comprised a bentonite clay, a Florida peat, and two other soils, which cover a wide range of solid organic matter (SOM) content (Table 7.19). The applied surfactant $X$ ranged from less than the CMC to two to three times the CMC. The

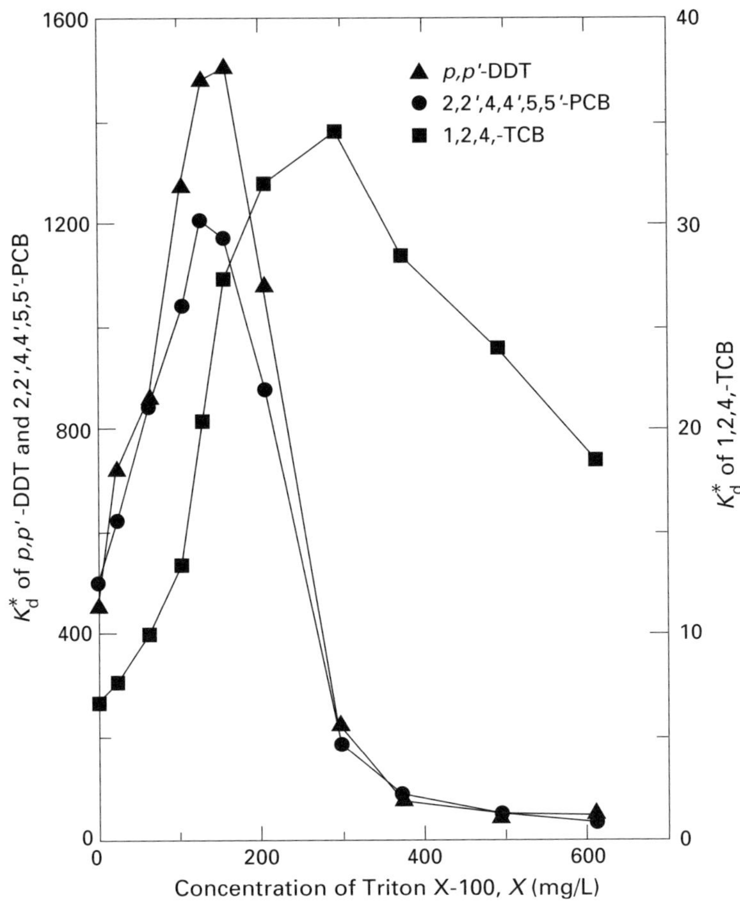


**Figure 7.35** Apparent $K_d^*$ values of $p,p'$-DDT, 2,2′,4,4′,5,5′-PCB, and 1,2,4-trichlorobenzene with a soil ($f_{om}$ = 0.0017) as a function of TX100 concentration in water at 23°C. [Data from Sun et al. (1995). Reproduced with permission.]

**TABLE 7.19. Properties of Natural Sorbents Chosen for Contaminant Sorption from Water with and without Triton X-100: SA = BET-N$_2$ Surface Area and $f_{om}$ = Fraction of Organic Matter in Sorbent**

| Sorbent | Abbreviation | SA (m²/g) | $f_{om} \times 100$ |
|---|---|---|---|
| Ca-montmorillonite | Bentonite | 76 | 0.03 |
| Taichung soil | TCS | 5.2 | 2.4 |
| Chihsing Mountain soil | CSMS | 32 | 14.8 |
| Florida peat | Peat | 1.3 | 86.4 |

*Source*: Data from Lee et al. (2000).

**TABLE 7.20. Distribution Coefficients of BTEX and Chlorinated Compounds in Solid–Water Systems with TX100 ($K_d^*$) and without TX100 ($K_d$) as a Function of TX100 Equilibrium Concentration in Water ($X$)**[a]

| Sorbent | Surfactant $X$ (mg/L)[b] | $K_d^*$ or $K_d$ | | | | |
| --- | --- | --- | --- | --- | --- | --- |
| | | Benzene | Toluene | $p$-Xylene | Lindane | HPOX |
| Bentonite | 0 | $\simeq 0$ | $\simeq 0$ | $\simeq 0$ | $\simeq 0$ | $\simeq 0$ |
| | 67 | — | — | — | 10.8 | 251 |
| | 84 | 8.95 | 18.0 | 22.5 | — | — |
| | 370 | — | — | — | 134 | 342 |
| | 381 | 14.2 | 37.8 | 91.4 | — | — |
| TCS | 0 | 0.29 | 0.79 | 2.27 | 12.6 | 121 |
| | 68 | 0.51 | 2.30 | 4.16 | 12.0 | 145 |
| | 312 | 1.64 | 6.20 | 15.4 | 18.3 | 82.4 |
| CSMS | 0 | 3.26 | 9.27 | 21.9 | 266 | 1840 |
| | 51 | 5.46 | 13.2 | 32.0 | 210 | 1400 |
| | 108 | 6.20 | 18.7 | 36.3 | 139 | 1370 |
| | 244 | 12.4 | 21.9 | 40.9 | 65.5 | 84.3 |
| Peat | 0 | 8.33 | 20.8 | 61.7 | 923 | 3980 |
| | 41 | 10.4 | 26.9 | 86.1 | 767 | 3120 |
| | 82 | 13.7 | 33.7 | 93.9 | 604 | 2410 |
| | 536 | 15.8 | 39.1 | 110 | 589 | 1860 |

*Source*: Data from Lee et al. (2000).

[a] $K_d$ values (i.e., at $X = 0$) are underlined.

[b] CMC = 158 mg/L.

experimental results of Lee et al. (2000) for benzene, toluene, $p$-xylene, lindane, and HPOX are given in Table 7.20. For the sorbents studied, the bentonite exhibits the highest uptake of TX100, followed in order by TCS/CSMS and peat (data not shown). For bentonite, which has a very low SOM content ($f_{om} = 0.00031$) and a high surface area (SA), the TX100 uptake should be principally by mineral adsorption. By contrast, the TX100 sorption to peat, which has a very high SOM content ($f_{om} = 0.864$) and a low SA, should be primarily by partition into SOM. For CSMS with a high SOM content ($f_{om} = 0.148$), TX100 partition to SOM is probably significant, although the soil has a moderately high SA. In contrast, the sorption of TX100 to TCS, which has a low $f_{om}$ (0.024), appears to occur more by adsorption than by partition. Since the TX100 uptake on a natural solid involves both adsorption on mineral matter and partition into SOM, the overall TX100 sorption is not closely related to the solid SA.

For all solutes on all soil solids, the $K_d$ values (at $X = 0$) are closely related to solid $f_{om}$, in reflection of solute partition to SOM being the primary process; the $K_d$ values are lowest (near zero) for bentonite and highest for peat. The observed $K_d$ values on a given solid, in the order benzene < toluene < ethylbenzene < $p$-xylene < lindane $\simeq$ α-BHC < HPOX, correlate inversely with solute water solubilities ($S_w$) and directly with their $K_{ow}$ values, as expected;

the correlation with $S_w$ for solid pesticide solutes is improved with the use of their supercooled-liquid $S_w$ values (not shown). For more water-soluble BTEX liquids, in which the $S_w$ ranges from 190 mg/L for $p$-xylene to 1780 mg/L for benzene, the $K_d^*/K_d$ ratios exceed 1 on all solids and increase with increasing $X$ up to two to three nominal CMCs. With solid $f_{om}$ decreasing from peat to CSMS, TCS, and bentonite, the $K_d^*/K_d$ ratios go up rapidly. Relative to BTEX data, the $K_d^*/K_d$ values for less soluble lindane ($S_w = 7.8$ mg/L), $\alpha$-BHC ($S_w = 2.0$ mg/L), and HPOX ($S_w = 0.2$ mg/L) are much more variable. On bentonite ($f_{om} \approx 0$), the $K_d^*/K_d$ ratios are >1 and increase rapidly with $X$ from below to above the CMC. On TCS ($f_{om} = 0.024$), the $K_d^*$ values at $X <$ CMC are about the same as $K_d$ for all three chlorinated compounds; when $X$ is about 2 CMCs, the $K_d^*/K_d$ values are about 1.5 for lindane and $\alpha$-BHC, while the $K_d^*/K_d$ value of HPOX drops to about 0.7. On CSMS ($f_{om} = 0.148$) and peat ($f_{om} = 0.864$), the $K_d^*/K_d$ values are <1 for all three solutes and decrease steadily with increasing $X$ from below to above the CMC.

In terms of Eq. (7.29), the observed $K_d^*/K_d$ values for solutes with TX100 on natural solids from all reports reveal the following trends: (1) the $(1 + f_{sf}K_{sf}/K_d)$ term is greater than the $(1 + X_{mn}K_{mn} + X_{mc}K_{mc})$ term for relatively water-soluble BTEX on all solids at low-to-moderate $X$, in which the ratio of $(1 + f_{sf}K_{sf}/K_d)$ to $(1 + X_{mn}K_{mn} + X_{mc}K_{mc})$ increases with decreasing solid $f_{om}$ and with increasing solute $S_w$; (2) if the solid has a very low $f_{om}$ value, then $(1 + f_{sf}K_{sf}/K_d) > (1 + X_{mn}K_{mn} + X_{mc}K_{mc})$ holds for practically all solutes at low $X$ ($\ll$CMC); (3) the reverse effect, that is, $(1 + f_{sf}K_{sf}/K_d) < (1 + X_{mn}K_{mn} + X_{mc}K_{mc})$ occurs at low $X$ only for low-$S_w$ solutes (e.g., lindane and HPOX) on high-$f_{om}$ solids (e.g., CSMS); and (4) the aromatic BTEX tend to exhibit higher $K_d^*/K_d$ than other relatively soluble nonaromatic compounds (e.g., TCE) at comparable $X$.

The diversity of the $K_d^*/K_d$ data manifests the sensitive balance between $(1 + f_{sf}K_{sf}/K_d)$ and $(1 + X_{mn}K_{mn} + X_{mc}K_{mc})$ for different solutes on different sorbents (solids) with TX100 at low to moderate $X$. In the water phase, we recall that the magnitude of $(1 + X_{mn}K_{mn} + X_{mc}K_{mc})$ increases sensitively with the $K_{ow}$ or $1/S_w$ of the solute (see the discussion on page 180); the TX100 applied at $X < 400$ mg/L is not expected to have a significant effect on $S_w$ for relatively soluble solutes, such as BTEX [i.e., the term $(1 + X_{mn}K_{mn} + X_{mc}K_{mc})$ should be close to 1]. By contrast, the applied surfactant at the same $X$ would significantly affect the solubility of the less-soluble chlorinated solutes; the estimated $(1 + X_{mn}K_{mn} + X_{mc}K_{mc})$ for lindane at $X = 200$ to 400 mg/L is about 1.6 to 3.0, and a higher value is anticipated for HPOX (Lee et al., 2000). To the solid sorbent, the effect of the sorbed surfactant on solute uptake (i.e., $f_{sf}K_{sf}/K_d$) is expected to be about the same for all solutes on a sorbent because $K_{sf}$ should be largely linearly related to $K_{om}$; the exception will be for certain solutes that exhibit a special affinity for a specific type of surfactant. However, as the solid $f_{om}$ increases, the amount of adsorbed surfactant on the mineral matter may decrease, which could then reduce the impact of the surfactant on solute uptake.

With the rationale above, the observation that the $K_d^*/K_d$ values of BTEX are all greater than 1 at the applied $X$ range should be a result of their small $(1 + X_{mn}K_{mn} + X_{mc}K_{mc})$ values, due to their high $S_w$; the increase in $K_d^*/K_d$ with decreasing $f_{om}$ is attributed to an increased mass of adsorbed and aggregated TX100 on mineral matter. In this case, the small increase in $K_d^*/K_d$ for BTEX with peat, to which a significant mass of TX100 is sorbed, may be realized on the account that most sorbed TX100 partitions to SOM and the extent of surface aggregation is small. The increase in $K_d^*/K_d$ from $p$-xylene to benzene on all solids is consistent with an increase in $S_w$. In light that $(1 + f_{sf}K_{sf}/K_d) > (1 + X_{mn}K_{mn} + X_{mc}K_{mc})$ for BTEX and TCE on low-$f_{om}$ solids with TX100, the finding that the $K_d^*/K_d$ values are greater for BTEX than for TCE may reflect an improved partition of more-aromatic BTEX to aggregated, partially aromatic TX100 (Lee et al., 2000). The finding that the $K_d^*$ value for TCE with a high-$f_{om}$ soil is about the same as $K_d$ over a large range of $X$ (Deitsch and Smith, 1995) is understood on the basis that the term $(1 + X_{mn}K_{mn} + X_{mc}K_{mc})$ should be small (i.e., close to 1) over the applied range of $X$ because TCE is relatively water soluble and that the adsorbed and aggregated TX100 on the solid surface should be small because of the high solid $f_{om}$, making the term $(1 + f_{sf}K_{sf}/K_d)$ close to 1.

For the less soluble chlorinated solutes, the variation in $K_d^*/K_d$ follows basically the same pattern as noted for BTEX, in which the $K_d^*/K_d$ decreases with increasing solid $f_{om}$ (or increasing $K_d$) and become eventually less than 1 for the solutes on high-$f_{om}$ solids (CSMS and peat). The transition in $K_d^*/K_d$, as found either for different solutes on a solid or for a given solute on different solids, escapes recognition in other studies (e.g., Deitsch and Smith, 1995; Sun et al., 1995) because these systems consisted mainly of either low-$f_{om}$ solids or relatively water-soluble solutes. The noted difference in $K_d^*/K_d$ between lindane on TCS ($f_{om} = 0.024$) (Table 7.20) and TCB on Oshtemo silt loam ($f_{om} = 0.0017$) (Figure 7.35), where the two solutes have similar $\log K_{ow}$ values, manifests the intimate effect of solid $f_{om}$ on solute $K_d^*$ when TX100 is applied. The impact of solid $f_{om}$ on the relative order of $(1 + f_{sf}K_{sf}/K_d)$ and $(1 + X_{mn}K_{mn} + X_{mc}K_{mc})$ for different solutes is detected more readily if the solid has a significant $f_{om}$. Thus, whereas the $K_d^*/K_d$ are much greater than 1 for DDT and 2,2',4,4',5,5'-PCB with Oshtemo silt loam at low $X$ (Sun et al., 1995), the reverse effect would most likely occur if the solid has a much higher SOM content. A more comprehensive analysis of all pertinent system parameters has been given by Lee et al. (2000).

Although the $K_d^*/K_d$ characteristics observed for contaminants with one surfactant do not suffice for generalization of the potential effects by all surfactants, the recognized major features as to the direction to which the $K_d^*/K_d$ ratio varies with soil/solid and contaminant properties should be applicable to most surfactants. As illustrated with TX100 surfactant and PSO microemulsion, the $K_d^*/K_d$ ratio increases with increasing solute water solubility ($S_w$) and decreases largely with increasing solid SOM content ($f_{om}$). This behavioral pattern should aid in our assessment of the contaminant distribution in natural environments and of the plausibility of certain surfactants or microemulsions for remediating soils and natural solids contaminated by specific organic compounds.

## 7.4  SORPTION FROM ORGANIC SOLVENTS

### 7.4.1  Effect of Solvent Polarity

There are relatively few studies reported on the soil sorption of organic compounds and pesticides from organic-solvent solutions. Results from such studies are useful for illustrating soil sorptive behavior in general and for establishing the theoretical basis for selection of suitable solvents to recover contaminants from soil, either for laboratory analysis or for remediation. Obviously, the best solvents to recover contaminants are those that effectively suppress the contaminant sorption to soil (or sediment). Based on the behaviors of soil minerals and organic matter as described earlier, one would anticipate that the partitioning of nonionic organic compounds from organic solvents into soil organic matter (SOM) should be weak because of the good solvency of the solution phase (Chiou et al., 1981; Chiou and Shoup, 1985). Thus the extent of soil uptake would be determined mainly by the efficiency of the compounds to compete with the solvent for adsorption on soil minerals. Because of the inherent polarity of inorganic minerals, the adsorptivity of a compound (solute or solvent) would depend strongly on its ability to engage in polar interactions with minerals.

Hance (1965) reported a set of experimental data which highlight the difference in the sorption of diuron from aqueous and petroleum solutions by an oxidized soil ($f_{om} \leq 0.015$) and an organic soil ($f_{om} = 0.76$). The sorption of diuron to the oxidized soil was remarkably greater from petroleum solution than from water solution, whereas the sorption to the organic soil was remarkably greater from water solution than from petroleum solution. The diuron sorption from water solution to both soils was largely linear, whereas the sorption from petroleum solution was noticeably nonlinear. The sorption isotherms from petroleum solution are presented in Figure 7.36. The solubility of diuron in petroleum is about 30 mg/L and in water about 36 mg/L at room temperature. From the data observed, Hance concluded that there was a competition between diuron and water for adsorption sites in soil under aqueous slurry conditions and that diuron competed more effectively for SOM than for soil mineral surfaces. The high uptake of diuron from the petroleum solution by the oxidized soil along with the noted isotherm curvature is illustrative of the strong (competitive) adsorption of the polar solute (diuron) on soil minerals and the relatively weak competition of a nonpolar solvent (petroleum spirit) for adsorption on minerals. The isotherm curvature with the organic soil in Figure 7.36 is not evident, as the overall uptake is very weak; this is because the soil has a low mineral content. On the other hand, the high

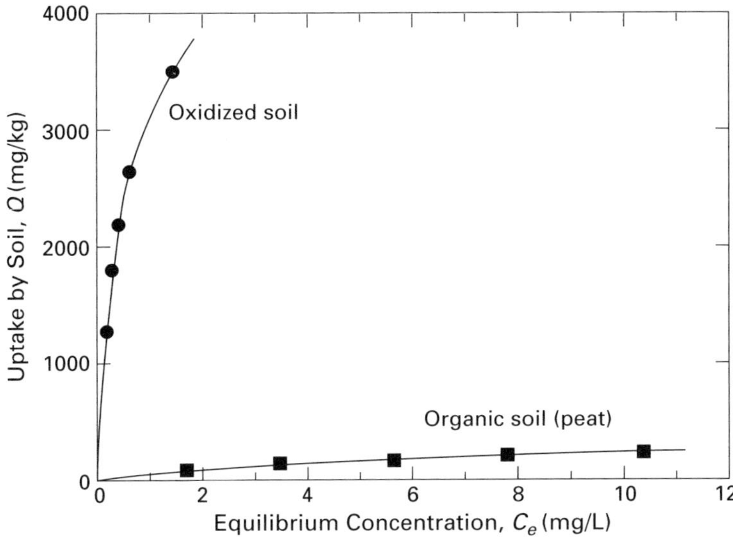


**Figure 7.36**   Sorption of diuron (DUN) from petroleum spirit solution on an oxidized soil ($f_{om} \leq 0.015$) and an organic soil ($f_{om} = 0.76$) at room temperature. [Adapted from the data of Hance (1965).]

uptake of diuron by the organic soil in water solution and the largely linear-sorption behavior are in keeping with the solute partition into SOM, with a concomitant suppression by water of adsorption on soil minerals.

Mills and Biggar (1969) found similar differences in the sorption of lindane by Venado clay (50% montmorillonite; $f_{om} = 0.06$) and Staten peat muck ($f_{om} = 0.22$) from aqueous and hexane solutions. The sorption capacities in aqueous systems are largely proportional to the SOM content in the soil samples, which is in accord with the partition model. In hexane solution, the sorption to the oven-dried Venado clay is considerably more enhanced than to dry peat muck. Moreover, whereas the equilibrium molar heats of lindane sorption in aqueous systems with both soil samples are less exothermic than the reverse heat of solution of lindane in water ($-\Delta \overline{H}_w$), the heats of sorption from hexane solution observed are significantly more exothermic than the reverse heat of solution in hexane. Once again, the presumably weak adsorption of relatively nonpolar hexane on minerals allows lindane to compete favorably for adsorption on mineral matter from hexane (while partition with organic matter is minimized), and consequently, the Venado clay shows a much greater sorption capacity.

Yaron and Saltzman (1972) studied the soil uptake of parathion from a wide range of organic solvents and the effect of soil water content on parathion uptake from these systems. Parathion shows a high uptake on dry soils from hexane, a lower uptake from benzene, and virtually no uptake from such polar solvents as methanol, ethanol, acetone, chloroform, ethyl acetate, and dioxane.

While the uptake from hexane by dry and nearly dry soils is remarkably higher than that from water, such uptake is strongly suppressed by humidity and approaches zero when the soils become water saturated (to be in contrast with the observation that parathion exhibits a definitive uptake on the same soils from water). The saturation water content ranges from about 2 wt% for the sandy Mivtahim soil (6% clay; $f_{om} = 0.003$) to 17% for the clay-rich Har Barqan soil (56% clay; $f_{om} = 0.019$). These observations led the authors to suggest that different mechanisms govern the parathion "adsorption" in aqueous soil suspensions and in hydrated soil–organic solvent systems. They assumed that in soil–water systems the solvent (water) is preferentially adsorbed but that small amounts of parathion diffuse through water films and get adsorbed as they approach the colloid surfaces.

It is evident from the parathion sorption data of Yaron and Saltzman (1972) that the dominant mechanism of the soil sorption in organic-solvent systems is different from that in aqueous systems. Results found in aqueous systems are reconcilable with the assumed solute partition into SOM. The high parathion uptake from hexane on dry soils is attributable to adsorption on soil minerals (mainly clays), on which the specific interactions of parathion's polar groups reduce the adsorptive competition of nonpolar hexane (while the partition of parathion into the organic matter is minimized by the good solvency of hexane). The parathion uptake from hexane would therefore be depressed by humidity because of the strong adsorptive competition of water for minerals (in this case, water is considered as a competing solute), which leads eventually to a nearly complete suppression of parathion uptake when the soils become fully water saturated. By comparison, parathion exhibits a definitive uptake on soil from aqueous solution because the poor solvency of water (in addition to its suppression of mineral adsorption) makes the partition of parathion into the SOM a favorable process. The failure of the soils to sorb parathion from polar organic solvents is due apparently to the fact that such solvents minimize solute (parathion) adsorption on minerals because of their polarity and reduce the solute partition into SOM by their good solvency. Therefore, polar solvents are much more effective than nonpolar solvents to recover nonionic organic contaminants from soil.

Since the presumed adsorption of organic solutes from an organic solvent onto a soil occurs by a competition of solute and solvent for relatively polar soil minerals, the amounts of adsorption for different solutes on a given soil will be closely related to the solute polarity. In light that solute adsorption would be most effective from a nonpolar solvent, as illustrated above, a polar solute should exhibit much higher adsorption than a nonpolar or weakly polar solute from a nonpolar solvent onto a soil. The sorption isotherms of parathion, lindane, 2,4′-PCB, and 1,2-dichlorobenzene from hexane onto Woodburn soil ($f_{om} = 0.019$), as shown in Figure 7.37, are in agreement with this expectation. Here the uptake of parathion is greatly enhanced over those of other relatively nonpolar solutes (lindane, 2,4′-PCB, and 1,2-dichlorobenzene) because parathion has a much higher polar-group content.

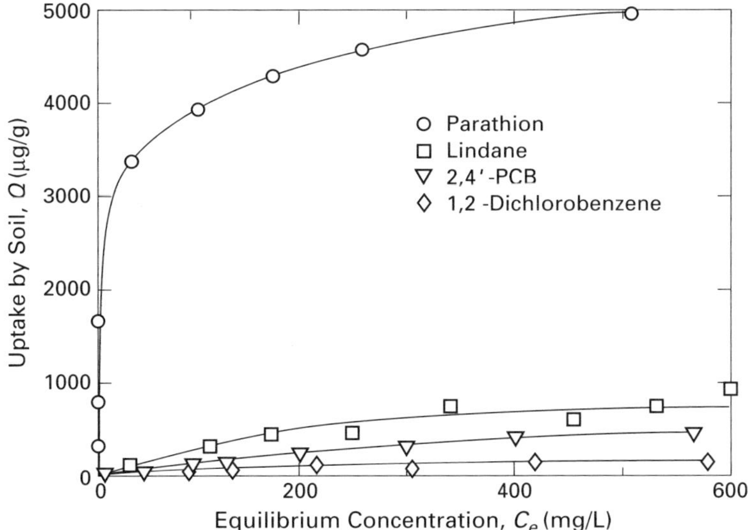


**Figure 7.37**   Sorption of parathion, lindane, 2,4′-PCB, and 1,2-dichlorobenzene from hexane on oven-dried Woodburn soil ($f_{om} = 0.019$) at 20°C. [Data of lindane, 2,4′-PCB, and 1,2-dichlorobenzene from C. T. Chiou, P. E. Porter, and D. W. Schmedding (unpublished research) and data of parathion from Chiou et al. (1985).]

Although the solute solubility in hexane must be considered in the comparison of isotherms, the solubility effect is relatively minor for these solutes because they exhibit comparable solubilities in hexane, as shown later for parathion and lindane.

### 7.4.2   Effects of Temperature, Moisture, and Contaminant Polarity

An interesting phenomenon observed by Yaron and Saltzman (1972) is that while parathion exhibited a significant sorption from hexane to certain partially hydrated soils, the sorption increased with increasing temperature. This finding is opposite to that of the parathion sorption from water, which decreased with increasing temperature as for most solutes. Such differences led Mingelgrin and Gerstl (1983) to suggest that the heat of adsorption of a solute from solution can be either exothermic or endothermic and consequently that the associated entropy change for solute adsorption can be either negative or positive.

The observed temperature dependence of parathion sorption from hexane with partially hydrated soils warrants careful consideration. As mentioned in Chapter 4, adsorption of single vapors or single solutes from a solution (to the extent that the solute concentration on adsorbent surfaces is enhanced significantly over that in the solution phase) should be accompanied by a high exothermic heat. Although the competition of a solute for adsorption against a strongly adsorbing solvent may lead to an endothermic or a small exothermic heat, the solute adsorption in this case must be either insignificant or weak. For example, as noted earlier, adsorption of phenanthrene from water onto various pure minerals (Huang and Weber, 1997) exhibits weak and linear uptake with small exothermic heats (i.e., less exothermic than the reverse heat of solute solution in water) because the solvent (water) is preferentially adsorbed. However, in certain binary-solute systems, the adsorptive competition of one solute against the other for adsorbent surfaces may give rise to an anomalous temperature effect, if their heats of adsorption per unit area as single solutes are relatively comparable and if the displacement of one solute by the other produces a large gain in entropy of the system.

In the parathion uptake from hexane on partially hydrated soils as reported by Yaron and Saltzman (1972), the amounts of water in soils and in hexane solution are below saturation. Since parathion contains many polar groups, it should have a relatively high heat of adsorption per unit area with dry soil minerals. In addition, parathion is considerably bigger in molecular size than water. Adsorption of 1 mole of parathion by displacing water from mineral surfaces of a partially hydrated soil (where there is enough water to cover all mineral sites) would release many moles of water into hexane solution. Although this displacement may be moderately endothermic for parathion, the gain in system entropy by releasing more water from surface sites to unsaturated hexane solution may be high enough to offset the enthalpic deficiency and thus to make the process favorable (Chiou, 1998). This entropic driving force diminishes as the hexane solution becomes more saturated with water. When the water content in soil exceeds saturation, there would be no more entropic driving force for parathion adsorption from hexane, making it difficult to detect the temperature effect. On the other hand, if the amount of water in soil is way below the monolayer capacity, the uptake of parathion from hexane should be exothermic because of its adsorption on abundant water-free mineral sites. Thus, although it is possible for the adsorption of a solute to be either exothermic or endothermic, as suggested by Mingelgrin and Gerstl (1983), the endothermic adsorption occurs only for special binary-solute systems in which the increase in total entropy is sufficient to compensate for the unfavorable enthalpic balance.

Chiou et al. (1985) further explored the sorption of parathion and lindane from aqueous and hexane solutions to substantiate the roles of soil minerals and organic matter in uptake by soil, using Woodburn soil (a mineral soil) and Lake Labish peat soil (an organic soil). The composition of Woodburn soil consists of 1.9% organic matter ($f_{om} = 0.019$), 68% silt, and 21% clay and that of Lake Labish peat soil of 51% organic matter ($f_{om} = 0.51$), 36% silt, and 3.5% clay. In aqueous systems, both parathion and lindane show linear isotherms, and there is no apparent sorptive competition between the two solutes; the results on Woodburn soil are illustrated in Figure 7.7. In oven-dried Woodburn soil–hexane systems, the sorption of parathion (and lindane) is nonlinear and much greater than the corresponding uptake from water. On air-dried Woodburn soil (with about 2.5% water), the uptake of parathion

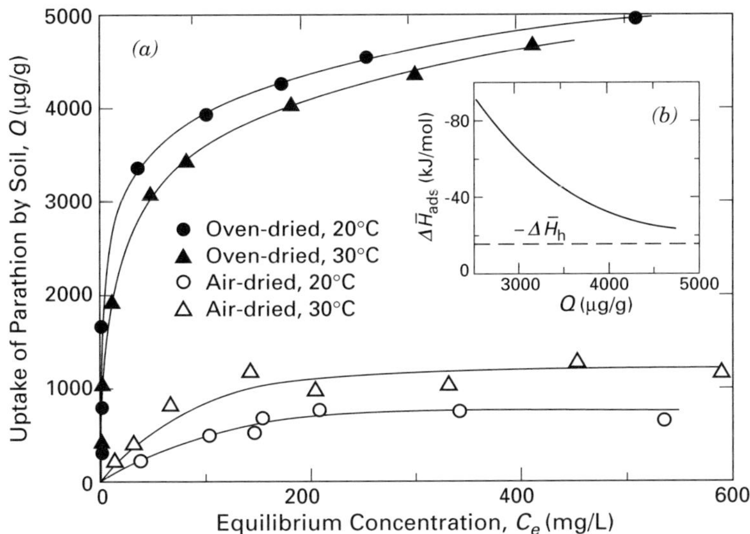


**Figure 7.38** (*a*) Sorption of parathion from hexane on oven- and air-dried Woodburn soil at 20 and 30°C. The air-dried soil contained about 2.5% moisture. (*b*) Isosteric heats of parathion sorption on Woodburn soil calculated from the oven-dried-soil isotherms [see Eq. (4.16)]. [Data from Chiou et al. (1985). Reproduced with permission.]

from hexane is significantly reduced; on water-saturated Woodburn soil (about 5% water), no detectable uptake is observed. The results are shown in Figure 7.38*a*. Parathion shows a similar endothermic uptake from hexane on air-dried Woodburn soil; the temperature effect is qualitative only because of a large scattering of the data.

As mentioned, comparison of the soil uptake of a solute from different solvents is more appropriately done on the basis of relative concentration (rather than absolute concentration), which corrects for differences in solubility of the solute in different solvents. Since the solubility of parathion in hexane is much higher ($5.74 \times 10^4$ mg/L at 20°C and $8.56 \times 10^4$ mg/L at 30°C) than in water (about 12 mg/L at 20°C) (Chiou et al., 1985), it is essential for the comparison to extend the parathion sorption from hexane to sufficiently high absolute concentrations. The data in Figure 7.38*a* show that over a relative concentration of parathion between 0 and 0.01 at 20°C, the isotherm exhibits a marked curvature, with capacities more than two orders of magnitude greater than in aqueous systems. Such curvature is not evident in the study of Yaron and Saltzman (1972) because their measurement was limited to very low relative concentrations, which fell within the Henry's law region. In the sorption of diuron from petroleum (Figure 7.36), a similar curvature arises when the relative concentration extends to 0.05 or so. Thus, the diuron and parathion isotherms, as depicted in Figures 7.36 and 7.38*a*, are mutually consistent.

The isosteric heat of parathion uptake determined from 20°C and 30°C isotherms with dry Woodburn soil is highly exothermic (more than the reverse heat of solution of parathion in hexane, $-\Delta\overline{H}_h$) and varies with the parathion loading, as shown in Figure 7.38$b$. These characteristics support the contention that parathion adsorption on soil minerals is the dominant sorption mechanism from a nonpolar solvent (hexane), as discussed previously. The high net exothermic heat per mole of parathion adsorption (up to 70 kJ/mol) comes presumably from the simultaneous interactions of parathion's many polar groups with relatively polar mineral surfaces. Thus, on a per mole basis, the heat of parathion adsorption can be greater than for water, whereas the heat of adsorption per unit mineral surface must be greater for water to account for the suppression of parathion adsorption by water.

The sorption data of lindane from hexane on dry and partially hydrated (Woodburn) soils exhibit essentially the same patterns as with parathion, except that the uptake of lindane is considerably lower at equal relative concentration and is more sharply reduced by the water content (Figure 7.39); the solubility of lindane in hexane at 20°C is $1.26 \times 10^4$ mg/L (relative to 7.8 mg/L in water at 25°C). With about 2.5% water in soil, the lindane uptake is reduced nearly 25 times relative to the capacity with dry soil, which is more intense than with parathion, and further addition of water to the saturation point suppresses the lindane uptake to a nondetectable level. These differences are

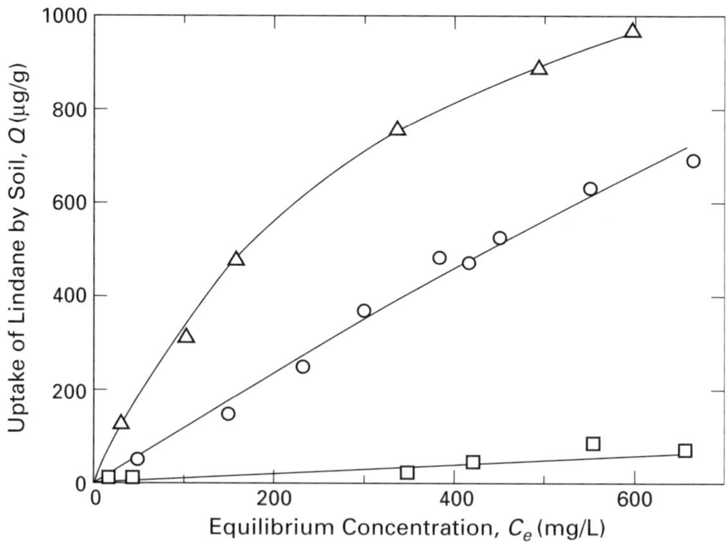

**Figure 7.39**  Sorption of lindane from hexane on oven-dried and partially hydrated Woodburn soil at 20°C: lindane only on oven-dried soil ($\triangle$): lindane on 5 mg of water per gram of soil ($\bigcirc$); lindane on 25 mg of water per gram of soil ($\square$). [Data from Chiou et al. (1985). Reproduced with permission.]

consistent with the lower polarity of lindane relative to parathion, making lindane a less potent adsorbate, and thus a much weaker competitor than parathion against water for mineral adsorption. Being a relatively nonpolar adsorbate, lindane must exhibit a considerably smaller heat of adsorption per unit area than water on dry soil minerals and consequently it cannot effectively displace water from a partially hydrated soil. Thus, when all mineral surfaces are essentially covered by water, as for Woodburn soil with about 2.5% water, the adsorption of lindane from hexane takes place presumably on less energetic and more uniform adsorbed-water surfaces, resulting in a weak and essentially linear uptake. This feature is in contrast to that found for parathion under the same system condition.

While the water content in soil suppresses the uptake of parathion and lindane from hexane, a similar competitive effect also occurs between parathion and lindane in their simultaneous sorption from hexane on dry Woodburn soil, as illustrated in Figure 7.40, where the uptake of lindane decreases with an increase of parathion uptake. Finally, both parathion and lindane show a considerably higher uptake from hexane on mineral-rich Woodburn soil than on organic-rich Lake Labish peat soil. This is the opposite of what is found in water solution, because of the dominance of solute adsorption on mineral matter from a nonpolar solvent and of the dominance of solute partition in SOM from water (Chiou et al., 1985).

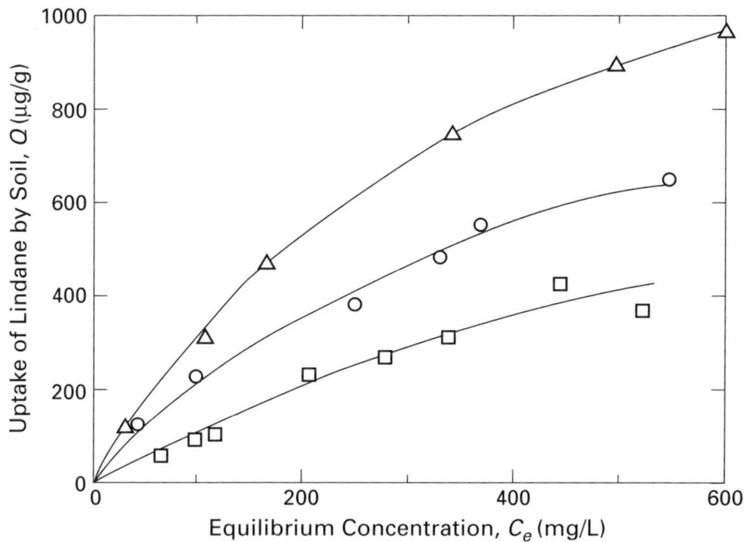

**Figure 7.40** Sorption of lindane from hexane on oven-dried Woodburn soil with and without parathion as the competing solute at 20°C: lindane only on oven-dried soil ($\triangle$): lindane on 0.5 mg of parathion per gram of soil ($\bigcirc$); lindane on 3 mg of parathion per gram of soil ($\square$). [Data from Chiou et al. (1985). Reproduced with permission.]

## 7.5  SORPTION FROM VAPOR PHASE

### 7.5.1  General Aspects of Vapor Sorption

Following our deliberations on the sorptive effects of SOM and minerals, we expect the sorption of organic vapors on relatively dry soils to consist of adsorption on soil minerals (and on HSACM, if present) and concurrent partition into SOM. For ordinary dry soils that are abundant in mineral content, the adsorptive contribution would undoubtedly predominate the overall soil uptake. Shipinov (1940) found that the vapor sorption of hydrogen cyanide on dry soils was nonlinear (BET type II shape). Stark (1948) showed similar nonlinear isotherms for chloropicrin vapor uptake on dry soils and found a close correlation between sorption capacities and soil clay contents; the sorption capacity decreased as the soil-moisture content increased. Hanson and Nex (1953) observed that at a soil moisture content substantially below the wilting point, ethylene dibromide (EDB) was strongly sorbed by the soil, but that the sorption decreased sharply to a minimum near the wilting point. Wade (1954) studied the EDB vapor sorption by three soils having very different clay versus SOM contents. A much greater EDB sorption was found on dry soils; moisture sharply suppressed the EDB sorption only before the soil-water content reached saturation, after which the sorption was unaffected by soil-water content and the EDB isotherms were essentially linear. On wet soils, the EDB sorption capacities were closely proportional to respective SOM contents, with no correlation to clay contents.

Jurinak (1957a,b) and Jurinak and Volman (1957) studied the vapor uptake of EDB and 1,2-dibromo-3-chloropropane on dehydrated clays (which still retained small amounts of water) in relation to clay types and exchanged cations. The extent of sorption was related to the surface areas of specific clays and the sorption data fit the BET adsorption model. The BET monolayer capacity of EDB increased from 17.5 g/kg for nonexpanding Ca-kaolinite (1.0% water) to 72.1 g/kg for expanding Ca-montmorillonite (2.3% water). Call (1957) studied the dependence of EDB vapor uptake on several soils and a Ca-montmorillonite on relative humidity (RH). On relatively dry Ca-montmorillonite and soils (RH < 20%), the EDB isotherms were type II shape, in which the sorption capacity increased with increasing clay content. An increase of RH from 0 to 50% progressively suppressed the EDB sorption on the soils, with a concomitant change of the isotherm shape toward linearity. On Ca-montmorillonite, however, the EDB sorption increased sharply when RH increased from 0 to 10%, which was attributed to the clay-layer expansion creating additional surfaces; however, with RH > 10%, the EDB sorption decreased with increasing RH, and at RH = 90% the sorption became very small relative to that with the dry clay. By these observations, Call suggested that sorption of EDB on clays and soils at low RH resulted from adsorption on mineral surfaces, whereas the sorption on wet soils and soils at high RH occurred by dissolution in soil water or by adsorption, as a Gibbs surface excess, onto the adsorbed water surface. We shall see later that the vapor uptake on wet soils is better explained in terms of the vapor partition into SOM.

Leistra (1970) presented results on the vapor uptake of *cis-* and *trans-*1,3-dichloropropene on three types of soils: humus sand ($f_{om}$ = 0.055), peat sand ($f_{om}$ = 0.18), and peat ($f_{om}$ = 0.95), with moisture contents of 17%, 41%, and 120% of the dry soil weight, respectively. The isotherms for all three soils were highly linear, with the soil-to-vapor distribution coefficients being proportional to the respective SOM contents; the SOM-normalized sorption coefficients ($K_{om}$) were largely independent of the soils for each vapor. The $K_{om}$ values of both compounds exhibited a small temperature dependence, with $\Delta \overline{H}$ being < 4 kJ/mol exothermic and nearly constant (Hamaker and Thompson, 1972). The data suggest that at these moisture contents the soils were fully water saturated. Note that the amount of water needed to saturate the soil in sorption is obviously much lower than the field water saturation capacity; the former seems to be close to the water content at the soil wilting point. Analogously, the moisture content in some conventional unsaturated zones (i.e., the vadose zones) may be well above the water-sorption saturation level. It is important that the water saturation level in soil sorption not be confused with the field water-saturation capacity.

From the vapor sorption data analyzed so far, it is evident that dry and slightly hydrated soil minerals (especially, clays) act as powerful adsorbents for organic vapors and that the contribution by clay adsorption greatly exceeds the concurrent vapor partition into the organic matter on most dry mineral soils. Apparently, at saturation-water contents, the adsorptive power of soil minerals for organic compounds is largely lost because of strong competitive adsorption of water (Chiou and Shoup, 1985), leaving the partition with SOM as the dominant mechanism. The fact that the organic-vapor uptake by dry soils is closely related to clay content rather than to SOM content suggests that dry clay is more powerful per unit weight in adsorption of organic compounds than is SOM per unit weight in uptake by partition; the reverse is true for the hydrated soils. The suppression of vapor uptake observed on soils by moisture is essentially the same as noted in the suppression by water of parathion and lindane uptake from a nonpolar solvent (hexane). The only difference is that the organic solvent also minimizes the solute partition in SOM, making the total solute uptake approach zero at full water saturation. Although the water content also affects the vapor partition in SOM, to be illustrated later, the partition uptake with water-saturated SOM remains substantial. More data on vapor sorption in relation to RH or soil-water content are presented later.

Relative to soil mineral adsorption, there have been few studies on the partition uptakes of different vapors by relatively dry SOM. Using the Florida peat ($f_{om}$ = 0.864) as a model for SOM, Rutherford and Chiou (1992) and Chiou and Kile (1994) measured the vapor partition to dry SOM at room temperature of some nonpolar and polar liquids: benzene, carbon tetrachloride

**TABLE 7.21. Limiting Partition Capacities ($Q_{om}^\circ$), Volume Fraction Solubilities ($\phi^\circ$) of Liquids, and Solubility Parameters ($\delta$) of Liquids in Peat Organic Matter**

| Liquid | $\delta$ (cal/cm$^3$)$^{0.5}$ | $Q_{om}^\circ$ (mg/g) | $\phi^\circ$ |
|---|---|---|---|
| *n*-Hexane | 7.3 | 28.2 | 0.053 |
| Carbon tetrachloride | 8.6 | 65.9 | 0.051 |
| Benzene | 9.2 | 38.9 | 0.054 |
| Trichloroethylene | 9.2 | 80.0 | 0.067 |
| 1,4-Dioxane | 10.0 | 80.2 | 0.092 |
| EGME | 10.5 | 190 | 0.21 |
| Acetone | 9.9 | 171 | 0.22 |
| Nitroethane | 11.1 | 272 | 0.25 |
| Acetonitrile | 11.9 | 344 | 0.36 |
| 1-Propanol | 11.9 | 313 | 0.34 |
| Ethanol | 12.7 | 396 | 0.40 |
| Methanol | 14.5 | 620 | 0.51 |
| Water | 23.4[a] | 370 | 0.33 |

*Source*: Data from Chiou and Kile (1994).

[a] Value uncertain.

(CT), trichloroethylene (TCE), *n*-hexane, 1,4-dioxane, EGME, nitroethane, acetone, acetonitrile, methanol, ethanol, 1-propanol, and water. Except for water vapor, the isotherms for the organic vapors are largely linear, reflecting the predominance of the partition effect. Water exhibits a large uptake and a unique isotherm, with a profound concave-downward curvature at low RH but a good linearity at moderate to high RH. It appears that water engages initially in a hydrate formation with SOM and then partitions subsequently into the hydrated SOM net work (Chiou and Kile, 1994). For the essentially linear uptakes of organic vapors, their limiting partition capacities (i.e., solubilities), $Q_{om}^\circ$, with the SOM may be obtained by extrapolating the respective isotherms to $P/P^\circ = 1$ and normalizing the capacities to $f_{om}$. The $Q_{om}^\circ$ values calculated in mg/g and the corresponding volume-fraction solubilities, $\phi^\circ$, of the liquids with dry SOM are given in Table 7.21.

We recall that SOM is a relatively polar phase. As a consequence, the $Q_{om}^\circ$ values of nonpolar liquids (e.g., hexane, CT, and benzene) are about an order of magnitude smaller than those of highly polar liquids (e.g., methanol and ethanol), in keeping with the solubility criterion. As seen, the $Q_{om}^\circ$ values of nonpolar liquids with dry SOM of the peat in Table 7.21 are comparable with the corresponding $S_{om}^\circ$ values of the liquids with water-saturated SOM in Table 7.2. In Table 7.21, one also notes that the nonpolar liquids display very consistent $\phi^\circ$ values, because the solubilities of solutes in a polymer or a macromolecular substance are usually better accounted for by their volume fractions (Flory, 1941). The much higher $Q_{om}^\circ$ values for polar liquids reflect their enhanced solubilities in relatively polar SOM, as assisted by more powerful polar and H-bonding forces. Since the solubility parameter ($\delta$) of a liquid is a

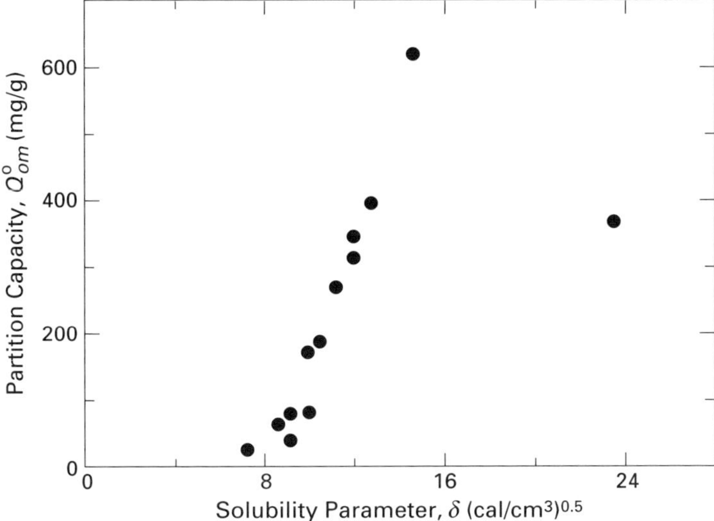

**Figure 7.41** Partition capacities of organic liquids and water in peat organic matter ($Q^{\circ}_{om}$) as a function of their respective solubility parameters ($\delta$) at room temperature. [Data from Chiou and Kile (1994).]

good index for its polarity, the $Q^{\circ}_{om}$ and $\delta$ values of the studied liquids are well correlated, as shown in Table 7.21. In the plot of $Q^{\circ}_{om}$ versus $\delta$ shown in Figure 7.41, which supposedly should yield a bell-shaped curve if sufficient data are available, the $\delta$ corresponding to the maximum $Q^{\circ}_{om}$ is taken as the best solubility parameter for dry SOM. Although there is a lack of data between $\delta = 14.5$ (methanol) and 23.4 (water), as there are few liquids with $\delta$ values falling into this range, the $\delta$ value for dry SOM should be more than 14.5, judging from the plot in Figure 7.41. The high $\delta$ value for SOM seems reasonable since it is relatively polar in nature. Finally, similar to the $S^{\circ}_{om}$ values in Table 7.2, the $Q^{\circ}_{om}$ value of a solid compound would be smaller than that of a similar liquid because of the melting-point effect.

## 7.5.2  Influence of Moisture on Vapor Sorption

Consider first the equilibrium vapor concentrations of lindane and dieldrin in a mineral soil at different soil-water contents, as determined by Spencer et al. (1969) and Spencer and Cliath (1970). At soil-water contents <2.2% on Gila silt loam (Typic Torrifluvent, $f_{om} = 0.006$), the measured equilibrium vapor densities of lindane at about 50 mg/kg soil and dieldrin at 100 mg/kg soil were substantially lower than the corresponding saturation vapor densities of the pure compounds. This suggests that the amounts of pesticides applied to the soil with water content <2.2% were much below the saturation limits. An increase in soil-water content to >3.9%, however, led to sharp increases of equilibrium

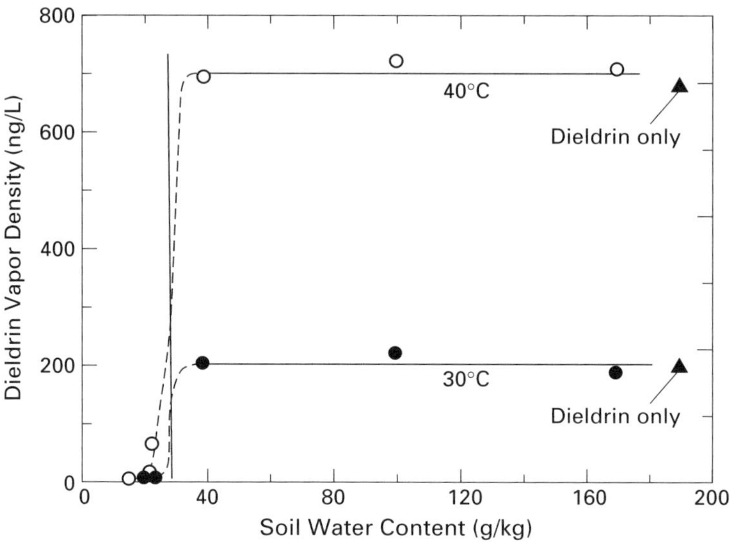


**Figure 7.42**    Influence of soil water content on the vapor density of dieldrin applied to Gila silt loam at 100 mg of dieldrin per kilogram of soil. [Data from Spencer et al. (1969). Reproduced with permission.]

vapor concentrations, which became equal to the saturation vapor concentrations of the pure compounds and stayed unchanged with water content up to the soil's field capacity (17%). Results with dieldrin at 30 and 40°C are shown in Figure 7.42.

The fact that applied dieldrin on Gila soil at 100 mg/kg soil and lindane at 50 mg/kg soil show subsaturation vapor concentrations when the soil-water content is low but display saturation vapor concentrations when the soil is wet is illustrative of the distinct sorption mechanisms of soil minerals and organic matter. The low vapor concentrations with the relatively dry soil are ascribed to strong mineral adsorption, which overrides the effect of partition with SOM. Upon wetting, water displaces most pesticide from soil minerals by adsorptive competition, and, as a result the amount of pesticide in soil becomes more than enough to saturate the SOM, and consequently, the vapor concentrations become saturated. At water saturation, one notes that 100 mg of dieldrin per kilogram of Gila soil (with $f_{om} = 0.006$) corresponds to 17 g of dieldrin per kilogram of SOM, and this loading is far greater than the solubility of solid dieldrin in SOM at 25°C; that is, $S_{om} = 1.5$ g of dieldrin per kilogram of SOM, calculated by Eq. (7.10) with $S_w = 0.20$ mg/L (Weil et al., 1974) and $K_{om} = 7400$ (Briggs, 1981). Similarly, at water saturation, a loading of 50 mg of lindane per kilogram of Gila soil corresponds to 8.3 g of lindane per kilogram of SOM, which is more than the calculated $S_{om} = 3.0$ g of lindane per kilogram of SOM for solid lindane at 25°C by Eq. (7.10) with $S_w = 7.8$ mg/L (Weil et al., 1974) and $K_{om} = 380$ (Chiou et al., 1985). Thus, for the amounts of dieldrin and lindane applied to water-saturated Gila soil, the pesticide vapor concentrations will remain unchanged at saturation with increasing soil-water content. This effect should persist until the water content becomes high enough to dilute the excess pesticides to below saturation.

The analysis above accounts further for the observation that the lindane sorption capacity on the hydrated Gila soil is largely the same as that on the same soil with 3.9% or 10% water (Spencer and Cliath, 1970). At 3.9% water on Gila soil, which corresponds to about the point of water saturation, the heat of lindane vapor sorption calculated from the temperature dependence of the isotherms was less exothermic than the reverse heat of lindane vaporization $(-\Delta \overline{H}_v)$, due presumably to the dominance of vapor partition in SOM of a water-saturated soil.

To substantiate the effect of humidity on the mechanism and capacity of soil sorption of organic compounds, Chiou and Shoup (1985) determined the vapor sorption of benzene, chlorobenzene, $m$-dichlorobenzene, $p$-dicholorobenzene, 1,2,4-trichlorobenzene, and water on dry Woodburn soil, and of benzene, $m$-dichlorobenzene, and 1,2,4-trichlorobenzene as functions of relative humidity (RH). Isotherms for all compounds as single vapors on dry soil in a normalized plot of $Q$ versus $P/P°$ (where $P$ and $P°$ are equilibrium and saturation vapor pressures, respectively) were distinctly nonlinear, with water showing the greatest capacity (Figure 7.43). The data observed closely fit the BET adsorption model. In addition to the isotherm nonlinearity, the sorption capacities of organic vapors on dry soil were about two orders of magnitude greater than those of the same compounds from water on the same soil shown in Figure 7.2 (when the data therein are expressed as $Q$ versus $C_e/S_w$). The considerably greater sorption on dry soil is attributed to strong adsorption on soil minerals, which predominates over the simultaneous uptake by partition into the SOM. The reason for this strong suppression by water of adsorption of organic compounds on soil minerals is elucidated in Chapter 6.

The sorption of benzene, $m$-dichlorobenzene, and 1,2,4-trichlorobenzene by initially dry Woodburn soil was depressed progressively by increasing RH (Chiou and Shoup, 1985). The results for $m$-dichlorobenzene and 1,2,4-trichlorobenzene are given in Figures 7.44 and 7.45, respectively. In addition to the reduced uptake in the presence of water vapor, the isotherms for these relatively nonpolar vapors at RH $\geq$ 50% assume a practically linear shape at $P/P° \leq 0.5$. At high RH before water saturation, the vapor uptake observed should consist largely of adsorption on adsorbed-water surfaces, as these nonpolar vapors cannot effectively displace the adsorbed water, and of a concomitant partition into the SOM, producing relatively linear isotherms up to moderate $P/P°$. At high $P/P°$ values of the vapor, the vapor isotherm may exhibit nonlinearity, as shown by Call (1957) for EDB, due to multilayer vapor adsorption on water surfaces, if a significant portion of the water-associated mineral surface area remains available (which should decrease with the

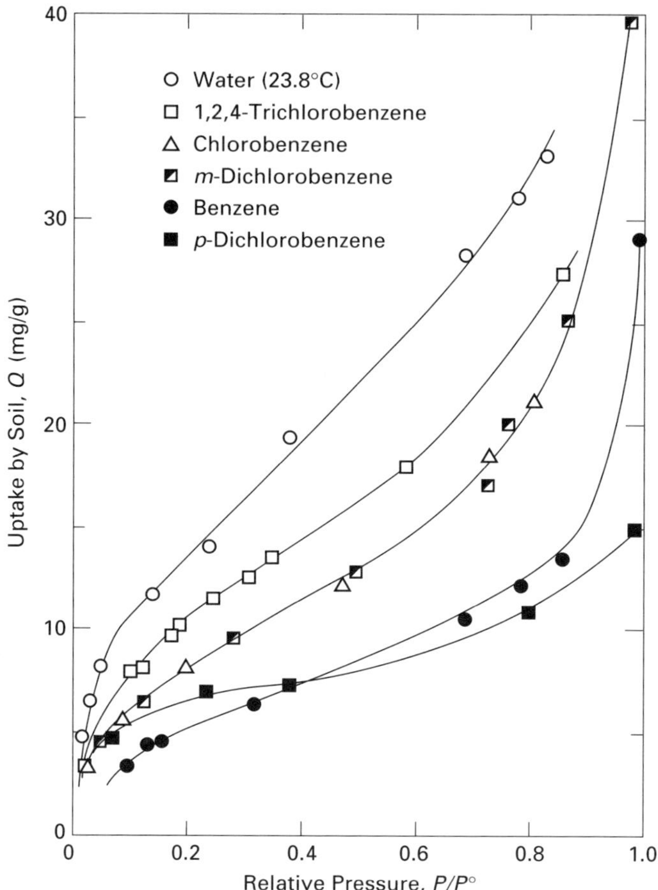



**Figure 7.43**  Vapor uptake of benzene, chlorobenzene, *m*-dichlorobenzene, *p*-dichlorobenzene, 1,2,4-trichlorobenzene, and water on dry Woodburn soil as a function of relative pressure at 20°C. [Data from Chiou and Shoup (1985). Reproduced with permission.]

amount of water adsorbed). As adsorption of the vapors on water surfaces is decreased further by increasing amounts of adsorbed water on minerals, the vapor partition into SOM becomes increasingly more important, and thus the isotherm linearity extends to higher $P/P°$, as noted similarly for the EDB vapor uptake in relation to RH (Call, 1957). At about 90% RH, the sorption capacities of the compounds fall into a range close to those on the water-saturated soil. Similar suppressions of the vapor sorption by RH and variations in vapor isotherm shape were found for trichloroethylene (TCE) on soil (Smith et al., 1990a), TCE on clays and oxides (Ong and Lion, 1991), *p*-xylene on soil and silica gel (Pennell et al., 1992), and chlorobenzene and toluene on soil (Thibaud et al., 1993).

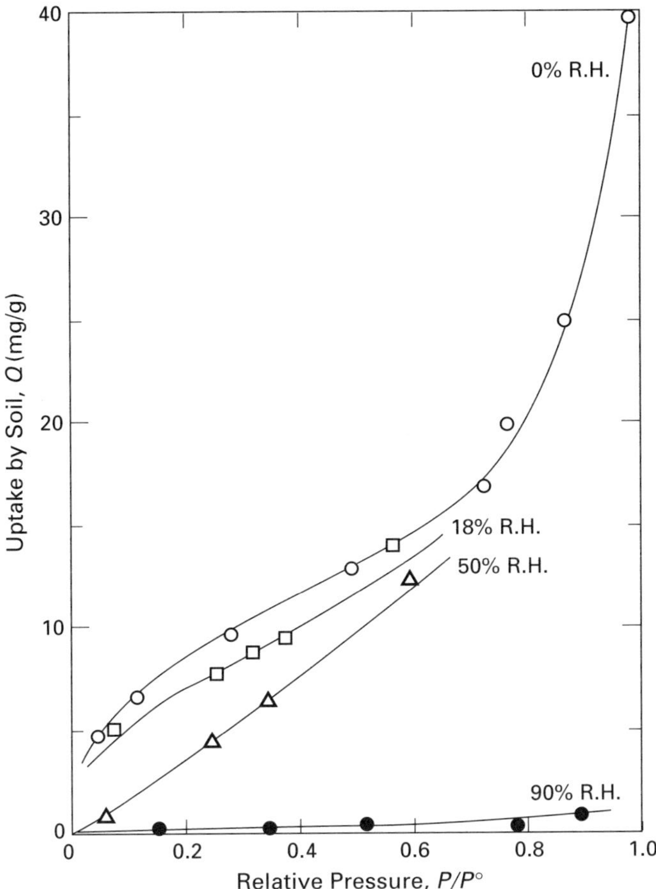



**Figure 7.44**   Vapor uptake of *m*-dichlorobenzene on dry Woodburn soil as a function of relative humidity at 20°C. [Data from Chiou and Shoup (1985). Reproduced with permission.]

Whereas the vapor sorption data with either relatively dry soils or with nearly water-saturated soils are well appreciated in terms of either mineral-dominated adsorption or of SOM-dominated partition, there are conceptually different views of how an organic vapor adsorbs on water-film-covered mineral surfaces. According to Chiou (1998) and Chiou and Shoup (1985), adsorption of a relatively nonpolar vapor onto the water film of a mineral is merely a consequence of the adsorptive competition between water and the less-adsorbing vapor, in which the more energetic adsorption of water forces the vapor to adsorb on top of the water film. Alternatively, Call (1957) and Pennell et al. (1992) viewed the organic-vapor adsorption as a Gibbs surface-excess effect of the adsorbed water film in which the vapor dissolves. It should be recognized, however, that the Gibbs equation applies only for solutes on

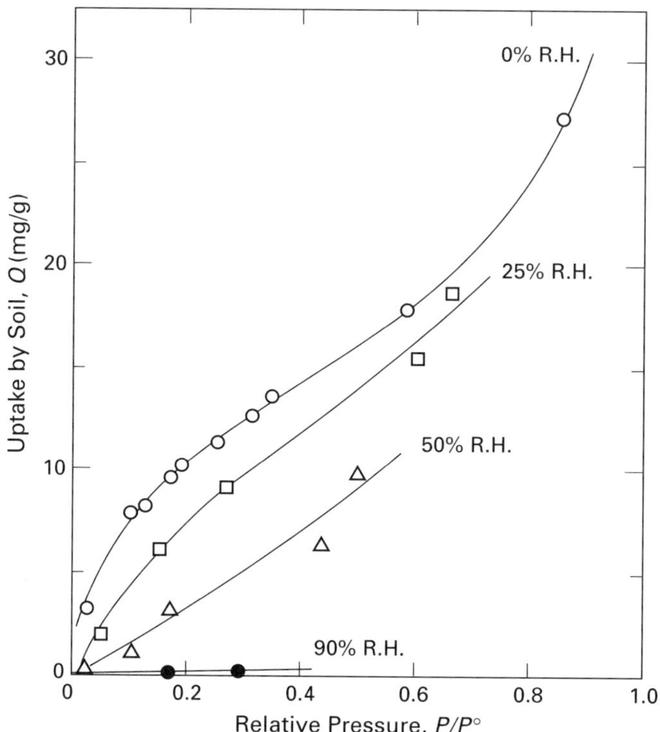



**Figure 7.45**   Vapor uptake of 1,2,4-dichlorobenzene on dry Woodburn soil as a function of relative humidity at 20°C. [Data from Chiou and Shoup (1985). Reproduced with permission.]

an energetically uniform bulk surface. As such, it is doubtfully applicable to a mineral-adsorbed water film, because it is not a bulk phase and may not be energetically uniform at a given film thickness. Moreover, the Gibbs surface excess calculated by assuming that the water film in minerals has the same surface area as the dry minerals (Pennell et al., 1992) would also overestimate the result. In all likelihood, the vapor adsorption by the Gibbs surface-excess effect is insignificant, since the vapor sorption normally decreases sensitively with increasing soil water content, which is not supported by the Gibbs theory. Incidentally, the suggested vapor dissolution into adsorbed water film (Call, 1957) is usually relatively insignificant, as shown by Pennell et al. (1992) and Thibaud et al. (1993).

In contrast to the observations with mineral soils, the organic-vapor sorption with a high-organic-content soil should be more linear and less dependent on RH because the vapor partition to SOM would be substantial relative to mineral adsorption. Rutherford and Chiou (1992) showed that vapor isotherms of benzene, carbon tetrachloride (CT), and trichloroethylene (TCE) on a peat ($f_{om} = 0.864$) and a muck ($f_{om} = 0.815$) are essentially linear. Results

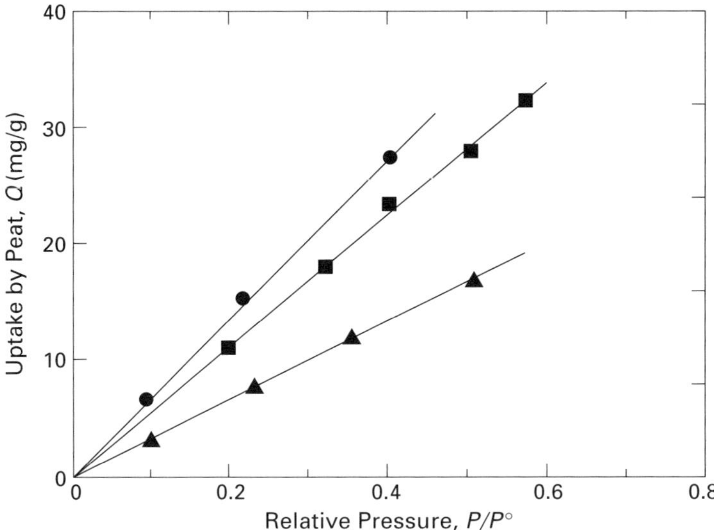

**Figure 7.46**   Vapor uptake of trichloroethylene (TCE) (●), carbon tetrachloride (CT) (■), and benzene (▲) on Florida peat as a function of relative pressure at room temperature. [Data from Rutherford and Chiou (1992).]

on dry peat are shown in Figure 7.46. The high isotherm linearity is indicative of the predominance of vapor partition to the organic matter. Note that the sorbed-vapor capacity on dry peat is orders of magnitude too high to be reconciled with the surface area of the sample ($1.4\,m^2/g$). The small mineral content in peat is likely covered by SOM and thus inaccessible to vapors. To determine the influence of water saturation in peat on vapor partition, one compares the sorption isotherms from vapor phase ($Q$ versus $P/P°$) with those from aqueous solution ($Q$ versus $C_e/S_w$), since $P/P° = C_e/S_w$. The data for benzene and CT on peat are shown in Figure 7.47. The saturation of organic matter by water reduces the vapor partition by some 40 to 50% on peat and 30 to 40% on muck (Rutherford and Chiou, 1992). This reduction results presumably from the increased polarity of the water-saturated SOM, which makes it less compatible with low-polarity compounds (due to a greater mismatch of their polarities). This effect is, however, small compared with the remarkable suppression by water of the adsorption of nonpolar vapors and solutes on minerals.

The strong dependence of vapor sorption by a mineral soil on RH (or soil-water content) provides a basis to explain the sharp variation in the activity of a soil-incorporated contaminant with soil-water content. As illustrated in Figure 7.44, the relative pressure $P/P°$ [i.e., the chemical activity; see Eq. (2.1)] of a contaminant in soil is affected not only by the level of contaminant in soil but also by RH because the limiting (saturation) capacity of a contaminant in soil is mediated by soil-water content. For $m$-dichlorobenzene on Woodburn

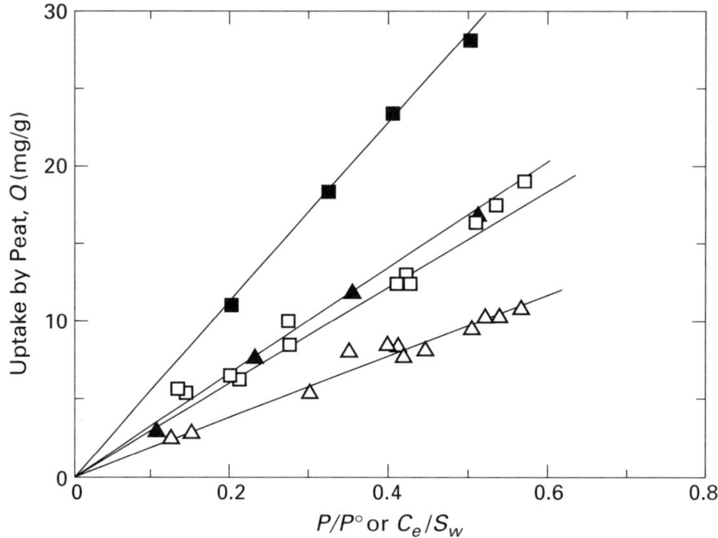

**Figure 7.47**    Comparison of the uptakes of CT (■,□) and benzene (▲,△) from vapor phase on dry Florida peat ($Q$ versus $P/P°$) and from water solution ($Q$ versus $C_e/S_w$) on wet Florida peat. Solid symbols are for vapor-phase data and open symbols for water-phase data. [Data from Rutherford and Chiou (1992).]

soil at, say, 0.4 mg per gram of soil, the $P/P°$ is <0.01 at RH ≤ 50% but increases to about 0.8 at RH = 90%, which amounts to nearly a 100-fold increase in vapor activity because of the change by water of the soil's sorption capacity. The effect of RH on vapor activity is intimately consistent with the results of Spencer and Cliath (1970) and Spencer et al. (1969) on the sharp variation of lindane and dieldrin vapor densities with soil-water content. In natural environments, the transport of organic contaminants across soil–air interfaces would hence be strongly influenced by a drastic change of ambient humidity or soil-water content.

## 7.6    INFLUENCE OF SORPTION ON CONTAMINANT ACTIVITY

In preceding sections we have elucidated the relation of soil/sediment sorption to contaminant properties, mineral–SOM composition, and ambient factors such as temperature and humidity. On the basis of the described mechanistic roles of SOM and minerals, and the effect of soil water, one can now gain a better perspective of the diverse environmental behavior of the contaminants. In aqueous systems, the governing factors for the sorption of nonpolar contaminants are the SOM content, the contaminant water solubility, the potential HSACM effect (at some special conditions), and for highly water-insoluble compounds, the amount and type of dissolved and suspended organic matters. The individual sorption coefficients of the polar contaminants may also be subject to change to a certain extent with their own concentrations and with the in situ concentrations of other compounds. In nonaqueous systems, the important factors are the mineral type and content, the ambient humidity (or soil-moisture content), and the contaminant and solvent (medium) polarities.

An important consequence of the suppression of mineral adsorption by water and the partition into SOM of contaminants in soil-water systems is that the contaminant mass sorbed to a soil (or sediment) should usually be only a small fraction of the SOM content, regardless of the contaminant level in water. This is because most contaminants of interest have a small solubility in SOM. For example, Choi and Chen (1976) found that concentrations of chlorinated hydrocarbons (DDT, DDE, and PCBs) in sediments of a marine site were well correlated with the organic carbon contents of the sediments; total chlorinated hydrocarbon concentrations ranged from 0.3 to 3.5 mg per kilogram of sediment, whereas total organic carbon contents ranged from 4.5 to 17 g per kilogram of sediment on a dry weight basis. Similarly, Goerlitz et al. (1985) found that sandy aquifer sediments, which contained trace amounts of organic matter, in a creosote-contaminated groundwater site exhibited little retention of substituted phenols and PAHs, despite the fact that the groundwater was contaminated with significant levels of these compounds. By this account, a more sensible criterion for evaluating the extent of sediment contamination or the sediment quality in an aquatic system should be based on the contaminant level normalized to the SOM content rather than on the level with respect to the whole sediment.

The contaminant activity in a terrestrial environment is expected to be influenced most sensitively by a change in soil-water content from above to below the saturation capacity in the field, because the drying–wetting cycle sharply affects the contaminant uptake by soils. The very top layer of surficial soils undergoes a drying–wetting transition, the extent of which depends on ambient humidity and field operation. Under relatively dry conditions, strong adsorption by soil minerals (particularly, high-surface-area clays) along with the partition into SOM lowers the activity (as measured by $P/P°$) of a contaminant in the soil. Upon wetting, the chemical activity would rise rapidly and sharply because of the displacement by water of those species previously adsorbed on soil minerals (Goring, 1967; Chiou and Shoup, 1985). Volatile compounds sorbed initially to dry soils and clays are thus readily released by wetting the soil and clay (Chisholm and Koblitsky, 1943; Stark, 1948; Hanson and Nex, 1953; Wade, 1954; Goring, 1967; Chiou and Shoup, 1985; Pennell et al., 1992; Thibaud et al., 1993). In fields to which pesticides have been applied, the rainfall event and the dew deposition on soil surfaces trigger large and sudden increases of volatile pesticides into the air (Glotfelty et al., 1984; Grover et al., 1988; Majewski et al., 1993). If the soil surface is relatively dry, the evaporative fluxes of pesticides are usually small. An example illustrating this phenomenon is given in Figure 7.48, where rainfall in a field brought about

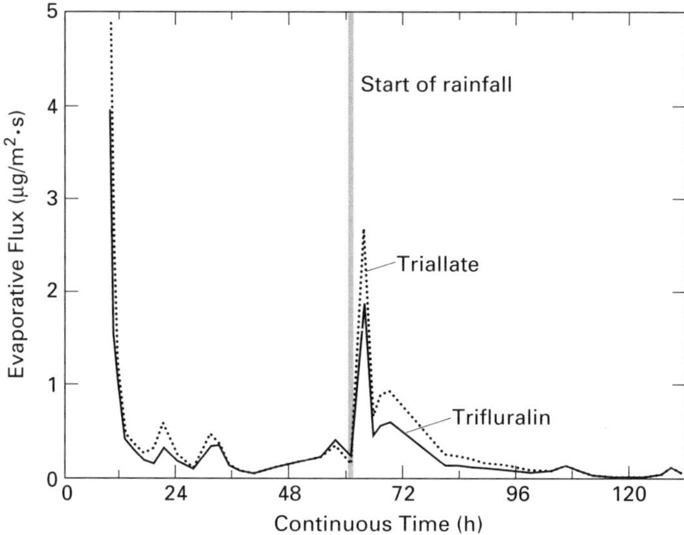


**Figure 7.48** Evaporative fluxes of triallate and trifluralin herbicides with time, showing the effect of a rainfall after the spray of herbicides to a field site in Ottawa, Canada. [Data from Majewski et al. (1993). Reproduced with permission.]

**TABLE 7.22. Influence of Soil Type and Soil Moisture on the Toxicity of Insecticides in Soils to First-Instar Nymphs of the Common Field Cricket [*Gryllus pennsylvanicus* (Burmeister)]**

| Insecticide | Soil Type[a] | $LD_{50}$ (mg/kg[1]) | |
| --- | --- | --- | --- |
| | | Moist | Dry |
| Heptachlor | Plainfield sand | 0.068 | 0.53 |
| | Muck | 4.19 | 5.39 |
| DDT | Plainfield sand | 1.75 | 17.3 |
| | Muck | 67.2 | 99.8 |
| Diazinon | Plainfield sand | 0.26 | 34.1 |
| | Muck | 17.0 | 11.5 |
| V-C 13 | Plainfield sand | 3.80 | 717 |
| | Muck | 279 | 165 |
| Parathion | Plainfield sand | 0.25 | 6.00 |
| | Muck | 22.6 | 9.10 |

*Source*: Data from Harris (1964).

[a] Plainfield sand (mixed, mesic Typic Udipsamment).

a sharp rise in the vapor fluxes of volatile pesticides (trifluralin and triallate) from the soil surfaces. Although the increased vapor fluxes may not be significant for nonvolatile chemicals with a similar soil wetting, their chemical activities should rise as sharply as those of the volatile chemicals as a result of their desorption from soils.

A change in the chemical activity of a soil-incorporated pesticide by soil moisture leads invariably to a change in the apparent pesticidal toxicity. Upchurch (1957) found that diuron was more toxic to cotton (*Gossypium hirsutum* L.) under moist than under dry soil conditions. Barlow and Hadaway (1955) observed that chlorinated insecticides (lindane, DDT, and dieldrin) were inactivated for mosquito control by dry clay but were reactivated under high humidities. Gerolt (1961) reported that the toxicity of dieldrin in a soil to insects increased sharply with an increase in ambient humidity. Harris (1964) observed correlations of the insecticide toxicities of heptachlor, DDT, diazinon, V-C 13 (dichlofenthion), and parathion on a sandy soil ($f_{om}$ = 0.0052) and a muck soil ($f_{om}$ = 0.65) with soil-moisture and organic-matter contents (Table 7.22). As shown, heptachlor was 7.8, DDT 9.9, parathion 24.4, diazinon 132, and V-C 13 189 times more toxic to crickets when the sandy soil was moist (5.5% water on a dry weight basis) than when it was dry. By contrast, moisture in the muck soil (162% water) had only a marginal effect on the insecticide toxicity. In moist soils all the insecticides were found to be strongly inactivated by the SOM content, with the extent of inactivation depending on the specific insecticide; in dry soils there was no obvious correlation between SOM content and pesticide toxicity. Harris thus concluded that inactivation of the insecticides in moist soils was proportional to the SOM content, while the inactivation with dry soils was related to the adsorptive capacity of the mineral fraction.

The sharp contrast of the effects of soil moisture on the apparent toxicity of pesticides with mineral and organic soils is closely related to the recognized effects of soil moisture on the contaminant adsorption on soil minerals and partition into soil organic matter. The remarkable effect of moisture to suppress the adsorption of organic compounds on soil minerals agrees vividly with our common experience that the air in the field has a pleasant fragrance following a rain shower that succeeds a long period of drought. This phenomenon does not occur if the field has been sufficiently wet prior to the rain shower. By such recognition, we now have a good perspective of the role of humidity in influencing the activity of soil-incorporated contaminants in terms of the distinct mechanisms through which contaminants are sorbed by soil organic matter and minerals.

# 8 Contaminant Uptake by Plants from Soil and Water

## 8.1 INTRODUCTION

The contamination of soils or water by pesticides and other substances leads to the subsequent contamination of plants grown in these soils. Many soil-incorporated pesticides are known to translocate into plants and crops following their applications. Although scientists began investigating plant uptake as early as the early 1950s, soon after insecticides were introduced for agricultural purposes, there has been only moderate progress in this area of research because of the complexity of the contaminant soil-to-plant transport process. Some key parameters affecting the system include the levels of contaminants in soil (or water), the contaminant physicochemical properties, the particular plant species, the soil type that sustains the plant, and the time of plant exposure. Because of insufficient understanding of the contaminant sorption to soils in earlier years (1950s through 1970s), the crucial link between contaminant levels in soils and plants could not be forged at that time. Even in a number of later simpler studies on plant uptake from water, the published work investigated only relatively simple systems, such as small, young plants or excised plant parts, in controlled laboratory systems usually not sufficiently representative of natural settings. As the plant-uptake process has yet to be further characterized with extensive data, we are especially interested in a simple physical model to guide future investigations. Knowledge of contaminant uptake by plants is of vital interest not only because it will improve our understanding and alleviation of the crop-contamination problem but also because it may provide us with a means to effectively bioremediate contaminated soil and groundwater sites by appropriate plantings. Some critical plant-uptake data and the proposed model concepts are presented in this chapter.

Plant uptake of contaminants is considered to occur by a passive and/or an active process, depending on the contaminant and plant type (Shone and Wood, 1974; Briggs et al., 1982). Passive transport proceeds in the direction of decreasing chemical potential, the same process that is primarily responsible for bioconcentration of nonionic compounds by fish; it also appears to be the primary process for plant uptake of these compounds. Active transport, on the other hand, takes place against the chemical potential gradient, requiring the expenditure of energy; it applies to certain plant nutrients and possibly to some other inorganic and organic ions. The passive plant uptake of contaminants may be treated as a series of partitions between plant water and plant organic constituents, similar to that in fish uptake, where the partition into fish lipids is largely responsible for bioconcentration of relatively water-insoluble compounds. However, the plant-uptake system differs from fish-uptake systems in two important respects: (1) the plant uptake is not restricted to that from external water (including soil water), since the plant is also exposed to atmosphere, although water is commonly the transport medium; and (2) the rate of water transport into plants, by which contaminants enter plants, is usually more limited than the high rate of water transport into fish. To the first effect, the atmosphere may act as a medium for both contaminant transport and dissipation. The second effect leads to the expectation that the contaminant level in plants at a given time may deviate profoundly from the equilibrium value with external water. In this chapter, we consider only the passive plant uptake from external water and soil water.

## 8.2  BACKGROUND IN PLANT-UPTAKE STUDIES

In early studies, Lichtenstein (1959) found that lindane in soil was taken up by root crops (e.g., carrots and potatoes) more readily from light mineral soils than from a muck soil. Similarly, Walker (1972) showed that the concentrations of atrazine in shoots of wheat plants growing in 12 different soils were inversely proportional to soil-organic-matter (SOM) contents. In a more specific study on the effect of soil type on crop uptake, Harris and Sans (1967) compared the levels of dieldrin accumulated by carrots, radishes, and other root crops from three well-characterized contaminated field plots in relation to the soil pesticide levels; the three soil types studied—a sandy soil, a clay loam, and a muck soil—differed widely in SOM content (1.4 to 66.5%) and other soil constituents. Plant dieldrin concentrations were much lower for crops from the muck soil than from sandy and clay soils; by contrast, soil dieldrin concentrations were considerably higher in the muck soil than in the two other soils.

For plant uptake of contaminants from soil-free nutrient solutions, Briggs et al. (1982) measured the uptake by barley roots of two series of organic compounds, $O$-methylcarbamoyloximes and substituted ureas, which vary widely in lipophilicity. They concluded that the root uptake of both types of compounds approached the equilibrium values in a relatively short time (24 to 48 h). However, the root concentration factors (RCFs), that is, the ratios of chemical concentrations in roots and in water, increased monotonically, but not proportionally, with the $K_{ow}$ values of the compounds. Similar empirical correlations for contaminants in plant roots and leaves were also observed (Trapp, 1995). In view of the influences of soil type and contaminant identity on plant uptake, we seek to relate the plant contaminant levels to physicochemical properties of the contaminants and to the properties and compositions of plants and soils.

Most current models for plant uptake of contaminants from soil, water, or air are formulated on a differential mass-balance basis in terms of the rates of contaminant interface transfer, plant growth and transpiration, and contaminant metabolism, along with some estimated transfer coefficients (Riederer, 1990; Trapp et al., 1990; Paterson et al., 1994; Trapp and Matthies, 1995; Tam et al., 1996). Although these models are intended primarily for delineating the rates of contaminant uptake by plants (or their specific parts) with time from given external source(s), the model calculations are very sensitive to the accuracy of assumed contaminant interface-transfer rates and coefficients. Alternatively, equilibrium models have been utilized in some studies to assess contaminant levels in plants (or their parts) after their exposure to chemicals in water over a certain period of time (Briggs et al., 1982; Trapp, 1995). However, as shown later, the actual state of a contaminant in plants may or may not be at equilibrium with the external source.

A quasiequilibrium partition model has recently been developed by Chiou et al. (2001) to account for the passive plant uptake of contaminants from their external sources in soil or water. The model takes explicit account of the plant contaminant level in relation to the source level and plant composition. Moreover, the model contains both equilibrium and kinetic features and sets the upper (equilibrium) limit for the level of a contaminant in a plant with respect to the external-source level, against which the actual approach to equilibrium of the contaminant in the plant at the time of analysis can then be estimated. Although in the initial model testing by Chiou et al. (2001) the partition coefficients of contaminants with certain plant components have had to be estimated, the observed consistency of the plant-uptake data with the conceived model parameters is stimulating to warrant further investigation. The essential features of the model are presented below.

## 8.3   THEORETICAL CONSIDERATIONS

Consider first the simpler case of a partition-limited model for the plant uptake of nonionic contaminants from a soil-free nutrient solution by passive transport through the plant vascular system. Here water is both the solvent for the contaminant and the medium that carries it to plant roots (and to other water-contacted surfaces) and eventually to other parts of the plant via the plant vascular system. The overall plant uptake process is driven by the external-water concentration and is considered to consist of a series of partition uptakes, with the understanding that the contaminant concentrations within the plant may or may not come to full equilibrium with the external water solution. On the other hand, for any given volume element inside the plant, local equilibrium is assumed for a contaminant between sap water and the various organic constituents within that volume element. With these considerations, the concentration of a contaminant either in the whole plant or in a specific part of the plant ($C_{pt}$), expressed as the mass of contaminant per unit wet mass of the plant, can be equated with the contaminant concentration in external water ($C_w$) at the time of sample analyses:

$$C_{pt} = \alpha_{pt} C_w [f_{pom} K_{pom} + f_{pw}] \tag{8.1}$$

in which $f_{pom} + f_{pw} = 1$ and

$$f_{pom} K_{pom} = \sum f_{pom}^i K_{pom}^i \qquad i = 1, 2, 3, \ldots, n \tag{8.2}$$

In Eq. (8.1), $K_{pom}$ is the contaminant partition coefficient between plant organic matter and water, $f_{pom}$ is the total weight fraction of the organic matter in the plant, and $f_{pw}$ is the weight fraction of water in the plant, either for the whole plant or for a specific part of it. In Eq. (8.2), the $f_{pom}K_{pom}$ term is expressed as the sum of contributions from all plant organic components according to their specific partition coefficients ($K_{pom}^i$) and weight fractions ($f_{pom}^i$). The term $\alpha_{pt}$ ($\leq 1$), called the *quasiequilibrium factor*, expresses the extent of approach to equilibrium of any absorbed contaminant in the plant (or in a part of it) with respect to the same contaminant in the external water phase. In this model it may be viewed as the ratio of the respective concentrations in plant water and external water. Thus, $\alpha_{pt} = 1$ denotes the attainment of equilibrium.

If passive transport is the dominant uptake process, $\alpha_{pt}$ should not exceed 1, except for highly unusual situations; if the uptake involves an active process, $\alpha_{pt}$ may however exceed 1. In principle, with passive transport, if the concentrations in whole plant and external water are at equilibrium, all parts of the plant must be at equilibrium. However, when equilibrium is not attained with the whole plant, the $\alpha_{pt}$ value may vary with the local composition of the plant and its proximity to the contaminant source. Further rationale for variations in $\alpha_{pt}$ will be presented later. The $\alpha_{pt}$ value for a contaminant with a plant (or a selected part of it) is thus determined with inputs of $C_{pt}$ and $C_w$ together with the overall (or local) plant compositions and the respective partition coefficients. Once the value of $\alpha_{pt}$ is known, it may then be used to predict $C_{pt}$ from $C_w$ and associated parameters.

We now extend the model formulation to the more typical case of plants in contaminated soils. For plant growth, the water content in soil must be well above the water content at the plant's wilting point. This means that with plants in soil, the soil interstitial (pore) space contains bulklike water with dissolved nutrients and contaminants that are available for plant-root uptake. Therefore, as before, water is the transfer medium. The approach to model the contaminant uptake from soil is in essence to relate the contaminant concentration in plants to the effective concentration in soil interstitial (or pore) water.

The concentration of a contaminant in soil interstitial water ($C_w$) can in principle be determined experimentally for a given loading of the contaminant in a soil ($C_s$), where $C_w$ is related to $C_s$ at equilibrium as

$$C_s = K_d C_w \tag{8.3}$$

where $K_d$ is the soil-water distribution coefficient specific to a contaminant on a given soil. $K_d$ is largely independent of $C_w$ when the ratio of $C_w$ to $S_w$ (water solubility) is moderately large, but may be a function of $C_w$ for certain contaminant–soil systems at very small $C_w/S_w$ (Chiou and Kile, 1998 and references therein). On the other hand, as rationalized below, reasonable and convenient estimates of $C_w$ can be achieved for relatively nonpolar contaminants on most soils from the observed effect of soil organic matter (SOM) on contaminant sorption to water-saturated soils.

As discussed in Chapter 7, an ordinary soil is a dual sorbent for nonionic organic compounds of limited water solubilities, in which the mineral matter acts as an adsorbent and the SOM as a partition medium. As described earlier, the soil uptake of relatively nonpolar contaminants from water occurs mainly by partition into SOM because of the suppression by water of their adsorption on mineral matter. Thus, the effective concentration of a contaminant in a hydrated soil is the concentration that is normalized to the SOM content of the soil, that is,

$$C_{som} = C_s / f_{som} \tag{8.4}$$

where $C_{som}$ is the SOM-normalized contaminant concentration in soil and $f_{som}$ is the weight fraction of the SOM in soil. Assuming local equilibrium for contaminants between soil particles and interstitial water, the contaminant $C_w$ in soil interstitial water, which will be designated as the driving force for contaminant transport, is related to $C_{som}$ as follows:

$$C_w = C_{som} / K_{som} \tag{8.5}$$

where $K_{som}$ is the contaminant partition coefficient between SOM and water (here the subscript *som* is used to replace the subscript *om* in earlier $K_{om}$ in order to contrast $K_{som}$ from $K_{pom}$); $K_{som}$ is practically concentration independent, with the possible exception at very low $C_w/S_w$ for some special soils, such as those containing a significant amount of high-surface-area carbonaceous material (HSACM), as elucidated in Chapter 7 (see section 7.3.7). The magnitude of $K_{som}$ is determined by the contaminant and SOM properties. We have seen that for contaminants of limited water solubilities, the $K_{som}$ values are usually much greater than 1. Since the SOM properties for soils from widely dispersed geographic sources have been found to be relatively comparable, as illustrated in Chapter 7 (see section 7.3.2), Eq. (8.5) should allow for a fairly general assessment of the influence of soil sorption on the contaminant uptake by plants. Thus, while the direct determination of $C_w$, or the evaluation of $C_w$ from the established relation with $C_s$, helps to capture the possible nonlinear-sorption effect, sufficiently accurate estimates of $C_w$ can readily be obtained for low-polarity contaminants from $C_{som}$ and $K_{som}$ for soils sufficiently high in SOM and low in HSACM if $C_w$ and $C_{som}$ are close to being at equilibrium. The latter approach is especially merited for analysis of earlier published studies where the $K_d$ data are usually unavailable.

With the foregoing considerations, the plant uptake of a relatively nonpolar contaminant from a soil can then be formulated to a good approximation by substituting $C_{som}/K_{som}$ for $C_w$ in Eq. (8.1):

$$C_{pt} = \alpha_{pt}(C_{som}/K_{som})(f_{pom}K_{pom} + f_{pw}) \tag{8.6}$$

Equation (8.6) is an alternative form of Eq. (8.1) that accounts for the effect of soil sorption on the contaminant concentration in soil interstitial water. The use of $C_{som}/K_{som}$ in Eq. (8.6) circumvents the need for the experimentally cumbersome determination of $C_w$ in soil interstitial water. The value of $\alpha_{pt}$ in Eq. (8.6) can be determined as before with the additional inputs of $C_{som}$ and $K_{som}$.

The water content and the organic composition of plants may vary considerably either between plant types or between the different parts of a plant. The partition limits for a given contaminant inside a plant from the water phase to different plant parts would therefore vary with the overall and local plant composition. Most root and leaf crops are composed of large amounts of water and polar organic constituents, such as carbohydrates, cellulose, and proteins, and lesser amounts of lipids. From the partition standpoint [Eq. (8.2)], the most striking differences in plant contamination level would probably occur for relatively nonpolar, lipid-soluble contaminants between plants that differ radically in their lipid contents. Such differences are anticipated because the partition coefficients of these contaminants are much higher with relatively nonpolar lipids than with polar organic matter (Chiou, 1985; Rutherford et al., 1992). For the more water-soluble solutes, the partition capacities from water to either nonpolar lipids or relatively polar carbohydrates and proteins should be small to moderate.

Because the value of $\alpha_{pt}$ characterizes the extent to equilibrium of a contaminant between external water and the plant (or a part of the plant) at the time of the analysis, it must depend in part on the contaminant partition capacity of the plant organic matter. For a plant with a given lipid content, a polar contaminant with a low $K_{lip}$(lipid-water) value would tend to exhibit a higher $\alpha_{pt}$ than a relatively nonpolar contaminant with a significantly higher $K_{lip}$ value, because the attainment of partition equilibrium for the latter requires a much greater volume of water transport within the plant. The magnitude of $\alpha_{pt}$ is therefore expected to be a function of the contaminant partition coefficient, the plant water-organic composition, and the plant water-transport rate. Finally, the contaminant uptake by plants through direct diffusion to the outer layers of plant roots should contribute to the rate of local and overall plant uptake and would therefore affect the local and/or overall $\alpha_{pt}$ value before the contaminant in plant and external water reaches equilibrium. The significance of contaminant uptake by diffusion relative to that by transport of external water into plant's vascular system depends on the plant system but should not in principle make $\alpha_{pt} > 1$.

We now evaluate the validity of the model against the pertinent literature data on crop/plant contamination. The present analysis is restricted to systems with relatively nonreactive nonionic compounds in water or soil. Ionizable compounds are excluded from consideration because of the possibility that their plant uptakes may involve active transport to certain plant organic constituents.

## 8.4   UPTAKE BY SMALL PLANT ROOTS FROM WATER

Starting with the simplest system, consider the uptake, by the roots of barley (*Hordeum vulgare* cv. Georgie), of various *O*-methylcarbamoyloximes and substituted ureas from nutrient water solution, as reported by Briggs et al. (1982). The experiments measured the root concentration factor (RCF = $C_{pt}/C_w$) for each of these $^{14}$C-labeled compounds individually in replicated laboratory systems after the 10-day-old barley plants were transferred to nutrient solution with the test compound for 24 and 48 h. The authors indicated that the RCF values for the parent compounds alone, or for the parent compounds plus their metabolites, were very similar after 24 and 48 h. The 24- to 48-h averaged RCF values of the parent compounds (with reported data uncertainties of ±5%) and their octanol–water partition coefficients ($K_{ow}$'s) are presented in Table 8.1.

The $K_{ow}$ values for many of the compounds in Table 8.1, such as the more lipophilic benzaldehyde *O*-methylcarbamoyloximes, were obtained either by indirect experimental methods or by empirical calculations. No information on the water–organic composition of the barley roots was provided. However, an approximate composition for barley roots (*Hordeum vulgare*) is given by Trapp et al. (1990); it comprises 87.5% water and 1% lipids by weight. It is assumed that the remainder consists mainly of carbohydrates and cellulose, with traces of proteins and nutrients, for a total of 11.5% by weight. We assume further that the partition coefficients of the compounds with the relatively polar carbohydrates, cellulose, and proteins are practically the same. Since octanol is known to mimic biological lipids closely in contaminant partition (see Table 5.5), the lipid–water partition coefficients ($K_{lip}$'s) are assumed to be the same as the corresponding $K_{ow}$'s.

On the premise of the assumptions above, the $f_{pom}K_{pom}$ term in Eq. (8.2) for barley roots can be simplified as the sum of the contributions by carbohydrates and lipids:

$$f_{pom}K_{pom} = f_{ch}K_{ch} + f_{lip}K_{lip} \tag{8.7}$$

where the subscripts "ch" and "lip" designate carbohydrates and lipids, respectively. Substituting Eq. (8.7) into Eq. 1 with the assumed barley root composition (Trapp et al., 1990) leads to

**TABLE 8.1. Root Concentration Factors (RCFs) of Pesticides and Related Compounds from Water into Barley Roots (*Hordeum vulgare* cv. Georgie) over a Period of 24 to 48 Hours (Briggs et al., 1982) and Calculated Quasiequilibrium Factors ($\alpha_{pt}$)**

| Compound | $\log K_{ow}$ | RCF | $\alpha_{pt}$ |
|---|---|---|---|
| *O*-Methylcarbamoyloximes | | | |
| Aldoxycarb | −0.57 | 0.66 | 0.74 |
| Oxamyl | −0.47 | 0.91 | 1.02 |
| Acetone *O*-methylcarbamoyloxime | −0.13 | 0.95 | 1.06 |
| Aldicarb | 1.08 | 0.94 | 0.90 |
| Benzaldehyde *O*-methylcarbamoyloxime | 1.49 | 1.48 | 1.19 |
| 4-Chlorobenzaldehyde *O*-methylcarbamoyloxime | 2.27 | 2.80 | 0.98 |
| 3,4-Dichlorobenzaldehyde *O*-methylcarbamoyloxime | 2.89 | 5.61 | 0.64 |
| 3-Phenylbenzaldehyde *O*-methylcarbamoyloxime | 3.12 | 8.72 | 0.61 |
| 3-(3,4-Dichlorophenoxy)benzaldehyde *O*-methylcarbamoyloxime | 4.6 | 81.1 | 0.20 |
| Substituted ureas | | | |
| 3-Methylphenylurea | −0.12 | 0.73 | 0.82 |
| Phenylurea | 0.80 | 1.20 | 1.25 |
| 4-Fluorophenylurea | 1.04 | 1.10 | 1.06 |
| 3-(Methylthio)phenylurea | 1.57 | 0.94 | 0.72 |
| 4-Chlorophenylurea | 1.80 | 2.00 | 1.28 |
| 4-Bromophenylurea | 1.98 | 3.17 | 1.63 |
| 3,4-Dichlorophenylurea | 2.64 | 5.86 | 1.09 |
| 4-Phenoxyphenylurea | 2.80 | 7.08 | 0.97 |
| 4-(4-Bromophenoxy)phenylurea | 3.7 | 34.9 | 0.68 |

*Source*: Data from Chiou et al. (2001).

$$C_{pt} = \alpha_{pt} C_w \left[ f_{pw} + f_{ch} K_{ch} + f_{lip} K_{lip} \right] \qquad (8.8)$$

or

$$\alpha_{pt} = (C_{pt}/C_w)/(0.875 + 0.115 K_{ch} + 0.01 K_{ow}) \qquad (8.9)$$

Calculations of the $\alpha_{pt}$ values for contaminants require values for the individual carbohydrate–water partition coefficients ($K_{ch}$'s); these are not available. However, the $K_{ch}$ values are expected to be small because of the high polarity of carbohydrates. For example, the cellulose–water partition coefficients for benzene ($\log K_{ow} = 2.13$) and carbon tetrachloride ($\log K_{ow} = 2.83$) are 0.56 and 1.75 (Rutherford et al., 1992), respectively; the cellulose is taken to be similar in composition to the carbohydrates. As an approximation, we thus assume $K_{ch}$ to be 0.1 for compounds with $\log K_{ow} < 0$; 0.2 for $\log K_{ow} = 0.1$ to 0.9; 0.5 for $\log K_{ow} = 1.0$ to 1.9; 1 for $\log K_{ow} = 2.0$ to 2.9; 2 for $\log K_{ow} = 3.0$ to 3.9; and 3 for $\log K_{ow} \geq 4.0$. With the assumed barley root composition, the partition contribution by carbohydrates becomes unimportant relative to that by the lipids for compounds with $\log K_{ow} > 3.0$.

The $\alpha_{pt}$ values calculated for the compounds are listed in the last column of Table 8.1. They are generally consistent with the overall hydrophilic-to-lipophilic trend of the solutes in that the water-soluble compounds have $\alpha_{pt}$ values close to 1 and that the $\alpha_{pt}$ values for lipophilic compounds (high $K_{ow}$ values) are less than 1. This relatively smooth transition is reflective of the passive transport of contaminants into the different plant-root organic matrices and of the spatial uniformity of the contaminant concentration in external water. As seen for both $O$-methylcarbamoyloximes and substituted ureas with $K_{ow} \leq 500$ or so, the $\alpha_{pt}$ values are, within the uncertainties from all sources, essentially 1, suggesting that the passive uptake of these relatively water-soluble compounds by barley roots comes close to equilibrium within 24 to 48 h. For compounds with increased lipophilicity (i.e., those with $K_{ow} > 1000$), the $\alpha_{pt}$ values are clearly below 1. With the assumed lipid content in barley roots (1%), the calculation shows that for compounds with $K_{ow} \leq 10$, the contaminant level in root water accounts for more than 85% of the total root uptake; for compounds with $K_{ow} = 100$, the uptake by the root water and the root lipid each contributes about 50% to the total uptake; for compounds with $K_{ow} > 1000$, the total root uptake is predominated by the lipid uptake.

The $\alpha_{pt}$ values calculated depend sensitively on the assumed lipid content and the accuracy of the $K_{ow}$ values for compounds having $K_{ow} > 100$; this sensitivity increases proportionately with increasing $K_{ow}$ of the compound. As for the $K_{ow}$, it is not uncommon for the reported value to be in error by a factor of 2 to 3, especially for compounds with large $K_{ow}$ values (Leo et al., 1971). Since the variation of the $\alpha_{pt}$ values in Table 8.1 is supportive of passive transport, one may verify the relative $K_{ow}$ values of the compounds in terms of their measured RCF (or $C_{pt}/C_w$) values. Here, for example, the RCF values for 4-bromophenylurea, 3,4-dichlorophenylurea, and 4-phenoxyphenylurea are 3.17, 5.86, and 7.08, respectively; the corresponding measured $K_{ow}$ values are 95, 436, and 631. Thus, based on the measured RCF values, the $K_{ow}$ for 4-bromophenylurea seems to be too low, by nearly a factor of 2, relative to the values for the other two compounds. The relatively high $\alpha_{pt} = 1.63$ value for 4-bromophenylurea, compared to the values for the other two compounds (where $\alpha_{pt} \approx 1$), may be an artifact of the calculation rather than a manifestation of active uptake. Similarly, the $\alpha_{pt}$ value for 3-(3,4-dichlorophenoxy)benzaldehyde $O$-methylcarbamoyloxime could be somewhat too low, as the estimated $K_{ow}$ value seems somewhat too high.

The less-than-1 $\alpha_{pt}$ values (i.e., the very small RCFs) for highly polar aldoxycarb (0.74) and 3-methylphenylurea (0.82) in Table 8.1 are noted with interest, since their uptake by root lipids and carbohydrates would be small relative to that by root water. Briggs et al. (1982) attributed the small RCFs of highly polar chemicals to difficulties in passing through the lipid membranes in the root, thus resulting in selective rejection of the chemicals at the membrane barriers. The low $\alpha_{pt}$ values could also result from the high ionic strength in root water that increases the activity (and thus decreases the solubility) of neutral compounds (Trapp, 2000). However, the expected high $\alpha_{pt}$ value (1.02) for equally polar and soluble oxamyl in Table 8.1 suggests that an effect other than membrane barrier and ionic strength may be relevant. Overall, the results indicate that the concentrations of compounds with $K_{ow} \leq 1000$ in barley roots are reasonably close to equilibrium with external water after 24 to 48 h; for compounds with $K_{ow} > 1000$, accurate $K_{lip}$ values are required to determine the approach to equilibrium.


## 8.5  UPTAKE BY PLANT SEEDLINGS FROM SOIL

Although there are numerous reports in the literature on contaminant uptake by plants or crops from soils, only a few provide the corresponding soil contaminant levels; even in studies where the levels in soil are reported, the requisite data on the SOM (or $K_d$) values are often not available. In other cases, the incorporation of contaminants into soils has not reached a stable condition, because either of insufficient time of incorporation or of instability of the compounds (e.g., large dissipation by vaporization). The following model calculations are performed for those systems that are considered to be relatively stable and where the data on SOM contents are reported.

Trapp et al. (1990) measured the concentrations of a herbicide (atrazine) and other chlorinated hydrocarbons in barley seedlings (*Hordeum vulgare*) germinated from a contaminated soil. The crop uptake was separately investigated for each of the $^{14}$C-labeled contaminants in a closed aerated laboratory system. After the barley seedlings had been in contact with the soil for 1 week, plant and soil samples were taken for analysis of the parent compound and its metabolites in both soil and plants; however, the metabolites were not identified. The soil contained 2.06% organic carbon, or approximately 3.5% in SOM, and was maintained at 20% water during the experiment. The measured concentrations of the parent compounds and their metabolites in both soil and plants (roots + shoots) are given in Table 8.2. For our model calculations, the concentrations in soil have also been normalized to the SOM.

For whole barley seedlings, the authors assumed the composition to be 87.5% water and 1% lipids. As before, we further assume that the remaining 11.5% consists essentially of carbohydrates and cellulose, on which the contaminants exhibit the same partition coefficients. With these assumptions together with Eq. (8.5), Eq. (8.6) may then be expressed as

$$C_{pt} = \alpha_{pt} C_{som} \left( f_{pw} + f_{ch} K_{ch} + f_{lip} K_{lip} \right) / K_{som} \tag{8.10}$$

or

$$\alpha_{pt} = K_{som} \left( C_{pt} / C_{som} \right) / \left( 0.875 + 0.115 K_{ch} + 0.01 K_{ow} \right) \tag{8.11}$$

**TABLE 8.2. Concentrations of Pesticides and Chlorinated Compounds in Soil ($C_s$) and Barley Plants ($C_{pt}$) after 1-Week Plant Uptake (Trapp et al., 1990) and Calculated SOM-Normalized Concentrations ($C_{som}$)[a], Soil-Interstitial-Water Concentrations ($C_w = C_{som}/K_{som}$), and Quasiequilibrium Factors ($\alpha_{pt}$)**

| Compound | $\log K_{ow}^{b}$ | $\log K_{som}^{c}$ | $C_s^{d}$ (ppm) | $C_{som}^{d}$ (ppm) | $C_w^{e}$ (ppm) | $C_{pt}^{d}$ (ppm) | $\alpha_{pt}^{e}$ |
|---|---|---|---|---|---|---|---|
| Atrazine | 2.71 | 2.17 | 0.84 (0.98) | 24 (28) | $1.6 \times 10^{-1}$ | 1.02 (2.53) | 1.0 |
| 1,2,4-Trichlorobenzene | 3.98 | 2.70 | 0.65 (0.75) | 19 (21) | $3.8 \times 10^{-2}$ | 0.70 (1.33) | 0.19 |
| 1,2,3,5-Tetrachlorobenzene | 4.65 | 3.42 | 1.13 (1.18) | 32 (34) | $1.2 \times 10^{-2}$ | 1.93 (3.54) | 0.35 |
| Dieldrin | 4.55 | 3.69 | 2.07 (2.08) | 59 (59) | $1.2 \times 10^{-2}$ | 1.14 (1.24) | 0.27 |
| Hexachlorobenzene | 5.50 | 4.19 | 1.86 (1.88) | 53 (54) | $3.4 \times 10^{-3}$ | 2.19 (2.21) | 0.20 |
| 2,4,6,2′,4′-PCB | 5.92 | 4.57 | 2.15 (2.16) | 61 (62) | $1.6 \times 10^{-3}$ | 2.54 (2.77) | 0.19 |
| DDT | 6.36 | 5.34 | 2.10 (2.16) | 60 (62) | $2.7 \times 10^{-4}$ | 0.70 (0.88) | 0.11 |

*Source*: Data from Chiou et al. (2001).

[a] The SOM content is 3.5% by weight.

[b] Values cited by Trapp et al. (1990) except for dieldrin from Brook et al. (1986) and hexachlorobenzene and DDT from Chiou et al. (1982b).

[c] The log $K_{som}$ for atrazine from Kenaga and Goring, (1980), 1,2,4-trichlorobenzene from Chiou et al. (1983), dieldrin from Felsot and Wilson (1980), and DDT from Shin et al. (1970); the log $K_{som}$ for 1,2,3,5-tetrachlorobenzene, hexachlorobenzene, and 2,4,6,2′,4′-PCB estimated from Eq. (7.14).

[d] Numbers outside the parentheses are for the parent compounds; numbers within the parentheses are sums of parents and metabolites.

[e] Calculated only for the parent compounds.

With the relatively high $K_{ow}$ values of the parent compounds in Table 8.2 and the assumed lipid content, the contributions to total barley uptake by plant water and cellulose would be quite small or negligible relative to that by lipids, as reasoned earlier. Although the levels of metabolites in soil and plants are a useful indicator of the contaminant fate, no calculations could be performed for the metabolites as their chemical identities are not known. The necessary $K_{som}$ values of the parent compounds in Table 8.2 are taken from the literature to complete the calculation of the $\alpha_{pt}$ values.

The $\alpha_{pt}$ values calculated for the compounds, except for 1,2,4-trichlorobenzene, are quite consistent with the expected countertrend between $\alpha_{pt}$ and $K_{ow}$, despite the fact that the calculated $\alpha_{pt}$ values for lipophilic compounds depend sensitively on the accuracy of the $K_{ow}$ and $K_{som}$ values. The noted results on $\alpha_{pt}$ are consistent with the model approach of substituting $C_{som}/K_{som}$ for $C_w$ in soil interstitial water. For 1,2,4-trichlorobenzene in soil, the system was recognized to be unstable because of its high volatility (Trapp et al., 1990); the total recovery of this compound and its metabolites from soil and plants was only 70%, whereas the recoveries for all other compounds exceeded 96%. Based on the model calculations, the uptake of atrazine by plant water and carbohydrates constitutes about 20% of the total, with the rest of the atrazine being taken up by lipids. The total uptakes of other less water-soluble contaminants are exclusively by the small amount of lipids, with the different $\alpha_{pt}$ values reflecting the relative efficiencies of the compounds inside the plants for approaching equilibrium with external soil water. One notes with interest that atrazine, with a moderate $\log K_{ow} = 2.71$, gives rise to $\alpha_{pt} = 1$ despite the fact that the amount of metabolites in the plant is more than that of the parent species. This suggests that the metabolic process or formation of metabolites in plants does not seem to retard the plant passive uptake of the parent compound.

Although the $\alpha_{pt}$ value of a contaminant is partly a function of the plant water uptake and transport, it could also be affected by other mechanisms. Consider here, for example, the $\alpha_{pt}$ value (about 0.1) for DDT (with $K_{lip} \simeq K_{ow} = 2.3 \times 10^6$) on barley seedlings containing about 1% lipids. Since the plant is considered to be about 90% water, the total plant mass is about the same as the plant–water mass. If all the DDT uptake by barley seedlings were to come only from absorption of the external soil water through the plant vascular system (i.e., as a consequence of the plant transpiration), the needed transport mass of water for DDT at $\alpha_{pt} = 0.1$ would exceed 2000 times the plant mass. Since this amount seems unreasonably high in a short-term experiment, the DDT uptake is more likely also facilitated by mass diffusion from interstitial water into or across the root surfaces of the plant. Moreover, had such a high water transport mass been involved, the $\alpha_{pt}$ values for contaminants with much lower $K_{ow}$ values (e.g., dieldrin and tetrachlorobenzene) would have been much closer to 1 than observed. This is because the transport masses required to saturate the lipid phase are much less for these contaminants.

The results in Table 8.2 reveal that although the $\alpha_{pt}$ values for compounds with high lipid-to-water solubility ratios (i.e., those with high $K_{ow}$ values) are significantly less than 1, their concentration factors from soil interstitial water ($C_w$ or $C_{som}/K_{som}$) to the plant ($C_{pt}$) (i.e., the $C_{pt}/C_w$ ratios) are markedly higher than for relatively water-soluble compounds (e.g., atrazine). As manifested in Table 8.2, the $C_{pt}/C_w$ ratio increases largely with increasing $K_{ow}$ because a net increase in $K_{ow}$ outweighs the resulting decrease in $\alpha_{pt}$, as exemplified, for example, by the data of dieldrin and DDT. As such, extremely water-insoluble DDT with a small $\alpha_{pt} = 0.11$ exhibits, nevertheless, a large concentration factor, about 2600, from soil interstitial water into the barley, based on the assumed lipid content (1%) and lipid–water partition coefficient ($K_{ow}$). Here the concentration of DDT inside the plant is only about 10% of the equilibrium value with respect to the soil-interstitial-water concentration; the theoretical concentration factor at equilibrium would be about 10 times greater than observed. The small $\alpha_{pt}$ values for DDT and other compounds with large lipid-to-water solubility ratios may be attributed to insufficient amounts of external-water transport into the plant circulatory system for these compounds to achieve the equilibrium partition capacities.

## 8.6  UPTAKE BY ROOT CROPS FROM DIFFERENT SOILS

A very instructive work on the effect of soil type on contaminant uptake by crops is that of Harris and Sans (1967), who measured the uptake of dieldrin and DDT by several root crops grown in three contaminated field plots of widely different soil types. Each of the three field plots maintained stable levels of dieldrin and DDT during the growing season. The soils studied were as follows: a Fox sandy loam (1.4% SOM), a clay soil (3.6% SOM), and a muck (66.5% SOM). The plots were seeded during mid-May; the growing season varied with the crop type and lasted, for example, about one month for radishes (Sparkler White Tip) and three months for carrots (Nantes). Soil insecticide-residue levels before seeding and after harvesting were measured and no significant changes were observed. The levels of dieldrin and DDT in crops and soils observed by Harris and Sans (1967) are given in Table 8.3. In addition to the original insecticide concentrations in soils ($C_s$), the SOM-normalized soil concentrations ($C_{som}$) are also presented.

Calculations of the $\alpha_{pt}$ values for dieldrin and DDT with each soil–crop system have been made using the $K_{ow}$ and $K_{som}$ values from Table 8.2 and the assumed compositions of the root crops. Among the root crops studied (carrots, radishes, turnips, and onions), carrots showed the highest uptake from the soils and radishes exhibited trace levels, while the uptake by turnips and onions was near or below the detection limit (<0.01 ppm). For DDT, the observed levels in these crops were mostly below the detection limit, and thus the data were more limited and less precise. The present analysis is confined largely to the dieldrin levels in carrots and radishes and to a lesser extent

**TABLE 8.3. Concentrations of Dieldrin and DDT in Soils ($C_s$) and Root Crops ($C_{pt}$) in Field Plots after a Growing Season (Harris and Sans, 1967) and Calculated SOM-Normalized Concentrations ($C_{som}$),[a] Soil-Interstitial-Water Concentrations ($C_w = C_{som}/K_{som}$), and Quasiequilibrium Factors ($\alpha_{pt}$)**

| System | $C_s$ (ppm) | $C_{som}$ (ppm) | $C_w$ (ppm) | $C_{pt}$ (ppm) | $\alpha_{pt}$ |
|---|---|---|---|---|---|
| Dieldrin | | | | | |
| Sandy soil/carrots | 0.48 | 34 | $6.9 \times 10^{-3}$ | 0.12 | 0.24 |
| Clay soil/carrots | 1.1 | 31 | $6.3 \times 10^{-3}$ | 0.11 | 0.24 |
| Muck/carrots | 3.9 | 5.9 | $1.2 \times 10^{-3}$ | 0.02 | 0.23 |
| Sandy soil/radishes | 0.48 | 34 | $6.9 \times 10^{-3}$ | 0.02 | 0.08 |
| Clay soil/radishes | 1.1 | 31 | $6.3 \times 10^{-3}$ | 0.05 | 0.22 |
| Muck/radishes | 3.9 | 5.9 | $1.2 \times 10^{-3}$ | 0.01 | 0.23 |
| DDT | | | | | |
| Clay soil/carrots | 0.34 | 9.4 | $4.3 \times 10^{-5}$ | <0.01 | <0.05 |
| Muck/carrots | 15 | 23 | $1.1 \times 10^{-4}$ | 0.01 | 0.02 |
| Clay soil/radishes | 0.34 | 9.4 | $4.3 \times 10^{-5}$ | <0.01 | <0.10 |
| Muck/radishes | 15 | 23 | $1.1 \times 10^{-4}$ | 0.01 | 0.04 |

*Source*: Data from Chiou et al. (2001).

[a] The SOM content is 1.4% for the sandy soil, 3.6% for the clay soil, and 66.5% for the muck.

to the DDT levels in these two crops. The inclusion of the DDT data for analysis is mainly to substantiate the relative order in $\alpha_{pt}$ between DDT and dieldrin.

According to the USDA Nutrient Database (*www.nal.usda.gov/fnic/foodcomp/*), a fully grown carrot (*Daucus carota*) has a lipid content of about 0.19%, whereas a baby carrot (*D. carota*) contains 0.53% lipids. The lipid content for the carrot (Nantes) used by Harris and Sans is unknown. As a working basis, we assume a lipid content of 0.2% for carrots. Similarly, according to the USDA Nutrient Database, the lipid content for radishes varies between the varieties, with most values around 0.1%, and no information is available for the species used by Harris and Sans (1967). We assume a lipid content of 0.1% to be representative of most radish varieties. With the $K_{ow}$ values for dieldrin and DDT, the crop uptake should be controlled predominantly by the lipid uptake. Contributions by carbohydrates and plant water are therefore ignored.

The calculated $\alpha_{pt}$ values for dieldrin and DDT follow the expected order. The $\alpha_{pt}$ values for dieldrin with carrots from three different soils are nearly constant and practically the same as the value found for dieldrin with barley seedlings (Table 8.2). This uniformity of $\alpha_{pt}$ values is more than anticipated, considering the differences in crop type and growth time and the potential nonuniformity in contaminant concentration and soil SOM content over the root-accessible soil zone. The use of $C_{som}/K_{som}$ for $C_w$ in Eq. (8.6) results in a practically linear relation between $C_{pt}$ and $C_{som}$ (as reflected by the consistent $\alpha_{pt}$ values) for dieldrin/carrots on the three soils of widely different properties. As may be seen, if the $C_s$ values are not normalized to the SOM contents, the $C_s/C_{pt}$ ratios vary radically, showing no apparent correlation. Similar results are noted for dieldrin with radishes and for DDT with carrots and radishes; however, the data here show more uncertainties where the contamination levels in the crops were either near or below detection limits (<0.01 ppm) (Harris and Sans, 1967).

## 8.7   EFFECT OF PLANT COMPOSITION

The generally lower uptake of dieldrin by radishes than by carrots appears to result largely from the difference in their lipid contents. This effect is illustrated by the finding that the $C_{pt}/C_w$ ratios for dieldrin with a given soil for radishes are about one-half the values for carrots, which are in accord with the presumed difference in their lipid contents. Moreover, the present dieldrin concentration factor ($C_{pt}/C_w$) with carrots correlates well with the dieldrin concentration factor with barley seedlings described earlier, in which a fivefold difference in crop lipids leads to about a fivefold difference in $C_{pt}/C_w$ values. Although the close agreement of the data in both cases seems a bit fortuitous for reasons stated earlier, the results suggest that plant lipids are the major factor for the observed difference in plant uptakes of the lipophilic contaminants. This view is further supported by the results of Lichtenstein (1960) on the uptakes of lipophilic aldrin, dieldrin, heptachlor, and heptachlor epoxide from a soil by carrots, radishes, beets, potatoes, onions, and lettuce. In this experiment, the field plots of a Carrington silt loam were treated with high insecticide levels on which various crops were seeded, and soil and crop contaminant levels were analyzed at harvest; however, information on the SOM content of the soil was not available. An essential portion of Lichtenstein's data on soil and crop contaminant levels is given in Table 8.4.

Data in Table 8.4 reveal that the whole-crop levels of these relatively water-insoluble pesticides in radishes (Early Scarlet Globe), beets (Detroit Dark Red), potatoes (Russet Sebago), onions (Yellow Globe Danvers), and lettuce (Great Lakes) were all significantly lower than in carrots (Red Cored Chantenay), despite significant variations in relative contaminant levels with different crops. Although the lipid contents for all these crops are very low, carrots appear to have a comparatively greater lipid content than the others according to the USDA Nutrient Database. Thus, the levels for each of the pesticides in these crops seem to be influenced most by the plant lipids. Significant variations in relative contaminant levels among crops may stem from various unspecified sources, such as the analytical sensitivity, the sample preparation loss with soil and crops, and other factors mentioned previously. Detailed evaluations of individual contaminant levels with these crops require the SOM content of the soil, accurate lipid data, and accurate $K_{ow}$ and $K_{som}$ values which are not readily available.

**TABLE 8.4. Concentrations of Aldrin, Dieldrin, Heptachlor, and Heptachlor Epoxide in Carrington Silt Loam and Crops in Field Plots after a Growing Season**

| | Concentrations in Soil and Crops (ppm) | | | |
| --- | --- | --- | --- | --- |
| Soil/Crop | Aldrin | Dieldrin | Heptachlor | Heptachlor Epoxide |
| Soil | 3.1 | 1.8 | 4.2 | 0.78 |
| Radishes | 0.05 | 0.13 | Trace | 0.12 |
| Beets | Trace | 0.17 | 0.03 | 0.12 |
| Potatoes (whole) | 0.11 | 0.31 | 0.29 | 0.49 |
| Potatoes (peels) | 0.63 | 1.82 | 3.03 | 2.35 |
| Potatoes (pulp) | Trace | 0.16 | Trace | 0.24 |
| Onions | Trace | 0.05 | NA[a] | NA |
| Carrots | 0.36 | 0.55 | 1.34 | NA |
| Lettuce | 0.03 | 0.17 | Trace | 0.05 |

*Source*: Data from Lichtenstein (1960).

[a] NA, data not available.

An interesting observation by Lichtenstein (1960) is that the contaminant concentrations in the peels (skin) of potatoes are considerably higher than those in either whole potatoes or their pulps (Table 8.4). Although the mass-diffusion effect would make the concentrations higher in the peels than in the pulps, it is not clear without the relevant lipid data whether this disparity is caused solely by the mass diffusion or in addition by the lipid-content difference. A similar but more pronounced disparity was later reported by Lichtenstein et al. (1965) for the uptake of dieldrin from soil by carrots and by Mattina et al. (2000) for the uptake of chlordane by carrots, potatoes, and beets. The exact cause of this phenomenon is yet to be uncovered.

## 8.8  CONTAMINANT LEVELS IN AQUATIC PLANTS AND SEDIMENTS

Contaminant levels in underwater plants of a polluted aquatic system should approach more closely the equilibrium values with their in situ bed sediments because of the enhanced plant uptake through roots and bulk-water diffusion. In other words, the $\alpha_{pt}$ values in Eq. (8.6) should be greater for underwater plants than for land-grown plants, if other parameters (e.g., plant composition and exposure time) are comparable. For relatively water-insoluble compounds such as DDT, PCBs, and some PAHs, where the contaminant levels in sediments and plants are controlled primarily by contaminant partition into SOM and plant lipids, respectively, Eq. (8.6) may be reduced to

$$C_{pt} \simeq \alpha_{pt}(C_{som}/K_{som})f_{\text{lip}}K_{\text{lip}} \tag{8.12}$$

or

$$C_{\text{lip}}/C_{som} \simeq \alpha_{pt}\, K_{\text{lip}}/K_{som} \simeq \alpha_{pt}\, K_{ow}/K_{som} \tag{8.13}$$

where $C_{\text{lip}} = C_{pt}/f_{\text{lip}}$ and $K_{ow} \simeq K_{\text{lip}}$. Equation (8.13) may be expressed alternatively as

$$C_{\text{lip}}/C_{oc} \simeq \alpha_{pt}\, K_{ow}/K_{oc} \tag{8.14}$$

where $C_{oc}$ is the contaminant level in sediment normalized to the sediment-organic-carbon content and $K_{oc}$ is, as defined earlier, the contaminant partition coefficient normalized to the sediment-organic-carbon content. Thus when the levels of a sparingly water-soluble contaminant in bed sediments and in situ aquatic plants reach equilibrium, the $C_{\text{lip}}/C_{som}$ (or $C_{\text{lip}}/C_{oc}$) value will be equal to $K_{ow}/K_{som}$ (or $K_{ow}/K_{oc}$), largely independent of sediment composition and plant species.

An intensive field study was carried out by Vanier et al. (2001) to compare the concentrations of 34 PCB congeners in bed sediments and shoots of submerged plants in Lake Saint-Francois (74°40′ W, 45°00′ N), one of the most PCB-contaminated areas in the Saint Lawrence River in Canada. Samples of macrophyte shoots (mostly of *Myriophyllum* sp. and *Elodea canadensis*) and in situ bed sediments were taken from three stations in the lake. The shoots at the time of sampling were about two months old. At each station, samples were collected from five locations around a circle of about 60 m in perimeter. Samples of plant shoots of all species were pooled. Concentrations of 34 PCB congeners in pooled plant shoots and in cored bed sediments (top 5 cm analyzed) were quantified individually to yield a total of 457 data points. PCB congeners with 2 to 5 chlorines comprised more than 85% of the total PCBs; the highest PCB levels in plants and sediments were found for congeners with 2 or 3 chlorines (Vanier et al., 1999). Levels of PCBs in sediments and in pooled plant shoots were normalized to sediment organic-matter contents ($f_{om}$) and plant-shoot lipid contents ($f_{\text{lip}}$), respectively. In plant shoots, levels of total PCBs ranged from 9.6 to 405 mg per kilogram of lipids. In sediments, the levels ranged from 4.4 to 256 mg per kilogram of organic matter. The relation between the resulting $C_{som}$ and $C_{\text{lip}}$ values were analyzed by the slope-range statistical method.

Analysis made by Vanier et al. (2001) for all PCB congeners led to the finding that

$$\log C_{\text{lip}} = 0.573(\text{SE}: 0.057) + 0.978(\pm 0.041)\log C_{som} \tag{8.15}$$

with $r^2 = 0.847$ and $n = 457$. Since the slope in Eq. (8.15) is not statistically different from 1, the relation between $C_{\text{lip}}$ and $C_{som}$ for all PCB congeners was determined only by the intercept in Eq. (8.15), giving $C_{\text{lip}} = 3.74\ (\pm 1.14)\ C_{som}$. This proportional factor (3.74) was considered by the authors to be within the reported range for PCB partition equilibrium between sediments and plants. To ascertain whether the $C_{lip}$ and $C_{som}$ values are truly at equilibrium, the observed $C_{lip}$–$C_{som}$ correlation may be evaluated against the model calculation using Eq. (8.13), as presented below.

To begin with, one recalls from Table 7.4 that the ratios of $K_{ow}$ to soil $K_{som}$ (i.e., $K_{om}$) for four PCB congeners (with 1 to 3 chlorines) are relatively comparable at $16.5 \pm 2.49$. Using these values as the approximate ratios for all PCBs, the estimated ratios of $K_{ow}$ to sediment $K_{som}$ would then be about $8.3 \pm 1.2$, because the sediment organic matter is about twice as efficient a partition medium as the soil organic matter for low-polarity solutes (see Chapter 7, section 7.3.2). Thus, with the correlation of Vanier et al. (2001) (i.e., $C_{lip} = 3.74\ C_{som}$ for all samples), the calculated average $\alpha_{pt}$ values would be $0.45 \pm 0.07$; that is, the PCB levels in plant shoots are about half the equilibrium capacities with respect to the PCB levels in in-situ bed sediments. For the entire set of data, in which $C_{lip}$ is 2.6 to 4.9 times $C_{som}$, the $\alpha_{pt}$ falls into the range 0.30 to 0.59. Thus, although the $C_{lip}$ values are reasonably close to their equilibrium values in some samples, the averaged values for all samples are only about one-half the equilibrium values. Despite these variations, practically all the PCB levels observed in shoots relative to their levels in bed sediments comply with the upper limits imposed by the partition-limited model, Eq. (8.6) or (8.13). In the calculation above, it is assumed that the bed-sediment samples from different geographic locations have comparable organic-matter compositions.

## 8.9   TIME DEPENDENCE OF CONTAMINANTS IN PLANTS

A subject of considerable interest to the plant uptake of a contaminant is the time dependence of the contaminant level in a specific part of the plant. In the practice of bioremediation by planting, this would determine the efficiency of a plant in taking up a target contaminant with time from an external source. Although not well established at this time, the preceding data analysis leads to the expectation that the change with time of the in-plant contaminant level should be a function of the contaminant partition capacity and the specific plant physiology (e.g., plant growth and water uptake rate). On a given plant species (or a specific part of it), one expects a major difference between a water-soluble compound and a lipid-soluble compound. For the water-soluble compound, not only will the water-to-plant concentration factor be low under any conditions, but the in-plant level should also approach an apparent steady-state value in a shorter time than is the case for a lipid-soluble compound. Thus, a plot of the in-plant contaminant level with time should display a small time-dependent region for a highly water-soluble contaminant and a pronounced time-dependent region for a lipid-soluble contaminant. This difference should especially be prominent with small plants, where the transport path of contaminants with water inside the plants is relatively short.

Hinman and Klaine (1992) studied the time dependence of the uptakes of atrazine, lindane, and chlordane from overlying water by small rooted aquatic vascular plants (*Hydrilla verticillata* Royle). It was found that the levels of more water-soluble atrazine ($\log K_{ow} = 2.71$) in shoots and roots of the plant approached apparent equilibrium within 1 and 2 h, while the levels of increasingly less water-soluble lindane ($\log K_{ow} = 3.75$) and chlordane ($\log K_{ow} = 5.58$) in roots and shoots reached apparent equilibrium within 24 and 144 h, respectively. The corresponding contaminant concentration factors with rooted hydrilla (i.e., $C_{pt}/C_w$) were 9.62, 38.2, and 1061 for atrazine, lindane, and chlordane. Although it is arguable whether all three contaminants in their plant–water systems have truly come to full equilibrium, because the plant composition was not available for calculating their theoretical $C_{pt}/C_w$ values, the time dependence and the order in $C_{pt}/C_w$ as observed are consistent with their relative uptake capacities according to their $K_{ow}$ values.

Consider now some likely scenarios in phytoremediation of contaminated soils or water. Let us start with highly water soluble compounds, in which the $K_{pom}$ values would be small with practically all plants. If the compounds are fairly resistant to biodegradation (i.e., if the metabolism is slow) and if they have low vapor pressures, there will be little driving force for the continuing plant uptake of these compounds from external water once the in-plant concentrations approach saturation values ($\alpha_{pt} \simeq 1$). In this case, the approach to saturation (or near saturation) should be relatively fast, as shown for solutes with small $K_{ow}$ values in Table 8.1. Thus, for such compounds, it would be nearly improbable to achieve effective remediation of polluted soil or water by plantings. Here, although the plant growth would in effect dilute the plant contaminant level, and thus create a driving force for continuing uptake, the uptake would be fairly limited. On the other hand, if the water-soluble compounds have instead high vapor pressures, such as methyl *t*-butyl ether (MTBE) and TCE, the high volatilization rate of the sorbed chemicals through plant leaves and surfaces should then exert a continuing driving force for the removal of these contaminants from the external water phase, despite the fact that the contaminant level in the plant may either approach saturation ($\alpha_{pt} \simeq 1$) or maintain a steady-state value. This expected consequence is supported by the finding that a sizable quantity of MTBE is removed from a groundwater site by planted poplar trees, which exhibit a very large transpiration rate (Hong et al., 2001). In a similar study of MTBE uptake by young poplar trees from a hydroponic solution, Rubin and Ramaswami (2001) found that the MTBE concentration in plant water ascends rapidly to about the same concentration as in external water (i.e., to approach the limit of $\alpha_{pt} = 1$).

For compounds with low water solubility (i.e., the lipid-soluble compounds), the situation is more straightforward. This is because the large partition coefficient with plants enables the system to maintain a high and continuing driving force for contaminant uptake. This is true whether or not the contaminant dissipation by either metabolism or volatilization is efficient and whether or not the plant has a high lipid content. Naturally, the rate of removal by plants should increase with the plant lipid content and water transpiration rate. As shown earlier, even for plants with a lipid content of 0.1% by weight, the potential toward accumulation by the plant lipid of a relatively water-insoluble compound (e.g., PCBs) is tremendously large.

Dissipation of a contaminant in the plant, which leads to a reduction in $\alpha_{pt}$, will in principle increase the driving force for plant uptake in situations where the contaminant level in plant could otherwise approach saturation (i.e., where the $\alpha_{pt}$ value is close to 1). If the dissipation loss is small, there will be a sharper dependence of the plant contaminant level with time. If the dissipation loss is high, an apparent steady-state trend may occur if the rate of contaminant uptake from external water happens to offset the rate of contaminant dissipation. The uptake of hexachlorobenzene, a relatively chemically stable, poorly volatile, and lipid-soluble compound, from a hydroponic solution by young rye grass shows a continuous rise of its level in plant over a long period of time (Li et al., 2001). By contrast, the uptake of tetrachloroethylene, a relatively volatile and far more water-soluble compound, by the same grass quickly approaches a steady-state level following a short period of exposure (Li et al., 2001). These results are consistent with the model expectation.

Based on the available plant-uptake data, the partition-limited model appears to give a satisfactory account of the passive transport of various contaminants from soil and water into thus far a small number of plants and crops. The plant uptake of nonionic contaminants through active transport does not appear to be significant for the systems examined. According to the model analysis, it may be generally concluded that for plants with a high water content, highly water-soluble contaminants occur mainly in the plant–water phase, as their uptakes by lipids and other plant matters are either small or insignificant. By contrast, the plant uptake of highly water-insoluble contaminants is predominated by partition into the plant–lipid phase, even though the lipid content may be very low. These conclusions, which appear to make good sense, serve as a useful guide to the problem of crop contamination and to the proper selection of plants in relation to contaminant type for intended bioremediation of contaminated soils and groundwater. More extensive experimental data are yet to be furnished to substantiate the range, and to define the limits, of the model applicability.

# BIBLIOGRAPHY


Acree, W. E., Jr., Department of Chemistry, University of North Texas, Denton, Texas, personal communication (1998).

Acree, W. E., Jr. and J. H. Rytting, "Solubility in binary-solvent systems. Part IV. Prediction of naphthalene solubilities using the UNIFAC group contribution model," *Int. J. Pharm.* 13, 197–204 (1983).

Adamson, A. W., *Physical Chemistry of Surfaces*, 2nd ed., Wiley, New York, 1967.

Aiken, G. R. and R. L. Malcolm, "Molecular weight of aquatic fulvic acids by vapor pressure osmometry," *Geochim. Cosmochim. Acta* 51, 2177–2184 (1987).

Allen-King, R. M., P. Grathwohl, and W. P. Ball, "New modeling paradigms for the sorption of hydrophobic organic chemicals to heterogeneous carbonaceous matter in soils, sediments, and rocks," *Adv. Water Resour.* (2002). (In press).

Bailey, G. W. and J. L. White, "Review of adsorption and desorption of organic pesticides by soil colloids, with implications concerning pesticide bioactivity," *J. Agric. Food Chem.* 12, 324–332 (1964).

Baldock, J. A., J. M. Oades, A. G. Waters, X. Peng, A. M. Vassallo, and M. A. Wilson, "Aspects of the chemical structure of soil organic materials as revealed by solid-state $^{13}C$ NMR spectroscopy," *Biogeochemistry* 16, 1–42 (1992).

Banerjee, S., R. H. Sugatt, and D. P. O'Grady, "A simple method for determining bioconcentration parameters of hydrophobic compounds," *Environ. Sci. Technol.* 18, 79–81 (1984).

Barlow, F. and A. B. Hadaway, "Studies on aqueous suspensions of insecticides. Part VI. Further notes on the sorption of insecticides by soils," *Bull. Entomol. Res.* 46, 547–559 (1955).

Barrer, R. M., *Zeolites and Clay Minerals as Sorbents and Molecular Sieves*, Academic Press, London, 1978, pp. 407–486.

Barton, A. F. M., "Solubility parameters," *Chem. Rev.* 75, 731–753 (1975).

Boucher, R. F. and G. F. Lee, "Adsorption of lindane and dieldrin pesticides on unconsolidated aquifer sands," *Environ. Sci. Technol.* 6, 538–543 (1972).

Bower. C. A. and F. B. Gschwend, "Ethylene glycol retention by soils as a measure of surface area and interlayer swelling," *Soil Sci. Soc. Am. Proc.* 16, 342–345 (1952).

Boyd, S. A., "Adsorption of substituted phenols by soil," *Soil Sci.* 134, 337–343 (1982).

Boyd, S. A., M. M. Mortland, and C. T. Chiou, "Sorption of organic compounds on hexadecyl-trimethylammonium-smectite," *Soil Sci. Soc. Am. J.* 52, 652–657 (1988).

Boyd, S. A., M. D. Mikesell, and J.-F. Lee, "Chlorophenols in soils," In *Reactions and Movement of Organic Chemicals in Soils*, Spec. Publ. 22, B. L. Sawhney and K. Brown, Eds., Soil Science Society of America, Madison, WI, 1989, pp. 209–228.

Brady, N. C., *The Nature and Properties of Soils*, 8th ed., Macmillan, New York, 1974, pp. 11–14.

Briggs, G. G., "Theoretical and experimental relationships between soil adsorption, octanol-water partition coefficients, water solubilities, bioconcentration factors, and the parachlor," *J. Agric. Food Chem.* 29, 1050–1059 (1981).

Briggs, G. G., R. H. Bromilow, and A. A. Evans, "Relationships between lipophilicity and root uptake and translocation of non-ionised chemicals by barley," *Pestic. Sci.* 13, 495–504 (1982).

Brook, D. N., A. J. Dobbs, and N. Williams, "Octanol–water partition coefficients ($P$): Measurement, estimation, and interrelation, particularly for chemicals with $P > E +$ 5," *Ecotoxicol. Environ. Saf.* 11, 257–260 (1986).

Browman, M. G. and G. Chesters, "The solid–water interface: Transfer of organic pollutants across the solid–water interface," In *Fate of Pollutants in the Air and Water Environments*, Part I, I. H. Suffet, Ed., Wiley, New York, 1977, pp. 49–105.

Brunauer, S., *Adsorption of Gases and Vapors*, Princeton University Press, Princeton, NJ, 1945.

Brunauer, S., P. H. Emmett, and E. Teller, "Adsorption of gases in multimolecular layers," *J. Am. Chem. Soc.* 60, 309–319 (1938).

Buffle, J., P. Deladoey, and W. Haerdi, "The use of ultrafiltration for the separation and fractionation of organic ligands in fresh water," *Anal. Chim. Acta* 101, 339–357 (1978).

Call, F., "The mechanism of sorption of ethylene dibromide on moist soils," *J. Sci. Food Agric.* 8, 630–639 (1957).

Carlson, H. C. and A. P. Colburn, "Vapor–liquid equilibria of nonideal solutions," *Ind. Eng. Chem.* 34, 581–589 (1942).

Caron, G., I. H. Suffet, and T. Belton, "Effect of dissolved organic carbon on the environmental distribution of nonpolar organic compounds," *Chemosphere* 14, 993–1000 (1985).

Carter, C. W. and I. H. Suffet, "Binding of DDT to dissolved humic materials," *Environ. Sci. Technol.* 16, 735–740 (1982).

Carter, D. L., M. D. Heilman, and C. L. Gonzalez, "Ethylene glycol monoethyl ether for determining surface area of silicate minerals," *Soil Sci.* 100, 356–360 (1965).

Chin, Y.-P., G. R. Aiken, and K. M. Danielsen, "Binding of pyrene to aquatic and commercial humic substances: the role of molecular weight and aromaticity," *Environ. Sci. Technol.* 31, 1630–1635 (1997).

Chiou, C. T., "Partition coefficient and water solubility in environmental chemistry," In *Hazard Assessment of Chemicals: Current Development*, Vol. 1, J. Saxena and F. Fisher, Eds., Academic Press, New York, 1981, pp. 117–153.

———, "Partition coefficients of organic compounds in lipid–water systems and correlations with fish bioconcentration factors," *Environ. Sci. Technol.* 19, 57–62 (1985).

———, "Comment on thermodynamics of organic chemical partition in soils," *Environ. Sci. Technol.* 29, 1421–1422 (1995).

———, "Soil sorption of organic pollutants and pesticides," In *Encyclopedia of Environmental Analysis and Remediation*, R. A. Meyers, Ed., Wiley, New York, 1998, pp. 4517–4554.

Chiou, C. T. and J. H. Block, "Parameters affecting the partition coefficients of organic compounds in solvent–water and lipid–water systems," In *Partition Coefficient: Determination and Estimation*, W. Dunn, J. Block, and R. Pearlman, Eds., Pergamon Press, New York, 1986, pp. 37–60.

Chiou, C. T. and D. E. Kile, "Effects of polar and nonpolar groups on the solubility of organic compounds in soil organic matter," *Environ. Sci. Technol.* 28, 1139–1144 (1994).

———, "Deviations from sorption linearity on soils of polar and nonpolar organic compounds at low relative concentrations," *Environ. Sci. Technol.* 32, 338–343 (1998).

Chiou, C. T. and M. Manes, "Application of the Polanyi adsorption potential theory to adsorption from solution on activated carbon. V. Adsorption from water of some solids and their melts, and a comparison of bulk and adsorbate melting points," *J. Phys. Chem.* 78, 622–626 (1974).

———, "Application of the Flory–Huggins theory to the solubility of solids in glyceryl trioleate," *J. Chem. Soc. Faraday Trans.* 1, 82, 243–246 (1986).

———, "Comment on temperature dependence of the aqueous solubilities of highly chlorinated dibenzo-*p*-dioxins," *Environ. Sci. Technol.* 24, 1755–1756 (1990).

Chiou, C. T. and D. W. Rutherford, "Effects of exchanged cation and layer charge on the sorption of water and EGME vapors on montmorillonite clays," *Clays Clay Miner.* 45, 867–880 (1997).

Chiou, C. T. and T. D. Shoup, "Soil sorption of organic vapors and effects of humidity on sorptive mechanism and capacity," *Environ. Sci. Technol.* 19, 1196–1200 (1985).

Chiou, C. T., V. H. Freed, D. W. Schmedding, and R. L. Kohnert, "Partition coefficient and bioaccumulation of selected organic chemicals," *Environ. Sci. Technol.* 11, 475–478 (1977).

Chiou, C. T., L. J. Peters, and V. H. Freed, "A physical concept of soil–water equilibria for nonionic organic compounds," *Science* 206, 831–832 (1979).

———, "Soil–water equilibria for nonionic organic compounds," *Science* 213, 683–684 (1981).

Chiou, C. T., J. H. Block, and M. Manes, "Substituent contribution to the partition coefficients of substituted benzenes in solvent-water mixtures," *J. Pharm. Sci.* 71, 1307–1309 (1982a).

Chiou, C. T., D. W. Schmedding, and M. Manes, "Partitioning of organic compounds in octanol–water systems," *Environ. Sci. Technol.* 18, 4–10 (1982b).

Chiou, C. T., P. E. Porter, and D. W. Schmedding, "Partition equilibria of nonionic organic compounds between soil organic matter and water," *Environ. Sci. Technol.* 17, 227–231 (1983).

Chiou, C. T., P. E. Porter, and T. D. Shoup, "Reply to comment on partition equilibria of nonionic organic compounds between soil organic matter and water," *Environ. Sci. Technol.* 18, 296–297 (1984).

Chiou, C. T., T. D. Shoup, and P. E. Porter, "Mechanistic roles of soil humus and minerals in the sorption of nonionic organic compounds from aqueous and organic solutions," *Org. Geochem.* 8, 9–14 (1985).

Chiou, C. T., R. L. Malcolm, T. I. Brinton, and D. E. Kile, "Water solubility enhancement of some organic pollutants and pesticides by dissolved humic and fulvic acids," *Environ. Sci. Technol.* 20, 502–508 (1986).

Chiou, C. T., D. E. Kile, T. I. Brinton, R. L. Malcolm, J. A. Leenheer, and P. MacCarthy, "A comparison of water solubility enhancements of organic solutes by aquatic humic materials and commercial humic acids," *Environ. Sci. Technol.* 21, 1231–1234 (1987).

Chiou, C. T., J.-F. Lee, and S. A. Boyd, "The surface area of soil organic matter," *Environ. Sci. Technol.* 24, 1164–1166 (1990).

Chiou, C. T., J.-F. Lee, and S. A. Boyd, "Reply to comment on the surface area of soil organic matter," *Environ. Sci. Technol.* 26, 404–406 (1992).

Chiou, C. T., D. W. Rutherford, and M. Manes, "Sorption of $N_2$ and EGME vapors on some soils, clays, and mineral oxides and determination of sample surface areas by use of sorption data," *Environ. Sci. Technol.* 27, 1587–1594 (1993).

Chiou, C. T., S. E. McGroddy, and D. E. Kile, "Partition characteristics of polycyclic aromatic hydrocarbons on soils and sediments," *Environ. Sci. Technol.* 32, 264–269 (1998)

Chiou, C. T., D. E. Kile, D. W. Rutherford, G. Sheng, and S. A. Boyd, "Sorption of selected organic compounds from water to a peat soil and its humic-acid and humin fractions: Potential sources of the sorption nonlinearity," *Environ. Sci. Technol.* 34, 1254–1258 (2000).

Chiou, C. T., G. Sheng, and M. Manes, "A partition-limited model for the plant uptake of organic contaminants from soil and water," *Environ. Sci. Technol.* 35, 1437–1444 (2001).

Chisholm, R. C. and L. Koblitsky, "Sorption of methyl bromide by soil in a fumigation chamber," *J. Econ. Entomol.* 36, 545–551 (1943).

Choi, W.-W. and K. Y. Chen, "Associations of chlorinated hydrocarbons with fine particles and humic substances in nearshore surficial sediments," *Environ. Sci. Technol.* 10, 782–786 (1976).

Collander, R., "The partition of organic compounds between higher alcohols and water," *Acta Chem. Scand.* 5, 774–780 (1951).

Cotton, F. A. and G. Wilkinson, *Advanced Inorganic Chemistry*, 2nd ed., Interscience, New York, 1966.

de Boer, J. H., B. C. Lippens, B. G. Linsen, J. C. P. Broekhoff, A. van den Heuvel, and Th. J. Osinga, "The *t*-curve of multilayer $N_2$-adsorption," *J. Colloid Interface Sci.* 21, 405–414 (1966).

Deitsch, J. J. and J. A. Smith, "Effect of Triton X-100 on the rate of trichloroethene desorption from soil to water," *Environ. Sci. Technol.* 29, 1069–1080 (1995).

Di Toro, D. M., "A particle interaction model of reversible organic chemical sorption," *Chemosphere* 14, 1503–1538 (1985).

Dobbs, A. J. and N. Williams, "Fat solubility—A property of environmental relevance?" *Chemosphere* 12, 97–104 (1983).

Dowdy, R. H. and M. M. Mortland, "Alcohol–water interactions on montmorillonite surfaces. I. Ethanol," *Clays Clay Miner.* 15, 259–271 (1967).

Dubinin, M. M. and D. P. Timofeyev, "Adsorption of vapours on active charcoals in relation to the properties of the adsorbate," *C. R. Acad. Sci. URSS* 54, 701–704 (1946).

Dyal, R. S. and S. B. Hendricks, "Total surface of clays in polar liquids as a characteristic index," *Soil Sci.* 69, 421–432 (1950).

Edwards, D. A., R. G. Luthy, and Z. Liu, "Solubilization of polycyclic aromatic hydrocarbons in micellar nonionic surfactant solutions," *Environ. Sci. Technol.* 25, 127–133 (1991).

Eichinger, B. E. and P. J. Flory, "Thermodynamics of polymer solutions. Part 1. Natural rubber and benzene," *Trans. Faraday Soc.* 64, 2035–2052 (1968a).

———, "Thermodynamics of polymer solutions. Part 2. Polyisobutylene and benzene," *Trans. Faraday Soc.* 64, 2053–2060 (1968b).

Eltantawy, I. M. and P. W. Arnold, "Reappraisal of ethylene glycol monoethyl ether (EGME) method for surface area estimations for clays," *J. Soil Sci.* 24, 232–238 (1973).

Felsot, A. and J. Wilson, "Adsorption of carbofuran and movement on soil thin layers," *J. Bull. Environ. Contam. Toxicol.* 24, 778–782 (1980).

Flory, P. J., "Thermodynamics of high polymer solutions," *J. Chem. Phys.* 10, 51–61 (1941).

———, *Principles of Polymer Chemistry*, Cornell University Press, Ithaca, NY, 1953.

———, "Thermodynamics of polymer solutions," *Discuss. Faraday Soc.* 49, 7–29 (1970).

Friesen, K. J. and G. R. B. Webster, "Temperature dependence of the aqueous solubilities of highly chlorinated dibenzo-*p*-dioxins," *Environ. Sci. Technol.* 24, 97–101 (1990).

Fujita, T., J. Iwasa, and C. Hansch, "A new substituent constant, $\pi$, derived from partition coefficients," *J. Am. Chem. Soc.* 86, 5175–5180 (1964).

Gauthier, T. D., W. R. Seitz, and C. L. Grant, "Effects of structural and compositional variations of dissolved humic materials on pyrene $K_{oc}$ values," *Environ. Sci. Technol.* 21, 243–247 (1987).

Gerolt, P., "Investigation into the problem of insecticide sorption by soils," *Bull. World Health Organ.* 24, 577–592 (1961).

Glotfelty, D. E., A. W. Taylor, B. C. Turner, and W. E. Zoller, "Volatilization of surface-applied pesticides from fallow soil," *J. Agric. Food Chem.* 32, 638–643 (1984).

Goerlitz, D. F., D. E. Troutman, E. M. Godsy, and B. J. Franks, "Migration of wood-preserving chemicals in contaminated groundwater in a sand aquifer at Pensacola, Florida," *Environ. Sci. Technol.* 19, 955–961 (1985).

Goring, C. A. I., "Physical aspects of soil in relation to the action of soil fungicides," *Annu. Rev. Phytopathol.* 5, 285–318 (1967).

Graham, D. P., "Physical adsorption on low-energy solids. III. Adsorption of ethane, *n*-butane, and *n*-octane on poly(tetrafluoroethylene)," *J. Phys. Chem.* 69, 4387–4391 (1965).

Gregg, S. J. and K. S. W. Sing, *Adsorption, Surface Area, and Porosity*, 2nd ed., Academic Press, London, 1982.

Griffin, J. J. and E. D. Goldberg, "Impact of fossil fuel combustion on sediments of Lake Michigan: A reprise," *Environ. Sci. Technol.* 17, 244–245 (1983).

Grover, R., A. E. Smith, S. R. Shewchuk, A. J. Cessna, J. H. Hunter, "Fate of trifluralin and triallate applied as a mixture to wheat field," *J. Environ. Qual.* 17, 543–550 (1988).

Gschwend, P. M. and S.-C. Wu, "On the constancy of sediment–water partition coefficients of hydrophobic organic pollutants," *Environ. Sci. Technol.* 19, 90–95 (1985).

Gu, T., B.-Y. Zhu, and H. Rupprecht, "Surfactant adsorption and surface micellization," *Prog. Colloid Polym. Sci.* 88, 74–85 (1992).

Gustaffson, Ö., F. Haghseta, C. Chan, J. MacFarland, and P. M. Gschwend, "Quantification of the dilute sedimentary soot phase: implications for PAH speciation and bioavailability," *Environ. Sci. Technol.* 31, 203–209 (1997).

Haderlein, S. B. and R. P. Schwarzenbach, "Adsorption of substituted nitrobenzenes to mineral surfaces," *Environ. Sci. Technol.* 27, 316–326 (1993).

Haderlein, S. B., K. W. Weissmahr, and R. P. Schwarzenbach, "Specific adsorption of nitroaromatic explosives and pesticides to clay minerals," *Environ. Sci. Technol.* 30, 612–622 (1996).

Hamaker, J. W. and J. M. Thompson, "Adsorption," In *Organic Chemicals in the Soil Environment*, Vol. 1, C. A. I. Goring and J. W. Hamaker, Eds., Marcel Dekker, New York, 1972, pp. 49–143.

Hance, R. J., "Observations on the relationship between the adsorption of diuron and the nature of the adsorbent," *Weed Res.* 5, 108–114 (1965).

Hansch, C., "A quantitative approach to biochemical structure–activity relationships," *Acc. Chem. Res.* 2, 232–239 (1969).

Hansch, C. and T. Fujita, "ρ-σ-π Analysis: A method for the correlation of biological activity and chemical structure," *J. Am. Chem. Soc.* 86, 1616–1626 (1964).

Hanson, W. J. and R. W. Nex, "Diffusion of ethylene dibromide in soils," *Soil Sci.* 76, 209–214 (1953).

Haque, R. and R. Sexton, "Kinetic and equilibrium study of the adsorption of 2,4-dichlorophenoxy acetic acid on some surfaces," *J. Colloid Interface Sci.* 27, 818–827 (1968).

Harris, C. R., "Influence of soil type and soil moisture on the toxicity of insecticides in soils to insects," *Nature (London)* 202, 724 (1964).

Harris, C. R. and W. W. Sans, "Absorption of organochlorine insecticide residues from agricultural soils by root crops," *J. Agric. Food Chem.* 15, 861–863 (1967).

Hassett, J. J., W. L. Banwart, S. G. Wood, and J. C. Means, "Sorption of α-naphthol: Implications concerning the limits of hydrophobic sorption," *Soil Sci. Soc. Am. J.* 45, 38–42 (1981).

Heilman, M. D., D. L. Carter, and C. L. Gonzalez, "The ethylene glycol monoethyl ether (EGME) technique for determining soil surface area," *Soil Sci.* 100, 409–413 (1965).

Hildebrand, J. H. and R. L. Scott, *Solubility of Nonelectrolytes*, Dover Publications, New York, 1964.

Hildebrand, J. H., E. T. Ellefson, and C. W. Beebe, "Solubilities of anthracene, anthraquinone, *para*-bromobenzene, phenanthrene, and iodine in various solvents," *J. Am. Chem. Soc.* 39, 2301–2302 (1917).

Hilbebrand, J. H., J. M. Praunitz, and R. L. Scott, *Regular and Related Solutions*, Van Nostrand Reinhold, New York, 1970.

Hinman, M. L. and S. J. Klaine, "Uptake and translocation of selected organic pesticides by the rooted aquatic plant *Hydrilla verticillata* Royle," *Environ. Sci. Technol.* 26, 609–613 (1992).

Hong, M., W. F. Farmayan, I. J. Dortch, C. Y. Chiang, S. K. McMillan, and J. L. Schnoor, "Phytoremediation of MTBE from a groundwater plume," *Environ. Sci. Technol.* 35, 1231–1239 (2001).

Huang, W. and W. J. Weber, Jr., "Thermodynamic considerations in the sorption of organic contaminants by soils. 1. The isosteric heat approach and its application to model inorganic sorbents," *Environ. Sci. Technol.* 31, 3238–3243 (1997).

Huang, W., M. A. Schlautman, and W. J. Weber, Jr., "A distributed reactivity model for sorption by soils and sediments. 5. The influence of near-surface characteristics in mineral domains," *Environ. Sci. Technol.* 30, 2993–3000 (1996).

Huggins, M. L., "Thermodynamic properties of solutions of long-chain compounds," *Ann. N.Y. Acad. Sci.* 43, 1–32 (1942).

Jafvert, C. T., P. Van Hoff, and J. K. Heath, "Solubilization of nonpolar compounds by nonionic surfactant micelles," *Water Res.* 28, 1009–1017 (1994).

Judy, C. L., N. M. Pontikos, and W. E. Acree, Jr., "Solubility of pyrene in binary solvent mixtures containing cyclohexane," *J. Chem. Eng. Data* 32, 60–62 (1987).

Jurinak, J. J., "The effect of clay minerals and exchangeable cations on the adsorption of ethylene dibromide vapor," *Soil Sci. Soc. Am. Proc.* 21, 599–602 (1957a).

———, "Adsorption of 1,2-dibromo-3-chloropropane vapor by soils," *J. Agric. Food Chem.* 598–601 (1957b).

Jurinak, J. J. and D. H. Volman, "Application of the Brunauer, Emmet, and Teller equation to ethylene dibromide adsorption," *Soil Sci.* 83, 487–496 (1957).

Karapanagioti, H. K., S. Kleineidam, D. A. Sabatini, P. Grathwohl, and B. Ligouis, "Impacts of heterogeneous organic matter on phenanthrene sorption: Equilibrium and kinetic studies with aquifer material," *Environ. Sci. Technol.* 34, 406–414 (2000).

Karapanagioti, H. K., J. Childs, and D. A. Sabatini, "Impacts of heterogeneous organic matter on phenanthrene sorption: Different soil and sediment samples," *Environ. Sci. Technol.* 35, 4684–4690 (2001).

Karickhoff, S. W., "Organic pollutant sorption in aqueous systems," *J. Hydraul. Eng.* 110, 707–735 (1984).

Karickhoff, S. W., D. S. Brown, and T. A. Scott, "Sorption of hydrophobic pollutants on natural sediments," *Water Res.* 13, 241–248 (1979).

Kenaga, E. E. and C. A. I. Goring, "Relationship between water solubility, soil sorption, octanol-water partitioning, and concentration of chemicals in biota," In *Aquatic Toxicology*, J. C. Eaton, P. R. Parrish, and A. C. Hendricks, Eds., American Society for Testing and Materials, Philadelphia, 1980, pp. 78–115.

Kile, D. E. and C. T. Chiou, "Water solubility enhancements of DDT and trichlorobenzene by some surfactants below and above the critical micelle concentration," *Environ. Sci. Technol.* 23, 832–838 (1989).

Kile, D. E., C. T. Chiou, and R. S. Helburn, "Effect of some petroleum sulfonate surfactants on the apparent water solubility of organic compounds," *Environ. Sci. Technol.* 24, 205–208 (1990).

Kile, D. E., C. T. Chiou, H. Zhou, H. Li, and O, Xu, "Partition of nonpolar organic pollutants from water to soil and sediment organic matters," *Environ. Sci. Technol.* 29, 1401–1406 (1995).

Kile, D. E., R. L. Wershaw, and C. T. Chiou, "Correlation of soil and sediment organic matter polarity to aqueous sorption of nonionic compounds," *Environ. Sci. Technol.* 33, 2053–2056 (1999).

Kleineidam, S., H. Rügner, B. Ligouis, and P. Grathwohl, "Organic matter facies and equilibrium sorption of phenanthrene," *Environ. Sci. Technol.* 33, 1637–1644 (1999).

Könemann, H. and K. van Leeuwen, "Toxickinetics in fish: Accumulation and elimination of six chlorobenzenes by guppies," *Chemosphere* 9, 3–19 (1980).

Laird, D. A., E. Barriuso, R. H. Dowdy, and W. C. Koskinen, "Adsorption of atrazine on smectites," *Soil Sci. Soc. Am. J.* 56, 62–67 (1992).

Langmuir, I., "The adsorption of gases on plane surfaces of glass, mica and platinum," *J. Am. Chem. Soc.* 40, 1361–1403 (1918).

Lee, J.-F., J. R. Crum, and S. A. Boyd, "Enhanced retention of organic contaminants by soils exchanged with organic cations," *Environ. Sci. Technol.* 23, 1365–1372 (1989).

Lee, J.-F., P.-M. Liao, C.-C. Kuo, H.-T. Yang, and C. T. Chiou, "Influence of a nonionic surfactant (Triton X-100) on contaminant distribution between water and several soil solids," *J. Colloid Interface Sci.* 229, 445–452 (2000).

Leistra, M., "Distribution of 1,3-dichloropropene over the phases in soil," *J. Agric. Food Chem.* 18, 1124–1126 (1970).

Leo, A. and C. Hansch, "Linear free-energy relationships between partitioning solvent systems," *J. Org. Chem.* 36, 1539–1544 (1971).

Leo, A., C. Hansch, and D. Elkins, "Partition coefficients and their uses," *Chem. Rev.* 71, 525–554 (1971).

Lewis, W. K., E. R. Gilliland, B. Chertow, and W. P. Cadogan, "Adsorption equilibria: Pure gas isotherms," *Ind. Eng. Chem.* 42, 1326–1332 (1950).

Li, H., O. Xu, C. T. Chiou, "Uptake of chlorinated hydrocarbons by rye grass from water," manuscript in preparation (2001).

Lichtenstein, E. P., "Absorption of some chlorinated hydrocarbon insecticides from soils into various crops," *J. Agric. Food Chem.* 7, 430–433 (1959).

———, "Insecticidal residues in various crops grown in soils treated with abnormal rates of aldrin and heptachlor," *J. Agric. Food Chem.* 8, 448–451 (1960).

Lichtenstein, E. P., G. R. Myrdal, and K. R. Schulz, "Absorption of insecticidal residues from contaminated soils into five carrot varieties," *J. Agric. Food Chem.* 13, 126–131 (1965).

Lu, P. Y. and R. L. Metcalf, "Environmental fate and biodegradability of benzene derivatives as studied in a model aquatic ecosystem," *Environ. Health Perspect.* 10, 269–284 (1975).

MacIntyre, W. G. and C. L. Smith, "Comment on partition equilibria of nonionic organic compounds between soil organic matter and water," *Environ. Sci. Technol.* 18, 295 (1984).

Majewski, M., R. Desjardine, P. Rochette, E. Pattey, J. Seiber, and D. Glotfelty, "Field comparison of an eddy accumulation and an aerodynamic-gradient system for measuring pesticide volatilization fluxes," *Environ. Sci. Technol.* 27, 121–128 (1993).

Malcolm, R. L. and P. MacCarthy, "Limitations in the use of commercial humic acids in water and soil research," *Environ. Sci. Technol.* 20, 904–911 (1986).

Manes, M., "Activated carbon adsorption fundamentals," In *Encyclopedia of Environmental Analysis and Remediation*, R. A. Myers, Ed., Wiley, New York, 1998, pp. 26–68.

Manes, M and L. J. E. Hofer, "Application of the Polanyi adsorption potential theory to adsorption from solution on activated carbon," *J. Phys. Chem.* 73, 584–590 (1969).

Masiello, C.A. and E. R. M. Druffel, "Black carbon in deep-sea sediments," *Science* 280, 1911–1913 (1998).

Mattina, M. J. I., W. Iannucci-Berger, and L. Dykas, "Chlordane uptake and its translocation in food crops," *J. Agric. Food Chem.* 48, 1909–1915 (2000).

McGroddy, S. E. and J. W. Farrington, "Sediment porewater partitioning of polycyclic aromatic hydrocarbons in three cores from Boston Harbor, Massachusetts," *Environ. Sci. Technol.* 29, 1542–1550 (1995).

McNeal, B. L., "Effect of exchangeable cations on glycol retention by clay minerals," *Soil Sci.* 97, 96–102 (1964).

Means, J. C., S. G. Wood, J. J. Hassett, and W. L. Banwart, "Sorption of polynuclear aromatic hydrocarbons by sediments and soils," *Environ. Sci. Technol.* 14, 1524–1528 (1980).

————, "Sorption of amino- and carboxyl-substituted polynuclear aromatic hydrocarbons by sediments and soils," *Environ. Sci. Technol.* 16, 93–98 (1982).

Meyer, H., "Zur Theorie der Alkohol-narkose. I. Welche Eigenschaft der Anesthetica bedingt ihre narkotische Wirkung," *Arch. Exp. Pathol. Pharmakol.* 42, 109–118 (1899).

Mills, A. C. and J. W. Biggar, "Solubility-temperature effects on the adsorption of gamma- and beta-BHC from aqueous and hexane solutions by soil materials," *Soil Sci. Soc. Am. Proc.* 33, 210–216 (1969).

Mingelgrin, U. and Z. Gerstl, "Reevaluation of partitioning as a mechanism of nonionic chemicals adsorption in soil," *J. Environ. Qual.* 12, 1–11 (1983).

Mooney, R. W., A. C. Keenan, and L. A. Wood, "Adsorption of water vapor by montmorillonite. I. Heats of desorption and application of BET theory," *J. Am. Chem. Soc.* 74, 1367–1374 (1952).

Nayyar, S. P., D. A. Sabatini, and J. H. Harwell, "Surfactant adsolubilization and modified admicellar sorption of nonpolar, polar, and ionizable organic compounds," *Environ. Sci. Technol.* 28, 1874–1881 (1994).

Neely, W. B., D. R. Branson, and G. E. Blau, "Partition coefficient to measure bioconcentration potential of organic chemicals to fish," *Environ. Sci. Technol.* 13, 1113–1115 (1974).

Oliver, B. G. and A. J. Nimii, "Bioconcentration of chlorobenzenes from water by rainbow trout: Correlation with partition coefficients and environmental residues," *Environ. Sci. Technol.* 17, 287–291 (1983).

Ong, S. K. and L. W. Lion, "Mechanism for trichloroethylene vapor sorption onto soil minerals," *J. Environ. Qual.* 20, 180–188 (1991).

Overton, E., *Studien uber die Narkose*, Fischer, Jena, Germany, 1901.

Paterson, S., D. Mackay, and C. McFarland, "A model of organic chemical uptake by plants from soil and the atmosphere," *Environ. Sci. Technol.* 28, 2259–2266 (1994).

Patton, J. S., B. Stone, C. Papa, R. Abramowitz, and S. H. Yalkowsky, "Solubility of fatty acids and other hydrophobic molecules in liquid trioleoylglycerol," *J. Lipid Res.* 25, 189–197 (1984).

Pennell, K. D. and P. S. C. Rao, "Comment on the surface area of soil organic matter," *Environ. Sci. Technol.* 26, 402–404 (1992).

Pennell, K. D., R. D. Rhue, P. S. C. Rao, and C. T. Johnston, "Vapor-phase sorption of *p*-xylene and water on soils and clay minerals," *Environ. Sci. Technol.* 26, 756–763 (1992).

Pennell, K. D., S. A. Boyd, and L. M. Abriola, "Surface area of soil organic matter re-examined," *Soil Sci. Soc. Am. J.* 59, 1012–1018 (1995).

Pereira, W. E., C. E. Rostad, C. T. Chiou, L. B. Barber, II., D. K. Demcheck, and C. R. Demas, "Contamination of estuarine water, biota, and sediment by halogenated organic compounds," *Environ. Sci. Technol.* 22, 772–778 (1988).

Pierce, R. H., C. E. Olney, and G. T. Felbeck, Jr., "*p,p'*-DDT adsorption to suspended particulate matter in sea water," *Geochim. Cosmochim. Acta* 38, 1061–1073 (1974).

Pignatello, J. J. and B. Xing, "Mechanism of slow sorption of organic chemicals to natural particles," *Environ. Sci. Technol.* 30, 1–11 (1996).

Plato, C. and A. R. Glasgow, "Differential scanning calorimetry as a general method for determining the purity and heat of fusion of high-purity chemicals. Application to 95 compounds," *Anal. Chem.* 41, 330–336 (1969).

Polanyi, M., "Adsorption of gases (vapors) by a solid nonvolatile adsorbent," *Verh. Dtsch. Phys. Ges.* 18, 55–80 (1916).

Quirk, J. P., "Significance of surface areas calculated from water vapor sorption isotherms by use of the B.E.T. equation," *Soil Sci.* 80, 423–430 (1955).

Raoult, F. M., "Loi générale des tensions de vapeur des dissolvants," *Compt. Rend.* 1430–1433 (1887).

———, "Über die Dampfdrucke ätherischer Lösungen," *Z. Phys. Chem.* 2, 353–373 (1888).

Reinert, R. E., "Pesticide concentrations in Great Lakes fish," *Pestic. Monit. J.* 3, 233–240 (1970).

Remy, M. J. and G. Poncelet, "A new approach to the determination of the external surface and micropore volume of zeolites from the nitrogen adsorption isotherm at 77 K," *J. Phys. Chem.* 99, 773–779 (1995).

Riederer, M., "Estimating partitioning and transport of organic chemicals in the foliage/atmosphere system: Discussion of a fugacity-based model," *Environ. Sci. Technol.* 24, 829–837 (1990).

Roberts, J. R., A. S. W. DeFrietas, and M. A. J. Gidney, "Influence of lipid pool size on bioaccumulation of the insecticide chlordane by northern redhorse suckers (*Moxostoma macrolepidotum*)," *J. Fish. Res. Board Can.* 34, 89–97 (1977).

Rosen, M. J., *Surfactants and Interfacial Phenomena*, Wiley, New York, 1978.

Rouse, J. D., D. A. Sabatini, and J. H. Harwell, "Minimizing surfactant losses using twin-head anionic surfactants in subsurface remediation," *Environ. Sci. Technol.* 27, 2072–2078 (1993).

Rubin, E. and A. Ramaswami, "The potential for phytoremediation of MTBE," *Water Res.* 35, 1348–1353 (2001).

Rutherford, D. W. and C. T. Chiou, "Effect of water saturation in soil organic matter on the partition of organic compounds," *Environ. Sci. Technol.* 26, 965–970 (1992).

Rutherford, D. W., C. T. Chiou, and D. E. Kile, "Influence of soil organic matter composition on the partition of organic compounds," *Environ. Sci. Technol.* 26, 336–340 (1992).

Rutherford, D. W., C. T. Chiou, and D. D. Eberl, "Effects of exchanged cation on the microporosity of montmorillonite," *Clays Clay Miner.* 45, 534–543 (1997).

Rutland, M. W. and T. J. Senden, "Adsorption of poly(oxyethylene) nonionic surfactant $C_{12}E_5$ to silica: A study using atomic force microscopy," *Langmuir* 9, 412–418 (1993).

Sabatini, D. A., R. C. Knox, and J. H. Harwell, Eds., *Surfactant Enhanced Subsurface Remediation: Emerging Technology.* ACS Symposium Series 594, American Chemical Society, Washington, DC, 1995.

Saltzman, S., L. Kliger, and B. Yaron, "Adsorption-desorption of parathion as affected by soil organic matter," *J. Agric. Food Chem.* 20, 1224–1226 (1972).

Schnitzer, M. and S. U. Kahn, *Humic Substances in the Environment*, Marcel Dekker, New York, 1972.

Schwarzenbach, R. P. and J. Westall, "Transport of nonpolar organic compounds from surface water to groundwater: Laboratory sorption studies," *Environ. Sci. Technol.* 15, 1360–1367 (1981).

Shin, Y.-O., J. J. Chodan, and A. R. Wolcott, "Adsorption of DDT by soils, and biological materials," *J. Agric. Food Chem.* 18, 1129–1133 (1970).

Shinoda, K. and J. H. Hildebrand, "The solubility and entropy of solution of iodine in octamethylcyclotetrasiloxane," *J. Phys. Chem.* 61, 789–791 (1957).

———, "The solubility and entropy of solution of iodine in $n$-$C_7F_{16}$, $c$-$C_6F_{11}CF_3$, $(C_3F_7COOCH_2)_4C$, $c$-$C_4Cl_2F_6$, $CCl_2FCClF_2$ and $CHBr_3$," *J. Phys. Chem.* 62, 292–294 (1958).

Shipinov, N. A., "Sorption of HCN on soils during their fumigation with cyanides," *Bull. Plant Prot. (U.S.S.R.)* No. 1–2, 192–199 (1940).

Shone, M. G. T. and A. V. Wood, "Uptake and translocation of herbicides. 1," *J. Exp. Bot.* 25, 390–400 (1974).

Sing, K. S. W., "Utilisation of adsorption data in the BET region," In *Surface Area Determination, Proc. Int. Symp.*, D. H. Everett and R. H. Ottewill, Eds., Butterworths, London, 1970, pp. 25–42.

Smith, D. W., J. J. Griffin, and E. D. Goldberg, "Elemental carbon in marine sediments: A baseline for burning," *Nature (London)* 241, 268–270 (1973).

Smith, J. A., P. J. Witkowski, and C. T. Chiou, "Partition of nonionic organic compounds in aquatic systems," *Rev. Environ. Contam. Toxicol.* 103, 127–151 (1988).

Smith, J. A., C. T. Chiou, J. A. Kammer, and D. E. Kile, "Effect of soil moisture on the sorption of trichloroethene vapor to vadose-zone soil at Picatinny Arsenal, New Jersey," *Environ. Sci. Technol.* 24, 676–683 (1990a).

Smith, J. A., P. R. Jaffé, and C. T. Chiou, "Effect of ten quaternary ammonium cations on tetrachloromethane sorption to clay from water," *Environ. Sci. Technol.* 24, 1167–1172 (1990b).

Spencer, W. F. and M. M. Cliath, "Desorption of lindane from soil as related to vapor pressure," *Soil Sci. Soc. Am. Proc.* 34, 574–578 (1970).

Spencer, W. F., M. M. Cliath, and W. J. Farmer, "Vapor density of soil-applied dieldrin as related to soil–water content, temperature, and dieldrin concentration," *Soil Sci. Soc. Am. Proc.* 33, 509–511 (1969).

Sposito, G., *The Surface Chemistry of Soils*, Oxford University Press, New York, 1984.

Spurlock, F. C. and J. W. Biggar, "Thermodynamics of organic chemical partition in soils. 2. Nonlinear partition of substituted phenylureas from aqueous solution," *Environ. Sci. Technol.* 28, 996–1002 (1994).

Stark, F. L., Jr., "Investigations of chloropicrin as a soil fumigant," *New York (Cornell) Agric. Exp. Stn. Mem.* 178, 1–61 (1948).

Stevenson, F. J., "Geochemistry of soil humic substances," In *Humic Substances in Soil, Sediment, and Water*, G. R. Aiken, D. M. McKnight, R. L. Wershaw, and P. MacCarthy, Eds., Wiley-Interscience, New York, 1985, pp. 13–29.

Sugiura, K., T. Washino, M. Hattori, E. Sato, and N. Goto, "Accumulation of organochlorine compounds in fishes. Difference of accumulation factors by fishes," *Chemosphere* 8, 359–364 (1979).

Sun, S. and S. A. Boyd, "Sorption of polychlorobiphenyl (PCB) congeners by residual PCB–oil phases in soils," *J. Environ. Qual.* 20, 557–561 (1991).

———, "Sorption of nonionic organic compounds in soil–water systems containing a petroleum sulfonate–oil surfactants," *Environ. Sci. Technol.* 27, 1340–1346 (1993).

Sun, S., W. P. Inskeep, and S. A. Boyd, "Sorption of nonionic organic compounds in soil–water systems containing a micelle-forming surfactant," *Environ. Sci. Technol.* 29, 903–913 (1995).

Swackhamer, D. L. and R. A. Hites, "Occurrence and bioaccumulation of organochlorine compounds in fishes from Siskiwit Lake, Isle Royale, Lake Superior," *Environ. Sci. Technol.* 22, 543–548 (1988).

Swoboda, A. R. and G. W. Thomas, "Movement of parathion in soil columns," *J. Agric. Food Chem.* 16, 923–927 (1968).

Tam, D. D., W.-Y. Shiu, K. Qiang, and D. Mackay, "Uptake of chlorobenzenes by tissues of the soybean plant: equilibria and kinetics," *Environ. Toxicol. Chem.* 15, 489–494 (1996).

Theng, B. K. G., D. J. Greenland, and J. P. Quirk, "Adsorption of alkylammonium cations by montmorillonite," *Clay Miner.* 7, 1–17 (1967).

Thibaud, C., C. Erkey, and A. Akgerman, "Investigation of the effect of moisture on the sorption and desorption of chlorobenzene and toluene from soil," *Environ. Sci. Technol.* 27, 2373–2380 (1993).

Thurman, E. M., R. L. Wershaw, R. L. Malcolm, and D. J. Pinckney, "Molecular size of aquatic humic substances," *Org. Geochem.* 4, 27–35 (1982).

Tiller, K. G. and L. H. Smith, "Limitations of EGME retention to estimate the surface area of soils," *Aust. J. Soil. Res.* 28, 1–26 (1990).

Trapp, S., "Model for uptake of xenobiotics into plants," In *Plant Contamination: Modeling and Simulation of Organic Chemical Processes*, S. Trapp and J. C. McFarland, Eds., Lewis Publishers, Boca Raton, FL, 1995.

———, "Modelling uptake into roots and subsequent translocation of neutral and ionisable organic compounds," *Pest. Manag. Sci.* 56, 767–778 (2000).

Trapp, S. and M. Matthies, "Generic one-compartment model for uptake of organic chemicals by foliar vegetation," *Environ. Sci. Technol.* 29, 2333–2338 (1995).

Trapp, S., M. Matthies, I. Scheunert, and E. M. Topp, "Modeling the bioconcentration of organic chemicals in plants," *Environ. Sci. Technol.* 24, 1246–1252 (1990).

Underdown, A. W., C. H. Langford, and D. S. Gamble, "Light scattering of a polydisperse fulvic acid," *Anal. Chem.* 53, 2139–2140 (1981).

Upchurch, R. P., "The influence of soil–moisture content on the response of cotton to herbicides," *Weeds* 5, 112–120 (1957).

Vanier, C., D. Planas, and M. Sylvestre, "Empirical relationships between polychlorinated biphenyls in sediments and submerged rooted macrophytes," *Can. Fish. Aquat. Sci.* 56, 1792–1800 (1999).

————, "Equilibrium partition theory applied to PCBs in macrophytes," *Environ. Sci. Technol.* 35, 4830–4833 (2001).

van Laar, J. J., "Über Dampfspannungen von binären Gemischen," *Z. Phys. Chem.* 72, 723–751 (1910).

————, "Zur Theorie der Dampfspannungen von binären Gemischen," *Z. Phys. Chem.* 83, 599–609 (1913).

Wade, P. I., "Soil fumigation. I. The sorption of ethylene dibromide by soils," *J. Sci. Food Agric.* 5, 184–192 (1954).

Walker, A., "Availability of atrazine to plants in different soils," *Pestic. Sci.* 3, 139–148 (1972).

Weber, W. J., Jr. and W. Huang, "A distributed reactivity model for sorption by soils and sediments. 4. Intraparticle heterogeneity and phase-distribution relationships under nonequilibrium conditions," *Environ. Sci. Technol.* 30, 881–888 (1996).

Weed, S. B. and J. B. Weber, "Pesticide-organic matter interactions," In *Pesticides in Soil and Water*, W. D. Guenzi, Ed., Soil Science Society of America, Madison, WI, 1974, pp. 223–256.

Weil, L., G. Duré, and K.-E. Quentin, "Wasserlöslichkeit von insektiziden chlorierten Kohlenwasserstoffen und polychlorierten Biphenylen im Hinblick auf eine Gewässerbelastung mit diesen Stoffen," *Z. Wasser Abwasser Forsch.* 7, 169–175 (1974).

Weissmahr, K. W., S. B. Haderlein, R. P. Schwarzenbach, R. Handy, and R. Nuesch, "In situ spectroscopic investigations of adsorption mechanisms of nitroaromatic compounds at clay minerals," *Environ. Sci. Technol.* 31, 240–247 (1997).

Wershaw, R. L., P. J. Burcar, and M. C. Glodberg, "Interaction of pesticides with natural organic material," *Environ. Sci. Technol.* 3, 271–273 (1969).

Westall, J. C., "Properties of organic compounds in relation to chemical binding." In *Biofilm Processes in Groundwater Research*, Symposium, Proceedings Stockholm, Sweden, 1983, pp. 65–90.

Whalen, J. W., "Adsorption on low-energy surfaces: Hexane and octane adsorption on polytetrafluoroethylene," *J. Colloid Interface Sci.* 28: 443–448 (1968).

Xia, G., "Sorption behavior of nonpolar organic chemicals on natural sorbents," Ph.D. dissertation, Johns Hopkins University, Baltimore, 1998, p. 297.

Xia, G. and W. P. Ball, "Adsorption-partition uptake of nine low-polarity organic chemicals on a natural sorbent," *Environ. Sci. Technol.* 33, 262–269 (1999).

————, "Polanyi-based models for the competitive sorption of low-polarity organic contaminants on a natural solid," *Environ. Sci. Technol.* 34, 1246–1253 (2000).

Xing, B. and J. J. Pignatello, "Dual-mode sorption of low-polarity compounds in glassy poly(vinylchloride) and soil organic matter," *Environ. Sci. Technol.* 31, 792–799 (1997).

Xing, B., J. J. Pignatello, and B. Gigliotti, "Competitive sorption between atrazine and other organic compounds in soils and model sorbents," *Environ. Sci. Technol.* 30, 2432–2440 (1996).

Xu, S. and S. A. Boyd, "Cation exchange chemistry of hexadecyltrimethylammonium in a subsoil containing vermiculite," *Soil Sci. Soc. Am. J.* 58, 1382–1391 (1994).

Yaron, B. and S. Saltzman, "Influence of water and temperature on adsorption of parathion by soils," *Soil Sci. Soc. Am. Proc.* 36, 583–586 (1972).

Yost, E. C. and M. A. Anderson, "Absence of phenol adsorption on goethite," *Environ. Sci. Technol.* 18, 101–109 (1984).

Young, T. M. and W. J. Weber, Jr., "A distributed reactivity model for sorption by soils and sediments. 3. Effects of diagenetic processes on sorption energetics," *Environ. Sci. Technol.* 29, 92–97 (1995).

# INDEX


Activated carbon
  benzene and water vapors on, 100, Fig. 6.13
  ethylene dibromide and chlorinated
    solvents from water on, Fig. 7.4
  $N_2$ vapor on, 87, Table 6.1, Fig. 6.3
  Polanyi theory for adsorption on, 47
  schematic pore structure of, Fig. 4.2
  surface area of, Table 6.1
Activity, definition, 9
  criterion for phase equilibrium, 30
  expression by Raoult's law, 15
  expression by Flory–Huggins model, 20
  for contaminants in soil, 210
  of pure solids, 17, 35
  reference states for gases, liquids, and
    solids, 10
  relation to chemical potential, 10
  relation to ideal solid solubility, 17, 69
  relation to supercooled-liquid solubility, 32
Activity coefficient, 15
  definition by Raoult's law, 15, Fig. 2.1
  Flory–Huggins version of, 20
  influence of solvent–water saturation on,
    31, 54, 55
  relation to concentration, 16, 21
  variation with temperature of, 33
  at infinite dilution, 18, 22
Adsorbate, definition, 39
  monolayer capacity on solid:
    by the BET model, 44
    by Langmuir model, 43
  molecular area estimation, 87
  requirement for surface area
    determination, 87
Adsorbent, definition, 39
Adsorption, definition, 39
  background in soil uptake, 108
  benzene and water vapors on minerals, 100,
    Figs. 6.7–6.12
  competition against solvent on solid, 47,
    108, 192
  competition between solutes on soil, 155,
    194, Figs. 7.16 to 7.21, Fig. 7.38

concept and models, 41–48
  enthalpy and entropy changes in, 6, 39, 198,
    Fig. 7.38*b*
  influence of moisture/humidity on vapor
    uptake, 203, Figs. 7.42, 7.44, and 7.45
  $N_2$ vapor on minerals and soils, 87, Tables
    6.1 to 6.3, Figs. 6.1 to 6.6
  nonpolar vapors on dry soils and minerals,
    99, 205, Table 6.3, Fig. 7.43
  solutes on soil from nonpolar solvents, 192,
    Figs. 7.36 to 7.38
Adsorption isotherm, definition, 39
  Brunauer's classification of, 40, Fig. 4.1
  conventional vs normalized plots, 39
  surface-area determination from, 44, 87, 96
Adsorption theory, *see* Adsorption, concept
  and models
Adsorption potential, 45
Aldrin
  absorption by crops of, 228, Table 8.4
Alumina
  adsorption of benzene and water vapors
    on, Fig. 6.8
  adsorption of EGME vapor on, Fig. 6.4
  surface area of, Tables 6.1 and 6.2
Athermal solubility, definition, 70
  values for solids in triolein, Table 5.4
Atrazine
  absorption from soil by barley of, Table 8.2
  adsorption by smectitic clays of, 166
  sorption by soil of, 150

Bed sediment
  $^{13}$C-NMR spectra of, Fig. 7.15
  contaminated by excess hydrocarbons, 145
  samples from wide sources, Table 7.3
Benzene
  limiting partition in SOM of, Table 7.2,
    Table 7.21
  partition into polymers of, 41
  $K_{hw}$, Table 5.2
  $K_{tw}$, Table 5.5
  $S_w$ and $K_{ow}$, Table 5.1

solubility of PAHs in, 141
sorption of parathion to soils from, 193
sorption to soil of, single and with
    surfactant, 113, 160, 187, Table 7.2,
    Table 7.20
vapor adsorption of, on minerals and
    activated carbon, 100, Figs. 6.7 to 6.13
vapor pressure, Table 1.1
as molecular probe for surface area,
    Table 6.3
BET equation, 43
analysis of vapor uptake by dry soil, 205
concept and derivation, 43
range of validity, 44
surface area determination with, 87, 92
Bioconcentration, background and concept,
    80
Bioconcentration factor (BCF)
correlation with $K_{tw}$ and $K_{ow}$, 81, 229
effect of fish lipids on, 80
effect of plant composition on, 217
factors on fish BCF, 83
field values on fish, Table 5.7
laboratory values on fish, Table 5.6
values on plants/crops, 217, Tables 8.1 to
    8.3
Brunauer–Emmett–Teller (BET) adsorption
    theory, 43
Butanol–water partition coefficient ($K_{bw}$), 62.
    *See also* Solvent–water partition

Carbamoyl oximes, 220
absorption by barley roots of, Table 8.1
Carbon tetrachloride
$K_{oc}$ with soils and bed sediments, 128,
    Table 7.3, Fig. 7.8, Fig. 7.10
$K_{oc}$ in relation to SOM composition, 132,
    Fig. 7.12
limiting partition in SOM of, Table 7.21
partition into polymers of, 41
vapor pressure, Table 1.1
Cation exchange, with clays and soils, 185
Cation solvation, 97, 104
Charcoal-like material, 149
identification, Fig. 7.22
model for nonlinear soil sorption, 157
Chemical potential, 6. *See also* Free energy;
    Free energy change
definition, 7
variation with pressure of, 8
variation with composition of, 23
as criterion for equilibrium, 8
solid vs supercooled liquid, 23
Chemisorption, definition, 39

Chlorinated benzenes
absorption by barley plants of, 223, Table 8.2
bioconcentration into fish of, 80, Tables 5.6
    and 5.7
interaction with DOM, 170, Table 7.15,
    Fig. 7.30, Fig. 7.31
limiting partition in SOM, Table 7.2
$S_w$ for supercooled liquids, 73
$S_w$, $K_{hw}$, $K_{ow}$, and $K_{tw}$, Tables 5.1, 5.2, and 5.5
soil $K_{om}$ values, Table 7.2, Table 7.4
sorption to sediment, 160
Chlorinated solvents
adsorption on activated carbon of, Fig. 7.4
$S_w$ and $K_{om}$, Table 7.1
sorption to soil of, Fig. 7.3
Clausius–Clapeyron equation, 11, 51, 116
schematic plot, Figs. 1.2 and 7.5
Cohesive energy density, definition, 27
Closed systems, definition, 2
Critical micelle concentration (CMC), 178.
    *See also* Surfactant

DDT
absorption by crops and plants of, 225,
    Table 8.2, Table 8.3
bioconcentration into fish of, 80
$K_{hw}$ and $K_{ow}$, Table 5.2
$K_{tw}$, Table 5.5
limiting partition in SOM, 120
melting point and heat of fusion, Table 5.4
$S_w$ as solid, 170
$S_w$ as supercooled liquid, 55, Table 5.1
$S_w$ enhancement by DOM, 170, Table 7.15,
    Figs. 7.28 to 7.31
$S_w$ enhancement by surfactants, 179,
    Figs. 7.32 and 7.33
soil sorption and heat of sorption, 119
soil sorption, impact of surfactant, 186,
    Fig. 7.35
solubility in triolein, Table 5.4
toxicity over soil, Table 7.22
Delta ($\Delta_X$) values, for substituents
concept and definition, 64
relation to $\pi_X$ (oct-w), Fig. 5.3
relation to $\pi_X$ (hep-w), Fig. 5.4
values, Table 5.3
Dibenzo-*p*-dioxins, chlorinated
$S_w$ and heats of solution of, 25, Table 2.1
1,2-Dibromoethane, *see* Ethylene dibromide
1,2-Dichlorobenzene (*o*-dichlorobenzene), *see
    also* Chlorinated benzenes
$K_{oc}$ with soils and bed sediments, 128, Table
    7.1, Table 7.3, Fig. 7.9, Fig. 7.11
limiting partition in SOM of, Table 7.2

1,3-Dichlorobenzene (*m*-dichlorobenzene),
   *see also* Chlorinated benzenes
   vapor sorption to soil of, effect of humidity,
      205, Fig. 7.44
   vapor pressure, Table 1.1
   as molecular probe for surface area,
      Table 6.3
1,4-Dichlorobenzene (*p*-dichlorobenzene), *see
   also* Chlorinated benzenes
   melting-point effect on $S_w$ of, 58
   limiting partition in SOM of, Table 7.2
Dieldrin
   absorption by crops/plants of, 225, 226,
      Tables 8.2 and 8.3
   $S_w$, $K_{om}$, and limiting partition in SOM,
      204
   toxicity over soil of, variation with
      humidity, 213
   vapor densities over dry and moist soils of,
      203, Fig. 7.42
Dipole-dipole forces, 27
Dispersion forces, *see* London forces
Dissolved organic matter, *see* DOM
Diuron, *see also* Substituted ureas
   sorption to soil from petroleum of, 192,
      Fig. 7.36
   sorption from water to soil of, 151, Table
      7.11, Table 7.13, Figs. 7.18 and 7.25
   sorption from water to soil HA of, 161,
      Table 7.13, Fig. 7.25
   toxicity over soil of, variation with
      moisture, 213
DOM, dissolved organic matter
   contaminant $K_{dom}/K_{doc}$ with, Table 7.15
   contaminant $K_{mn}$ and $K_{mc}$ with, Table 7.17
   as dissolved surfactant/microemulsion,
      Table 7.16, Table 7.18
   derived from aquatic and soil sources,
      Table 7.14
   impact on contaminant partition, 169, 184,
      185
   impact on contaminant $S_w$, 169, 179, 183
   polarity and size on contaminant solubility,
      170

EGME, ethylene glycol monoethyl ether
   apparent BET monolayers on solids of,
      Table 6.2
   cation solvation by, 97
   molecular area of, 96
   $N_2$-equiivalent BET monolayer of, 96
   partition to peat of, 98, Table 7.21
   sorption to minerals and soils of, 93
Enthalpy, definition, 3

Enthalpy change, *see also* Heat of adsorption;
      Heat of partition; Heat of solution
   at constant pressure, 3
   in phase transition, 11
Entropy, definition, 3
Entropy change
   for isolated systems, 4
   for solute partition, 119
   in adsorption, 6, 39, 196
   in phase transition, 11, 35
   in solubilization, 24
   reversible and spontaneous processes, 3
Ethylene dibromide (EDB)
   adsorption on activated carbon of, 115,
      Fig. 7.4
   impact of humidity on vapor sorption to
      soils and clays of, 200
   limiting partition in SOM of, Table 7.1
   nonlinear sorption to soil of, 151, Table
      7.11, Table 7.13, Fig. 7.17, Fig. 7.20
   sorption to soil HA, 160, Table 7.13,
      Fig. 7.23
   sorption to soil of, 115, Table 7.1, Fig. 7.3
   sorption to soil humin of, 160, Fig. 7.24
   $S_w$ and $K_{om}$, Table 7.1, Table 7.11
   as molecular probe for surface area,
      Table 6.3
Evaporative flux, pesticides over soil, 211,
      Fig. 7.48
Excess heat of mixing, 34
Extensive properties, definition, 6
External surface, 49, 92. *See also* Surface area

First law, thermodynamics, 2
Fish, bioconcentration studies, 80
   BCF, 80, Table 5.6, Table 5.7
   correlation with partition coefficients, 81,
      Figs. 5.7 and 5.8
   lipid effect, 80, 81
   estimation, 85
Flory–Huggins model
   application to lipids, 70, 73, Tables 5.4 and
      5.5
   application to SOM, 133, Fig. 7.13
   theory, 19–21
Free energy, definition, 4
Free-energy change
   as criterion for equilibrium, 4, 8
   at constant temperature, 5
   function of state variables, 7
   in closed systems, 4
   in open systems, 7
   solid to supercooled liquid, 23
   solute to solution, 23

Freundlich equation, 43
    analysis of sorption linearity, Table 7.13
Fugacity, 9
    reference states for gases, liquids, and
        solids, 10
    relation to chemical potential, 10

Gibbs free energy, *see* Free energy
Goethite
    adsorption of benzene and water vapors
        on, Fig. 6.9
    surface area of, Table 6.1
Group contribution, to partition coefficient,
    63
    $\pi_X$ (oct-w), $\pi_X$ (hep-w), and $\Delta_X$ values,
        Table 5.3

Heat of adsorption, 6, 39
    isosteric heat, 50, 198, Fig. 7.38*b*
    net and total heats, 47, 198
    relation to heat of solution, 48, 51, 122
    solutes from solution, 48, 193
    vapors, 47
Heat of desorption, 51, 118
Heat of fusion, 11
    impact on heat of solution, 25, 34
    relation to solid activity, 35
    relation to solid solubility, 69
Heat of partition, 34
    relation to heats of solution, 35, 118
    solutes between phases, 117
    vapors, 117, 201, 205
Heat of solution, 24
    excess heat of mixing, 34
    liquids vs solids, 25, 34
    relation to heat of adsorption, 48, 51,
        122
    relation to heat of partition, 35, 117, 119
    value for parathion in hexane, Fig. 7.38*b*
    values for polychlorinated dibenzo-*p*-
        dioxins in water, 25
Heat of vaporization, 11
    relation to heat of adsorption, 47, 51
    relation to heat of partition, 117, 205
Henry's law, 18, 115, 197
Heptachlor
    absorption by crops of, 228, Table 8.4
    toxicity over soil of, Table 7.22
Heptachlor epoxide
    absorption by crops of, 228, Table 8.4
    sorption to soil of, impact of surfactant, 187,
        Table 7.20
Heptane-water partition coefficient ($K_{hw}$), 59.
    *See also* Solvent–water partition

polarity effect on, 60
substituent effect on, 64
values, Tables 5.2 and 5.3, Figs. 5.2 and 5.4
Hexachlorobenzene
    absorption by barley of, Table 8.2
    bioconcentration into fish of, Table 5.6
    $K_{hw}$ and $K_{ow}$, Table 5.2
    $K_{tw}$, Table 5.5
    $S_w$, Table 5.1
    $S_w$ enhancement by surfactants, 179
Hexane
    partition limit in SOM of, Table 7.21
    solubility parameter, Table 2.2
    sorption to soil of lindane from, 193, 196
    sorption to soil of parathion from, 193, 196
    sorption to soil of PCBs from, 194, Fig. 7.37
    vapor pressure of, Table 1.1
Humic acid (HA)
    peat HA as sorbent, 160
    elemental contents of soil and aquatic
        humic acids, Table 7.14
    surface area of peat HA, Table 7.12
Humic substance
    composition, 106
    $^{13}$C-NMR spectra of, soil and sediment
        samples, Fig. 7.15
Humidity, *see also* Soil water content
    effect on vapor sorption to soil, 200, 205,
        Figs. 7.44 and 7.45
    effect on contaminant toxicity over soil, 213
Humin (HM)
    peat HM as sorbent, 160
    surface area of peat HM, Table 7.12

Ideal solubility, 17, 69. *See also* Athermal
    solubility
    solids in triolein, Table 5.4
Ideal solution
    definition, 15
    graphic illustration, Figs. 2.1 and 2.2
Illite, sorption of EGME and $N_2$ on, Table 6.2
    sorption of nitroaromatic solutes from
        water on, 166, Fig. 7.27
Induced dipole-induced dipole forces, *see*
    London forces
Intensive properties, definition, 6
Internal energy, definition, 2
    change in closed systems, 2, 4
    change in open systems, 7
    relation to cohesive energy density, 27
    relation to enthalpy, 3
Internal surface, 49, 92, 109. *See also* Surface
    area
Isolated system, definition, 4

INDEX   253

Isosteric heat
  concept and equation, 50, 116
  graphic illustration, Figs. 4.3 and 7.5
  values, Fig. 7.38*b*

$K_d$, solid-water distribution coefficient, 113, 169, 184, 218. *See also* Soil–water distribution
$K_{hw}$, *see* Heptane–water partition coefficient
$K_{oc}$ and $K_{om}$, *see* SOM–water partition coefficient
$K_{ow}$, *see* Octanol–water partition coefficient
$K_{tw}$, *see* Triolein–water partition coefficient
Kaolinite
  adsorption of benzene and water vapors on, Fig. 6.10
  adsorption of EGME on, Fig. 6.4
  sorption of nitroaromatic solutes from water on, 166, Fig. 7.27
  surface area of, Table 6.1, Table 6.3

Langmuir adsorption model, 41
Lindane
  absorption by aquatic plants of, 232
  interaction with DOM, 170, Table 7.15
  soil sorption from hexane of, single and with parathion, 193, 196, Fig. 7.40
  soil sorption from hexane of, impact of moisture; 198, Fig. 7.39
  soil sorption from water of, single and with parathion, 121, 154, Fig. 7.7
  soil sorption of, impact of surfactant, 187, Table 7.20
  $S_w$ and $K_{om}$ values, 204
  solubility in triolein, Table 5.4
  toxicity over soil of, effect of humidity, 213
  vapor density over soil, 203
Linear free-energy relationship, 78, 119
Lipids, 68, *see also* Triolein
  analysis of solute solubility in, 69
    by Raoult's law, 69
    by Flory–Huggins model, 70
  impact on fish bioconcentration, 81
  impact on plant uptake, 220, 225, 228
  solubility of solids in, 5–6, Table 5.4
Lipid–water partition coefficient ($K_{lip}$), *see* Triolein–water partition coefficient ($K_{tw}$)
Lipophilicity, concept, 58
London forces, 27

Medium, definition, 1
Melting-point effect
  on solute $S_w$, 32, 73, Tables 5.1, 5.5, and 7.4

on log $K_{ow}$-log $S_w$ correlation, 57, Fig. 5.1
on log $K_{om}$-log $S_w$ correlation, 135, Fig. 7.13
Micelle, 179. *See also* Surfactant
  contaminant partition to, 180, Table 7.17, Figs. 7.32 and 7.33
  influence on $K_d$, 184, Fig. 7.35
Microemulsion, 182
  contaminant partition to, 183
  examples, Table 7.18
  influence on $K_d$, 185, Fig. 7.34
Micropore volume, 90, Table 6.1
  analysis by $t$ plot, 90
  analysis by $\alpha_s$ plot, 91
Minerals
  adsorption of water and benzene vapors on, 100, Figs. 6.7 to 6.12
  composition, 106
  role in soil sorption, 110
Molar volume, 9, 20
  as scale factor for adsorption potential, 47
  as scale factor for cohesive energy, 27
  impact on partition by Raoult's law, 32, 72
  impact on partition by Flory–Huggins model, 33, 73, 133
  relation to component activity, 20, 21
  relation to athermal solubility, 70
  relation to ideal solubility, 69
  values, Tables 5.4–5.5
Molecular area
  calculation for adsorbates, 87
  EGME, 96
  nitrogen, 87
Molecular forces, types for neutral molecules, 27
Monolayer capacity, 44, 87, 96, Table 6.1, Table 6.2
Montmorillonite, $Ca^{+2}$ form
  sorption of atrazine from water on, 166
  sorption of benzene and water vapors on, Fig. 6.11
  sorption of nitroaromatic compounds from water on, 166
  sorption of EGME vapor on, Fig. 6.5
  surface area of, Table 6.1, Table 6.2
Montmorillonite, $K^{+1}$ form
  sorption of benzene and water vapors on, Fig. 6.12
  sorption of EGME on, 93, Table 6.2
  sorption of nitroaromatic compounds from water on, 166, Fig. 7.27
  surface area of, Tables 6.1 and 6.2

Naphthalene, *see also* Polycyclic aromatic
    hydrocarbons (PAHs)
    molecular properties, Table 7.7
    $S_w$ enhancement by surfactants of, 179
    solubility in triolein of, Table 5.4
    solubility parameter of, Table 2.2
    sorption to soil and sediment of, 138, 140,
        Table 7.6, Table 7.8
Nitroaromatic compounds
    adsorption from water on clays of, 166,
        Fig. 7.27
Nitrogen ($N_2$)
    adsorption on minerals and soils of, 87,
        Tables 6.1 to 6.3, Figs. 6.1 to 6.6
    adsorption data for $t$-plot and $\alpha_s$ plots, 90,
        91
    liquid density, 93
    molecular area, 87
Nonlinear sorption, with soil, 149
    capacity, 154, Table 7.11, Table 7.13
    hypotheses, 149
    HSACM model, 157
    range, Table 7.11, Table 7.13
    nonpolar solutes, 157, Figs. 7.16 to 7.17,
        Fig. 20, Figs. 7.23 to 7.24
    polar solutes, 158, Figs. 7.18–7.19, Fig. 7.21,
        Figs. 7.25 to 7.27
    specific-interaction model, 150, 158

$n$-Octanol, 54
    unique solvent power, 59
    model for lipids, 58
Octanol–water partition coefficient ($K_{ow}$), 54.
        *See also* Solvent–water partition
    correlation with $S_w$, 55, Fig. 5.1
    correlation with fish BCF, 82
    ideal line, 55
    relation to $K_{tw}$, 75
    substituent effect on, 63, Table 5.3
    values, Tables 5.1 to 5.3, Figs. 5.3 and 5.5

Parathion
    sorption to soil from organic solvents of,
        193
    sorption to soil from hexane of, effects of
        moisture and temperature, 195, 196,
        Fig. 7.38
    sorption to soil from water of, single and
        with lindane, 121, Fig. 7.7
    toxicity over soil of, Table 7.22
Partition theory, 30–33
Partition isotherm
    general equations, 31, 33
    isotherm shape, 36, Figs. 3.1 and 3.2

macromolecular phase-water mixtures, 33
    solvent–water mixtures, 32
    temperature dependence, 33
Partition-like interaction
    contaminants with DOM, 173, 179
Passive uptake by plants, 214. *See also*
        Plant–water partition
Peat
    HA and humin prepared from, 160,
        Table 7.12
    sorption of EGME vapor on, Fig. 6.6
    sorption of contaminants from water on,
        150, 151, 187
    surface area, Table 6.2
    vapor partition to, 201, Table 7.21
Phase, definition, 1
Phase transition, 5
    enthalpy and entropy changes, 10
Phenanthrene, *see also* Polycyclic aromatic
        hydrocarbons (PAHs)
    adsorption on mineral oxides of, 122
    melting-point effect on $S_w$, 57
    molecular properties, Table 7.7
    $S_w$ enhancement by surfactants of, 179
    solubility in triolein of, Table 5.4
    solubility parameter of, Table 2.2
    sorption to soil and sediment of, 138, 140,
        Table 7.6, Table 7.8
    soil sorption of, impact of emulsified
        material, 186
Phenols
    sorption to soil of, 122, 136, 151, 211, Table
        7.5, Table 7.13, Fig. 7.19, Fig. 7.21
    sorption to soil HA and humin of, 163,
        Fig. 7.26
Physical adsorption, definition, 39
Pi ($\pi_X$) values, for substituents
    concept and definition, 64
    $\pi_X$ (oct-w) and $\pi_X$ (hep-w) values,
        Table 5.3
    relation to $\Delta_X$ values, 64, Figs. 5.3 and 5.4
Plant–water partition, 215
    analysis, 220, Tables 8.1 to 8.4
    lipid effect, 228, 229
    model and equations, 216–219
    molecular diffusion, 225
    root concentration factor, 220
    time dependence, 231
Polanyi adsorption potential theory, 45
    analysis of nonlinear solute sorption, 160
Polar contaminants, 65, 112
    $K_{om}$ and $K_{ow}$ for selected compounds,
        Table 7.5
    limiting partition in SOM, Table 7.21

Polar groups, 65
Polychlorinated biphenyls (PCBs)
　absorption by barley of, Table 8.2
　absorption by underwater plants of,
　　230
　bioconcentration into fish of, 80
　limiting partition in SOM of, Table 7.2
　$S_w$, $K_{hw}$, $K_{ow}$, and $K_{tw}$, Tables 5.1, 5.2,
　　and 5.5
　$S_w$ enhancement by DOM of, 170,
　　Table 7.15, Figs. 7.29 and 7.31
　soil $K_{om}$ values, Tables 7.2 and 7.4
　soil sorption from hexane of, 194,
　　Fig. 7.37
　soil sorption from water of, impact of
　　surfactant and microemulsion, 186,
　　Figs. 7.34 and 7.35
Polychlorinated dibenzo-*p*-dioxins, *see*
　　Dibenzo-*p*-dioxins
Polycyclic aromatic hydrocarbons (PAHs)
　melting-point effect on $S_w$ of, 58
　molecular properties of, Table 7.7
　$S_w$ and $K_{ow}$, Tables 5.1 and 5.5
　$S_w$ enhancement by surfactants, 179
　soil/sediment sorption of, 138, 160, Tables
　　7.2, 7.4, 7.6, and 7.8
　solubility in triolein of, Table 5.4
　solubility parameters of, Table 2.2
Pore size, classification, 90
Pyrene, *see also* Polycyclic aromatic
　　hydrocarbons (PAHs)
　melting-point effect on $S_w$, 57
　molecular properties, Table 7.7
　$S_w$ enhancement by DOM of, 177
　$S_w$ enhancement by surfactants of, 179
　sorption to soil and sediment of, 138,
　　Table 7.8

Raoult's law, theory, 14
　for heptane–water mixtures, 59
　for octanol–water mixtures, 54
　graphic illustration, Figs. 2.1 and 2.2
　limitation, 20, 69, 72
Reference state, for substances, 10
Relative humidity, *see* Humidity
Reversible process, 3

$S_w$, *see* Water solubility
Second law, thermodynamics, 3
Silica
　adsorption of benzene and water vapors
　　on, Fig. 6.7
　BET surface area of, Table 6.1
Soil, *see also* Bed sediment

definition and formation, 106
composition, 106
as dual sorbent, 109
contaminated by excess PCB oil, 147,
　Table 7.9
samples from wide sources, Table 7.3
Soil organic matter, *see* SOM
Soil sorption, from vapor phase, 200
　influence of soil water and humidity, 203,
　　Fig. 7.42, Figs. 7.44 to 7.46
Soil sorption, from water solution, *see*
　　Soil–water distribution
Soil-to-bed sediment transformation, 131
Soil water content
　effect on pesticide toxicity over soil, 213,
　　Table 7.22
　effect on soil sorption from hexane, 194,
　　196, Figs. 7.38 to 7.40
　effect on soil sorption of vapors, 201
　effect on vapor density over soil, 203,
　　Fig. 7.42
Soil–water distribution, *see also* SOM–water
　　partition coefficient
　background, 107
　general characteristics, 112
　dependence on SOM content, 108
　equilibrium heat, 115
　isotherm shape, 112, 149, Fig. 7.1
　limiting capacity with SOM, 119,
　　Table 7.2
　$K_{oc}$ values for nonpolar contaminants, 145
　single vs binary solutes, 121, 151–165, Fig.
　　7.7, Figs. 7.16 to 7.21, Figs. 7.23, 7.25,
　　and 7.26
Solubility, *see also* Water solubility
　models, 14, 19
　solid and supercooled liquid, Fig. 2.2
　values in lipid triolein, 68, Table 5.4
　values in SOM, 119, 201, Tables 7.2 and
　　7.21, Fig. 7.41
　values in water, Tables 5.1, 5.3, and 5.5
Solubility parameter ($\delta$), definition, 29
　relation to contaminant partition to SOM,
　　143
　values, Table 2.2, and Table 7.21
Solution theory, 14. *See also* Activity
　　coefficient
　Raoult's law, 14
　Henry's law, 18
　Flory–Huggins model, 19
Solvent–water partition, 31. *See also*
　　Butanol–water; Heptane–water;
　　Octanol–water; Triolein–water
　　partition coefficient

concentration dependence of, 36
correlation equation, 77
polar-group effect on, 66
solvent–water mutual saturation, 31, 55, 62
substituent/group effect on, 63
temperature dependence of, 33
theory, 30–33
SOM, soil/sediment organic matter
composition, 106
functional-group analysis by $^{13}$C-NMR, 143,
Fig. 7.15
role as a partition medium for
contaminants, 110, Fig. 7.1
soil vs sediment source, 124
surface area of, 98, 109, 163, Table 7.12
SOM–vapor partition, 201, Table 7.21,
Figs. 7.41, 7.46 and 7.47
SOM–water partition coefficient ($K_{om}$, $K_{oc}$),
112
correlation with $K_{ow}$:
nonpolar contaminants, 138
PAHs, 138
polar contaminants, 136
effect of SOM composition on, 132,
Fig. 7.12
impact of DOM on, 169, 184, Table 7.20,
Figs. 7.34 and 7.35
$K_{om}$–$K_{oc}$ conversion, 135
with previously contaminated soils, 146
relation to $S_w$, 135
values for soils vs sediments, Tables 7.3 and
7.8
Sorption, definition, 107. *See also*
Adsorption; Partition between phases;
Plant–water partition, Soil–water
distribution
Specific interaction (SI), polar contaminants
with SOM, 150, 158
State function, thermodynamic, 2
Substituent constant, in partition coefficient,
*see* Group contribution
Substituted ureas, herbicides
absorption by barley roots of, 220, Table 8.1
$S_w$, $K_{ow}$, and $K_{om}$, Table 7.5, Table 7.10,
Table 8.1
Surface area, definition, 48
external and internal surfaces, 49, 92
measurement, 92
nonporous surface, 90, Table 6.1
theoretical basis, 44
values for soils and minerals, Tables 6.1 to
6.3, Table 7.3
values for organic matter, Table 7.12

Surfactant, 178
adsorption onto minerals of, 184
contaminant solubilization by, 179,
Table 7.17
CMC, 179, Table 7.16
examples, Table 7.16
influence on $K_d$, 184, 189, Table 7.20,
Fig. 7.35
Suspended solids, source and behavior, 132,
Table 7.3
System, definition, 1

TCE, trichloroethylene, *see also* Chlorinated
solvents
$K_{oc}$, Table 7.11
nonlinear soil sorption of, 151, Table 7.11,
Fig. 7.16
sorption from water to soil of, 150
$S_w$ and $K_{ow}$, Table 7.10
vapor partition to SOM of, 202, Table 7.21,
Fig. 7.46
vapor sorption to soil and minerals of, 206
vapor pressure, Table 1.1
Thermodynamics
extensive and intensive properties, 6
first law, 2
reversible and spontaneous processes, 3
second law, 3
state functions, 2
Toluene
$S_w$, $K_{hw}$, $K_{ow}$, and $K_{tw}$ values, Tables 5.1, 5.2,
and 5.5
sorption to soil of, single and with
surfactant, 189, Table 7.20
as molecular probe for surface area,
Table 6.3
Toxicity, pesticides, relation to soil type and
soil water content, 213, Table 7.22
1,2,4-trichlorobenzene, *see also* Chlorinated
benzenes
absorption by barley of, 225, Table 8.2
$K_{hw}$, Table 5.2
$K_{ow}$ and $S_w$, Table 5.1
soil $K_{om}$, Table 7.4
vapor sorption to soil of, effect of humidity,
205
Trichloroethylene, *see* TCE
Triolein, *see also* Lipids
as a model biological lipid, 69
solubility of solids in, Table 5.4
Triolein–water partition coefficient ($K_{tw}$). *See
also* Solvent–water partition
analysis by Raoult's law, 72

analysis by Flory–Huggins model, 73
correlation with fish BCF, 81
correlation with $S_w$ and $K_{ow}$, 75
values, Table 5.5

van der Waals forces, *see* Molecular forces
van Laar equation, 21
van't Hoff equation, 25
Vapor pressure
  of contaminants on soils, impacted by
      humidity, 203, 205, Figs. 7.42, 7.44, and
      7.45
  effect on chemical potential, 9
  function of temperature, Table 1.1
  relation to activity, 15, 209, 211
  relation to chemical potential, 23
  solid vs supercooled liquid, 12, Fig. 1.2
Volatilization, effect of soil wetting, 203, 211,
      Fig. 7.42, Fig. 7.48

Water solubility ($S_w$), *see also* Solubility
  correlation with $K_{om}$, 135
  correlation with $K_{ow}$, 57
  correlation with $K_{tw}$, 76

relation to $\Delta_X$ and $\pi_X$, 64
solid vs supercooled liquid, 32
temperature effect on, 34
values, Tables 5.1, 5.3, and 5.5
Water solubility enhancement, 32, 168
  by DOM, 169, Table 7.15, Figs. 7.28 and
      7.29
  by phenylethanoic acids, 174, Fig. 7.30
  by polyacrylic acid, 175, Fig. 7.31
  by surfactant and microemulsion, 178,
      Figs. 7.32 and 7.33
  correlation with $K_{ow}$, Table 7.15
  impact on solid–water distribution, 169, 184
  relation to contaminant $S_w$, 176
  theory, 169, 179, 182

Xylenes
  $S_w$, $K_{hw}$, and $K_{ow}$ values, Tables 5.1 and 5.2
  sorption to soil of, single and with
      surfactant, 187, Table 7.20
  vapor partition to polymers of, 41
  vapor pressure, Table 1.1
  as molecular probe for surface area,
      Table 6.3